AF361843

Smart Grid Analytics for Sustainability and Urbanization in Big Data

Smart Grid Analytics for Sustainability and Urbanization in Big Data

Editors

Sheraz Aslam
Herodotos Herodotou
Nouman Ashraf

Basel • Beijing • Wuhan • Barcelona • Belgrade • Novi Sad • Cluj • Manchester

Editors

Sheraz Aslam
Department of Electrical
Engineering, Computer
Engineering and Informatics
Cyprus University of Technology
Limassol
Cyprus

Herodotos Herodotou
Department of Electrical
Engineering, Computer
Engineering and Informatics
Cyprus University of Technology
Limassol
Cyprus

Nouman Ashraf
School of Electrical and
Electronic Engineering
Technological University
Dublin
Ireland

Editorial Office
MDPI
St. Alban-Anlage 66
4052 Basel, Switzerland

This is a reprint of articles from the Special Issue published online in the open access journal *Sustainability* (ISSN 2071-1050) (available at: www.mdpi.com/journal/sustainability/special_issues/Smart_Data).

For citation purposes, cite each article independently as indicated on the article page online and as indicated below:

Lastname, A.A.; Lastname, B.B. Article Title. *Journal Name* **Year**, *Volume Number*, Page Range.

ISBN 978-3-0365-9173-5 (Hbk)
ISBN 978-3-0365-9172-8 (PDF)
doi.org/10.3390/books978-3-0365-9172-8

Contents

About the Editors

Sheraz Aslam

Sheraz Aslam received a B.S. degree in Computer Science from the Bahauddin Zakariya University (BZU), Multan, Pakistan, in 2015, and a M.S. degree in Computer Science with a specialization in energy optimization in the smart grid from the COMSATS University Islamabad (CUI), Islamabad, Pakistan, in 2018. He also worked as a Research Associate with Dr. Nadeem Javaid during his M.S. period at the same University. Mr. Sheraz recently finished his Ph.D in Computer Engineering and Informatics from Cyprus University of Technology, Limassol, Cyprus, under the supervision of Dr. Herodotos Herodotou. Currently, Dr. Sheraz is working as a Post-doc Researcher at DICL Research Lab., Cyprus University of Technology (CUT), Limassol, Cyprus, where he is also a part of an European Union funded research project named as aerOS. He also working as senior lecturer (part-time) at CTL Eurocollege, Limassol, Cyprus. Previously, he worked on several EU funded research projects, i.e., STEAM and MARI-Sense. He has authored more than 80 research publications in ISI-indexed international journals and conferences, including IEEE Internet of Things Journal, Renewable & Sustainable Energy Reviews, and Electric Power System Research. Mr. Sheraz also served as a TPC member and invited reviewer of international journals and conferences. His research interests include data analytics, generative adversarial networks, wireless networks, smart grid, and cloud computing. He is constantly looking for collaboration opportunities with professors and students from different universities around the globe.

Herodotos Herodotou

Herodotos Herodotou is an Assistant Professor at the Department of Electrical Engineering and Computer and Informatics Engineering of the Cyprus University of Technology, where he is the Scientific Manager of the Data Intensive Computing Research Lab. He received his Ph.D. in Computer Science from Duke University in May 2012. His PhD thesis received the honors "SIGMOD Jim Gray Doctoral Dissertation Award Honorable Mention" and "Outstanding Ph.D. Dissertation Award in Computer Science at Duke". His research interests include large-scale data processing systems, database systems, and storage systems. In particular, his work focuses on the simplification of the use, manageability, and automatic tuning of centralized as well as distributed large-scale data analysis systems. In addition, he is interested in the application of database and machine learning techniques in other domains such as energy, environment, maritime, tourism, and social computing. In the past, Dr. Herodotou worked as a Senior Research SDE in the "Data Management, Exploration and Mining (DMX)" team of Microsoft Research. He has been engaged in various research projects related to cloud computing, covering compute, storage, and networking layers in global data centers. His professional experience includes additional research positions at Yahoo! Labs and Aster Data, as well as programming jobs at Microsoft and RWD Technologies.

Nouman Ashraf

Nouman Ashraf received B.S. and M.S. degrees in electrical engineering from COMSATS University Islamabad, Pakistan, and a Ph.D. degree in electrical engineering from Frederick University, Cyprus, under the Erasmus Mundus Scholarship Program. He has worked with Turku University of Applied Sciences, Finland, the TSSG, Waterford Institute of Technology, Ireland and with the University of Cyprus. Currently, he is working with Technological University Dublin, Ireland. His research interests include the application of control theory for management of emerging networks with applications in the Internet of Things, 5G and beyond communication networks, electric vehicles, metamaterial-based beam steering, wireless sensor networks, smart grid, networks resilience, and power load modeling.

Preface

Recently, microgrids have become a fundamental element within the framework of a smart grid. They bring together distributed renewable energy sources (RESs), prediction of RESs, energy storage units, and load control to enhance the reliability of the power system, promote sustainable growth, and decrease carbon emissions. Simultaneously, the swift progress in sensor and metering technologies, wireless and network communication, IoT-based technologies, as well as cloud and fog computing, is resulting in the gathering and storage of substantial volumes of data, such as device status information, energy generation statistics, and consumption data.

Furthermore, IoT devices are found in various parts of the smart grid, such as smart appliances, smart meters, and substations. These IoT devices generate petabytes of data, which are known to be one of the most scalable properties of a smart grid. Without smart grid analytics, it is difficult to make efficient use of data and to make sustainable decisions related to smart grid operations. With the energy system of the developing world heading towards smart grids, there needs to be a forum for analytics that can collect and interpret data from multiple endpoints. Data analytics platforms can analyze data to produce invaluable results that lead to many advantages, such as operational efficiency and cost savings. In addition, proper forecasting of energy generation from RESs and energy theft detection help a lot while maintaining smart and sustainable energy systems. This reprint comprises a variety of noteworthy and original research contributions that pertain to smart grid analytics for sustainability and urbanization in big data. It also plays a fundamental part in sharing and promoting novel ideas within this field.

Sheraz Aslam, Herodotos Herodotou, and Nouman Ashraf
Editors

sustainability

Article

Solar and Wind Energy Forecasting for Green and Intelligent Migration of Traditional Energy Sources

Syed Muhammad Mohsin [1,2,*], Tahir Maqsood [3] and Sajjad Ahmed Madani [1]

[1] Department of Computer Science, COMSATS University Islamabad, Islamabad 45550, Pakistan
[2] College of Intellectual Novitiates (COIN), Virtual University of Pakistan, Lahore 55150, Pakistan
[3] Department of Computer Science, COMSATS University Islamabad, Abbottabad 22060, Pakistan
* Correspondence: syedmmohsin9@yahoo.com

Abstract: Fossil-fuel-based power generation leads to higher energy costs and environmental impacts. Solar and wind energy are abundant important renewable energy sources (RES) that make the largest contribution to replacing fossil-fuel-based energy consumption. However, the uncertain solar radiation and highly fluctuating weather parameters of solar and wind energy require an accurate and reliable forecasting mechanism for effective and efficient load management, cost reduction, green environment, and grid stability. From the existing literature, artificial neural networks (ANN) are a better means for prediction, but the ANN-based renewable energy forecasting techniques lose prediction accuracy due to the high uncertainty of input data and random determination of initial weights among different layers of ANN. Therefore, the objective of this study is to develop a harmony search algorithm (HSA)-optimized ANN model for reliable and accurate prediction of solar and wind energy. In this study, we combined ANN with HSA and provided ANN feedback for its weights adjustment to HSA, instead of ANN. Then, the HSA optimized weights were assigned to the edges of ANN instead of random weights, and this completes the training of ANN. Extensive simulations were carried out and our proposed HSA-optimized ANN model for solar irradiation forecast achieved the values of MSE = 0.04754, MAE = 0.18546, MAPE = 0.32430%, and RMSE = 0.21805, whereas our proposed HSA-optimized ANN model for wind speed prediction achieved the values of MSE = 0.30944, MAE = 0.47172, MAPE = 0.12896%, and RMSE = 0.55627. Simulation results prove the supremacy of our proposed HSA-optimized ANN models compared to state-of-the-art solar and wind energy forecasting techniques.

Keywords: renewable energy; forecasting; machine learning; energy efficiency; sustainability; low carbon emission

check for updates

Citation: Mohsin, S.M.; Maqsood, T.; Madani, S.A. Solar and Wind Energy Forecasting for Green and Intelligent Migration of Traditional Energy Sources. *Sustainability* **2022**, *14*, 16317. https://doi.org/10.3390/su142316317

Academic Editor: Elena Cristina Rada

Received: 1 November 2022
Accepted: 30 November 2022
Published: 6 December 2022

Publisher's Note: MDPI stays neutral with regard to jurisdictional claims in published maps and institutional affiliations.

1. Introduction

Increased energy consumption leads to higher fossil fuel consumption. Brown energy is produced using expensive and environmentally damaging fossil fuels including coal, natural gas, and oil. On the other hand, green energy is produced by inexpensive and widely available renewable energy sources (RESs), such as solar and wind energy. When compared to brown energy sources, RESs have a substantially lower carbon emission rate (CER) [1]. In this context, governments and the scientific community face major difficulties related to lowering electricity costs and ensuring environmental sustainability. Due to the long-term effects of carbon emissions from traditional power plants, some countries have imposed large taxes on carbon emissions [2–4]. Therefore, the research community has made tremendous research efforts to reduce electricity costs and carbon emissions [5–8].

State-of-the-art literature recommends integration of RESs with brown energy sources [9–19] to meet users' energy demand in an environmental friendly and cost efficient manner. Solar and wind energy are the abundantly available main sources of renewable energy. However, both sources are inherently highly variable due to volatile weather conditions. Figures 1 and 2 show how intermittent energy is generated from solar and wind energy sources. Weather data are taken from the measurement and instrumentation data centre at the national renewable energy laboratory [20], as these data are sufficiently accurate [21]. The intermittency of solar and wind generation creates significant difficulties in seamless integration of solar and wind power into the existing power system [22].

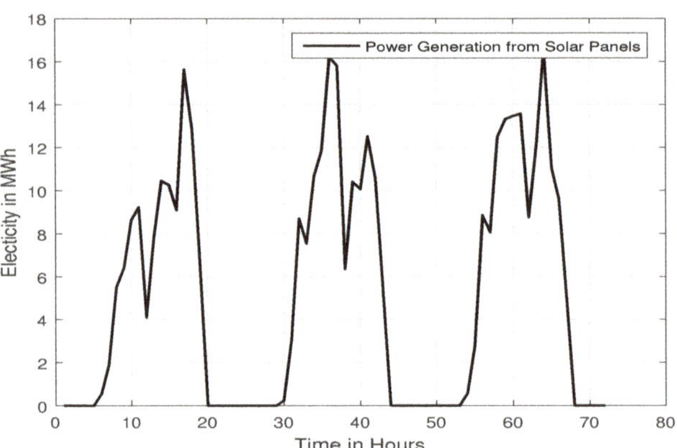

Figure 1. Pattern of solar power generation.

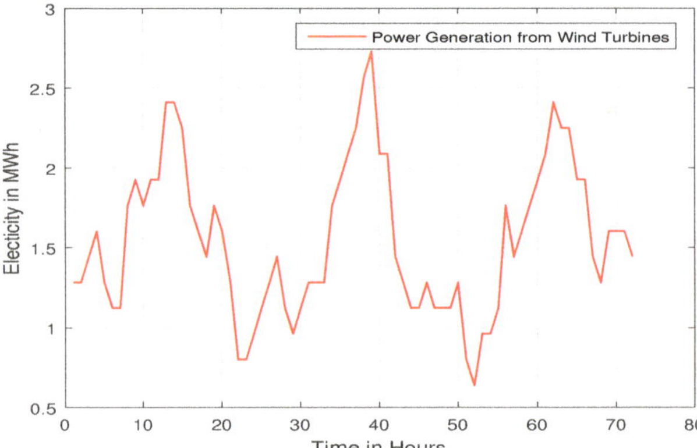

Figure 2. Pattern of wind power generation.

Meteorological conditions affect the green energy generation, and green energy production is different in various time zones [23–27]. Therefore, green energy produced by RESs is intermittent in nature. That is a big challenge, especially, while utilizing RESs as the only power source. Consequently, integration of RESs with brown energy is mostly studied and recommended in the literature to cope up with the intermittent nature of RESs. As the renewable energy generation is dependent upon wind direction, wind speed, solar radiance, temperature, humidity, and other weather conditions, it is unreliable [28]. Hence, accurate forecasting of energy production by RESs is necessary to:

1. Minimize the carbon emission,
2. Decrease operational cost of the grid,
3. Trustworthy and safe operations of the power grid
4. Minimize the gap between electricity demand and supply,
5. Reduce the use of electricity reserves through improved scheduling of generation.

Given the importance of renewable energy forecasting, DeepMind, a subsidiary of Alphabet and Google, has chosen machine learning (ML) for 36-h wind energy forecasts to ensure the availability of clean and carbon-free wind energy [29]. Machine learning, which encompasses a variety of other areas, such as data mining, image and speech recognition, optimization, virtual personal assistants, fraud detection, product recommendations, self-driving cars, and artificial intelligence, can be used to process extensive historical Big Data to solve data-driven problems [30,31]. During training, machine learning approaches search for relations between input and output data and make predictions based on the input data. Model generalisation and feature extraction are two areas where machine learning outperforms traditional statistical predictive models.

Machine learning can help make smarter, faster, data-driven estimates about how electricity generation can meet electricity demand [32]. ML can be used for a variety of energy-related tasks, including demand-side management, energy theft detection, demand forecasting, energy generation forecasting, energy price forecasting, predictive maintenance and control, prediction of weather phenomena and optimised energy storage operation that could impact energy forecasting and build energy management. All forms of renewable energy, including hydro, marine, wind, solar, geothermal, bio, hydrogen, and hybrid, can be harnessed with AI models [33].

In the literature, a number of ML forecasting methods for renewable energy have been put forth, and many patents have been registered in this regard. The inventors of US patent US 2015/0186904 A1 at [34] have invented a system for managing and predicting solar and wind energy. They have proposed a current–voltage curve of a solar cell, a diagram to illustrate energy management and use of energy generated by renewable energy sources. US patent 2005/0039787 A1 [35] presents a tool to help grid operators plan and allocate generation resources in a power grid with solar generation capacity on an hourly basis and a week in advance. Tools are also provided to help entities involved in the generation, distribution, and sale of electric energy.

Another invention, patent WO 2011/124226 A1 at [36], discloses a forecasting technique for wind power generation. The invention discusses establishment of a forecasting site at a given location and collecting power generation data from a series of wind turbines, with the first wind turbine at a first location and the second wind turbine at a second location. The estimation is based on the performance data of the group of wind turbines extrapolated into the future or used in conjunction with the location of the forecast site.

Artificial neural networks (ANN) and time series methods, such as autoregressive integrated moving average (ARIMA), are among the most popular ML-based forecasting techniques [37]. The authors of [38] found that time series techniques such as ARIMA lose accuracy when dealing with noisy data and are less accurate than ML techniques. However, ANN may also lose its prediction accuracy due to the high uncertainty of the input data and the random determination of the initial weights between different layers [38].

In this study, we consider reliable and accurate forecasting of solar and wind energy. Previously, the weights to the edges of ANNs were randomly assigned for solar and wind energy forecasting, while in this study we use a meta-heuristic optimization algorithm called harmony search algorithm (HSA) [39] for optimal weights assignment to the edges of ANN for improved forecasting. The main contributions of this study are given below.

- We summarize the state-of-the-art literature on solar and wind energy forecasting.
- During the literature review, we find out that artificial neural networks lose prediction accuracy when dealing with highly intermittent data, such as solar irradiance and wind speed.
- We propose a reliable solar irradiance and wind speed forecasting algorithm named HSA-optimized ANN.
- In our proposed forecasting model, we use HSA for assignment of optimized weights to the edges of ANN.

The rest of the article is organized as follows. State-of-the-art literature review is presented in Section 2. Motivation and problem statement are described in Section 3, and

Section 4 provides a thorough description of an artificial neural network, the harmony search algorithm, and our proposed system model. The simulation setup and methodology for this study are covered in Section 5, while Section 6 presents the simulation findings and pertinent discussions. The study is finally concluded along with future research directions in Section 7.

2. Literature Review

Prominent machine learning based solar and wind energy forecasting models [40–48] and [49–58] are briefly explained in Sections 2.1 and 2.2, respectively. The summary of the literature review on solar and wind energy forecasting studies is given in Tables 1 and 2, respectively.

2.1. State-of-the-Art Literature on Solar Energy Forecasting

An efficient and effective building energy management system (EMS) can be developed with a reliable energy supply system. Photovoltaic (PV) generation is intermittent; hence, its reliable and accurate forecasting is very important in the development of an efficient EMS. Authors of [40] have proposed a probabilistic day-ahead PV generation forecasting model. A clear sky model is transformed into temperature and shading, and then its deviation is used to train a bagging regression tree for point forecasting of PV energy. A proposed probabilistic forecast model was tested for one year in Munich, Germany, and results proved its accuracy in point forecasting for energy management system applications.

The paper [41] developed and evaluated a daily global solar radiation model from the European centre for medium range weather forecasting by using an ANN-based machine learning model. They compared the ANN model with other models, namely support vector regression, genetic programming, and gaussian process machine learning. Mean absolute error (MAE) and root mean square error (RMSE) were implemented for benchmarking. Results concluded that the ANN-based prediction model was better than other data-driven prediction models.

Solar irradiance is affected by meteorological factors, such as temperature, humidity, cloud cover, dust in desert locations, and sunshine intensity. As a result, solar output varies. Authors of [42] used aerosol optical depth and angstrom data for an hour-ahead solar irradiance forecasting. The proposed forecasting model was compared with different data-driven forecasting models, namely k nearest neighbors, multilayer perception, and support vector regression model. The proposed model was tested on Saudi Arabia data, and it was concluded that it is superior to compared forecasting models, especially for desert areas.

Photovoltaic cells produce electric power when exposed to sun rays. The relationship between energy supply and demand needs to be optimized by reliable solar energy forecasting. The authors of [43] proposed a multi-variant neural network ensemble framework trained on meteorological data. After combining the results with Bayesian model averaging, the proposed technique was compared with real-time solar PV data from the University of Queensland. Validation of the proposed framework was performed by one-day ahead forecasting. Results prove that the proposed multi-variant neural network ensemble framework helps improve the accuracy of PV power output.

Hourly solar irradiation for the following day was predicted by the authors of [44] using LSTM. Inputs include data from the weather forecast for the following day, which includes information on temperature, humidity, sky coverage, wind speed, and precipitation. The model was trained with data from multiple locations of Korea Meteorological Administration. It was found that the proposed model had strong forecasting capabilities with RMSE of 30 W/m^2.

Table 1. Summary of literature on solar energy forecasting.

Paper	Energy Source	RES Forecasting	Implementation Strategy	Objective(s)	Dataset Type	Performance/Result
[40]	Solar energy	Yes	Probabilistic day ahead PV forecast	PV energy forecasting	Historical	Continuous ranked probability skill score = 90.94%
[41]	Solar energy	Yes	Global solar radiance prediction	Daily global solar radiance	Historical	ANN RMSE = 1.613, ANN MAE = 1.146
[42]	Solar energy	Yes	Aerosol optical depth (AOD) and angstrom data for solar irradiance forecasting	One hour solar irradiance prediction	Historical	MLP RMSE = 32.75 (W/m^2)
[43]	Solar energy	Yes	Multivariate neural network ensemble framework	PV output power forecast	Historical	MAPE = 3.1
[44]	Solar energy	Yes	LSTM model for solar irradiance forecasting	Accurate forecast of solar irradiance	Historical	RMSE = 30 W/m^2
[45]	Solar energy	Yes	LSTM model for solar irradiance forecasting	Forecasting of solar irradiance	Historical	3.2% improvement in nRMSE over the SVR model
[46]	Solar energy	Yes	FFNN and LSTM	Accurate forecast of solar irradiance	Historical	Combination of MM and MO performed better
[47]	Solar energy	Yes	Image-based dataset and LSTM model	Solar irradiance forecasting	Historical	Pearson Product-Moment Correlation Coefficient (PCCs) is used in this study
[48]	Solar energy	Yes	GRU, LSTM, RNN, SVR, and FFNN	Accurate forecast of solar irradiance	Historical	GRU is better than LSTM

Using an LSTM model, the authors of [45] predicted hourly solar radiation for the city of Johannesburg. Solar radiation, temperature, daylight hours and relative humidity were used as training inputs for the LSTM network. Model was build using solar radiation data of National Oceanic and Atmospheric Administration from 2009 to 2019. The simulation results showed that the proposed LSTM network had a 3.2% improvement in normalised RMSE over the SVR model.

To anticipate solar radiation over a multilevel horizon in northern Italy, the authors of [46] used two different neural network types, FFNN and LSTM. The proposed models used a variety of methods, including multi-model (MM) and multi-output (MO), to build their predictive models. Six years of weather data from 2014 to 2019 was collected from the Italian weather station used in this study. Historical solar radiation data were used to train the model. Comparative results of the study proved that the proposed models performed better by combining the techniques of MM and MO.

The authors of [47] proposed a new method of solar irradiance prediction using an image-based dataset and LSTM model. The developed model can predict solar radiation up to 60 min in advance. The LSTM model was introduced with two different methods based on the input variables. Prediction results of the second model were better. Authors of [48] used the historical data from Korea Department of Meteorological Administration SURFRA system to analyze different deep learning and machine learning solar irradiance prediction algorithms. Simulation results showed that performance of deep learning models was better.

2.2. State-of-the-Art Literature on Wind Energy Forecasting

Due to its wide availability and limitless supply, wind energy is a particularly popular source of energy. Production of wind energy is hampered by uncertainty of air flow/pressure, among other things. The authors of [49] performed probabilistic wind speed forecasting through an ensemble model. The proposed ensemble model is composed of a recurrent neural network, wavelet threshold de-noising (WTD), and an adaptive neural fuzzy inference system (ANFIS). Sub-model variance is used to calculate wind speed forecasting, which is then confirmed for one hour wind speed prediction. Accuracy of the proposed model over its counterparts was proved by simulation results.

Wind energy is highly dependent on wind speed, wind direction, weather temperature, and weather pressure that make it unpredictable, hence unreliable. In [50], the authors exploited ANN to measure different local meteorological conditions that affect wind flow. In order to predict the wind speed, MAE and RMSE were determined. Reliable wind speed forecasting is needed to plan, develop, and monitor an intelligent power system. As the wind energy relies upon wind speed, pressure, temperature, and wind direction, its forecasting mechanism was proposed by the authors of [51]. Raw data is decomposed using an empirical wavelet transformation in a deep-learning-based hybrid wind speed forecasting model. The proposed model was validated in a way that simulation results show highly accurate wind speed prediction.

The authors of [52] stated that wind energy generation, conversion, and optimal control are dependent on reliable wind speed prediction. They proposed EnsembleLSTM using non-linear learning to predict the wind speed. Long short-term memory (LSTM) neural network neurons and numerous hidden layers are used in the suggested method to help reliable wind speed prediction. Later on, the wind speed forecasting process involves the usage of support vector regression machines and external optimization methods. The proposed method was compared with two cases of Inner Mongolia, China, for 10 min ahead and one hour ahead forecasting. Results proved the efficacy of the proposed method.

Wind energy has economical and environmental advantages, so it has garnered much attention of policy makers and the research community. However, uncertainty in wind power generation is unacceptable and a challenging task to overcome. A deep-learning-based ensemble solution was proposed by the authors of [53] for probabilistic wind power forecasting. In order to deal with uncertainties, this study proposed an enhanced point forecasting technique based on wavelet processing and convolutional neural networks (CNN) for wind energy forecasting. The non-linearity of each frequency also increased predicting accuracy. Results show that the suggested technique outperforms its competitors.

Wind power generation is economical and environment friendly; however, irregular wind power generation leads to peak load pressure and frequency regulation issues at grid stations. Wind power forecasting can make its supply steady. Therefore, for accurate wind power prediction, the authors of [54] suggested a long short-term memory improved forget gate network model. Results demonstrate significant improvement in prediction accuracy and speed up in the convergence process.

The authors of [55] presented the SSA-EMD-CNNSVM model, which uses singular spectrum analysis (SSA) for noise reduction and trend extraction from actual data. Time empirical mode decomposition (EMD), as the name suggests, is used to separate time series of wind speed into sublayers. Following that, a convolutional support vector machine (CSVM) is used to forecast wind speed. The proposed prediction model was compared with other wind speed prediction models, including the CNNSVM, EMD-BP, SVM, and EMD-Elman models. Results demonstrated the superiority of the proposed model.

Table 2. Summary of literature on wind energy forecasting.

Paper	Energy Source	RES Forecasting	Implementation Strategy	Objective(s)	Dataset Type	Performance/Result
[49]	Wind energy	Yes	WTD–RNN–ANFIS based ensemble model	Reliable wind speed forecasting	Historical	MAE ANN = 0.929, MAE SVM = 0.963, RMSE ANN = 1.293, RMSE SVM = 1.349
[50]	Wind energy	Yes	Wind speed forecasting through ANN	Forecasting of wind speed	Historical	RMSE = 0.675, MAE = 0.536
[51]	Wind energy	Yes	Wind speed prediction through wavelet transformation and recurrent neural networks	Wind speed is predicted	Historical	Wind speed series 1, MAPE ARIMA = 7.17, MAE ARIMS = 0.93, RMSE ARIMA = 1.21
[52]	Wind energy	Yes	Ensemble LSTM using non-linear learning to predict the wind energy	Wind speed forecasting	Historical	MAE EnsemL-STM = 0.574, RMSE EnsemL-STM = 0.755, MAPE EnsemLSTM = 5.41
[53]	Wind energy	Yes	Deep learning based ensemble approach for probabilistic wind power forecasting	Wind power forecasting	Historical	Performance improvemney by 48.42%, 45.02%, and 45.10% as compared to three benchmarks
[54]	Wind energy	Yes	Long short-term memory enhanced forget gate network model for reliable wind power prediction	Wind power forecasting	Historical	18.3% rise in accuracy
[55]	Wind energy	Yes	Convolutional support vector machine (CNNSVM)	Wind speed forecasting	Historical	RMSE = 39.25%, MAE = 39.21% , MAPE = 42.85%
[56]	Wind energy	Yes	SVM-based prediction and MLP are used	Wind speed forecasting	Historical	MSE SVM = 0.78%, MSE MLP = 0.9%
[57]	Wind energy	Yes	Wavelet transform	Wind speed forecasting	Historical	MAPE increased from 14.79% to 22.64%
[58]	Wind energy	Yes	ARIMA and ANN	Wind speed forecasting	Historical	MAPE = 6.97%

The support vector machine method is used in [56] for wind prediction. The regression analysis is performed after mapping the time series data for any variable into a higher dimensional space (e.g., Hilbert space), according to the procedure. In addition, the results of SVM-based prediction and multilayer perceptron (MLP) models were compared. MSE and RMSE were the performance measures used in [56]. Since the SVM had a mean square error of 0.78% compared to the MLP of 0.9%, it was found that the SVM performed better than the MLP.

Using a wavelet transform, the authors of [57] deconstructed a wind series. The selection of input parameters for the SVM was supported by the genetic algorithm approach. The input must be improved to select the best forecast candidates. According to the results, persistence increased MAPE from 14.79% to 22.64%, while WT-SVM-GA did not. The NNWT technique, in which the prediction for the next three hours is made using the historical data of the last twelve hours, was compared with ARIMA (1,2,1) and NN by [58]. The MAPE value was found to be 6.97% using the NNWT technique.

The authors of [59] discussed the benefits and limitations of solar energy in detail. A brief description along with benefits and limitations of solar energy and wind energy are briefly described in Table 3.

Table 3. Short description and merits/limitations of solar energy and wind energy.

Type	Short Description	Benefits	Limitations
Solar energy	Energy of sunrays is transformed into electricity with the help of PV cells	• Inexhaustible energy source • Pollution-free energy • Directly exploitable and widely available • Renewable energy • Being labor intensive industry, improves job opportunities • Reduces electricity cost	• Huge initial installation cost • Dependent over climate and weather • Intermittent in nature • Performance issues of batteries and inverters, etc. • Shortage of skilled manpower
Wind energy	Produced by kinetic energy caused by flow of air on the surface of Earth	• Clean energy • Carbon free generation • Minimizes dependence over fossil fuels	• Intermittent in nature • Dependent over air dynamics such air flow, pressure, direction, humidity etc.

3. Motivation and the Problem Statement

3.1. Motivation

Many researchers have considered the energy optimization and environmental implications caused by excessive brown energy usage. A few have proposed using both energy sources, while some have recommended using RESs only. Recent research articles cited at [40–58] have focused on solar energy and wind energy forecasting, respectively. The literature review motivated us towards accurate and reliable RES forecasting because it is very important for effective and efficient grid management. Moreover, it is helpful in minimizing user energy cost, reducing carbon emissions, overcoming energy imbalances, decreasing dependence upon electricity reserves, and better scheduling of different energy sources.

3.2. Problem Statement

Accurate renewable energy forecasting is important for the minimization of user energy cost and carbon emission. ANN-based renewable energy forecasting techniques lose prediction accuracy due to uncertainty of input data and random determination of initial weights between different layers of the ANN. Therefore, the objective of this work is to develop "a harmony search algorithm optimized artificial neural network model for reliable and accurate solar and wind energy forecasting".

4. Proposed System Model

4.1. Artificial Neural Network

A collection of linked nodes referred to as artificial neurons makes up an artificial neural network, which functions similarly to the human nervous system. Artificial neural nodes are connected to each other through edges, and the edges carry signal or output to the next artificial neuron where some logical action is performed [37]. This signal is possibly sent to the next neuron for further processing or final output. ANN may have a single hidden layer or multiple hidden layers, and each layer may have a different number of nodes. The quantity of hidden layers, learning rate, and iterations are the main governing factors of an ANN. The activation functions that have an impact on ANN processing include softmax, sigmoid, gaussian error linear units, exponential linear units and swish, among others. Figure 3 depicts the basic architecture of ANN.

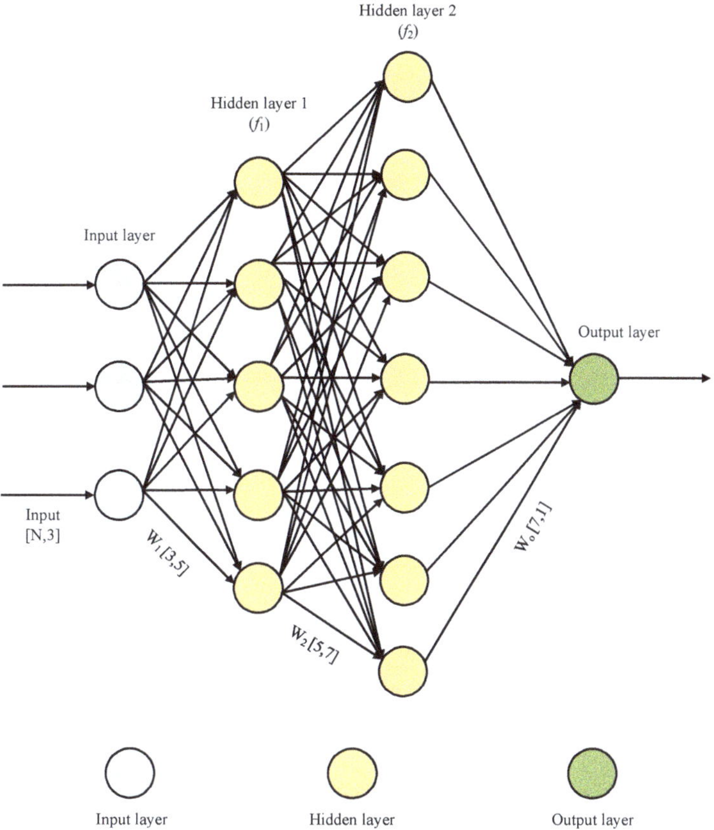

Figure 3. Basic architecture of artificial neural network.

4.2. Harmony Search Algorithm

Harmony search algorithm (HSA) is a nature inspired evolutionary meta-heuristic optimization algorithm [39], proposed by Zong Wo Geem et al. [60] in 2001. Harmony improvisation is the term used to describe the process by which artists apply this algorithm to enhance their harmony. Every time a musical band where each performer plays a different instrument completes the harmony improvisation process. Each member of the musical ensemble serves as a decision variable in this situation, and each musical instrument has a different pitch. Successful musical harmony is achieved throughout the process of improvising harmony, and this successful harmony is then updated in the harmony memory (HM). HM contains the top solution vectors. The HSA process is illustrated in Figure 4, and Table 4 displays manually chosen HSA control parameters in the context of Equations (1)–(5). According to Table 4, *HMS, NI, HMCR, PAR, PAP$_{max}$*, and *PAI$_{max}$* stand for harmony memory size, number of iterations, harmony memory consideration rate, pitch adjustment rate, maximum pitch adjustment proportion (used for continuous variables), and maximum pitch adjustment index (used for discrete variables), respectively.

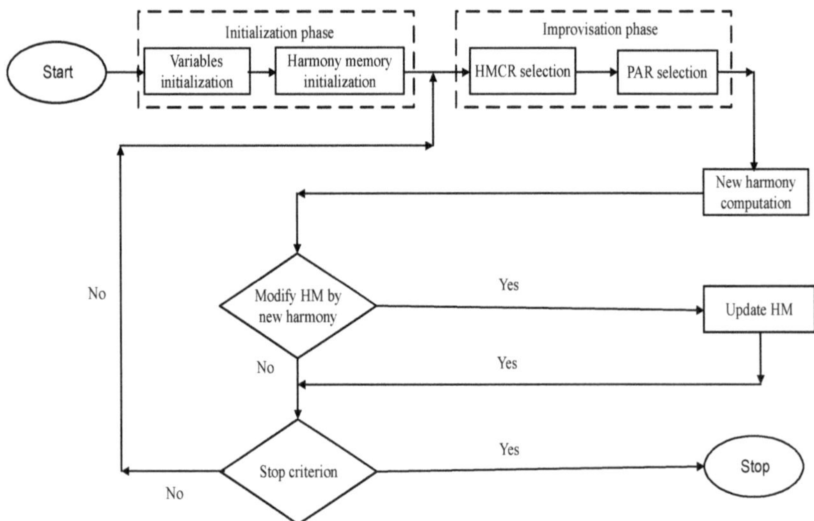

Figure 4. Procedural steps of harmony search algorithm.

Table 4. HSA control parameters.

Control Parameter	Value
HMS	3
NI	10
$HMCR$	0.9
PAR	0.5
PAP_{max}	0.25
PAI_{max}	2

A brief description of each procedural step of HSA is given in the following.

1. Variables initialization
    Values limit of different variables used in HSA are defined during this step.

    - Upper and lower limit of variables

    $$x_i^L \leq x_i \leq x_i^U \tag{1}$$

    - Harmony memory size (HMS)

    $$10 \leq HMS \leq 3 \tag{2}$$

    - Harmony memory consideration rate ($HMCR$)

    $$0.0 \leq HMCR \leq 1.0 \tag{3}$$

    - Pitch adjustment rate (PAR)

    $$0.0 \leq PAR \leq 1.0 \tag{4}$$

    - Maximum number of iterations (NI), i.e., stopping criteria

    $$0 \leq NI \leq 10 \tag{5}$$

2. HM initialization

   The harmony memory matrix is randomly initialized during the HM initialization step using following equation.

$$x_{(i,j)} = l_j + rand().U_j - l_j \tag{6}$$

   The *j*th element of the initial harmony memory in Equation (6) is indicated by $x_{(i,j)}$, whereas the rand() function is used to generate random values between zero and one. U_j and l_j in Equation (6) represent upper and lower bounds of variables, respectively.

3. *HMCR* selection

   A random number between zero and one is created during the *HMCR* selection phase of the HSA improvisation phase using the rand() function, as indicated in Equation (7). The value for that specific place is chosen if the randomly generated value is smaller than the *HMCR*; otherwise, a new random number is generated.

$$V_{i,j} = \begin{cases} x(randj) & if\,randb()\,is < HMCR \\ l_j + rand().U_j - l_j & else \end{cases} \tag{7}$$

4. *PAR* selection

   PAR selection is another sub-part of the the HSA improvisation phase in which the memory elements selected during the *HMCR* step are further improved in the *PAR* selection step. The *PAR* selection step works on the basis of Equation (8); *bw* in Equation (8) represents bandwidth which plays an important role in pitch adjustment.

$$V_{i,j} = \begin{cases} V_i^j rand().bwj & if\,rand()\,is < PAR \\ V_i^j & else \end{cases} \tag{8}$$

5. Update HM

   Upon successful completion of the new harmony selection process, it is updated in the HM by replacing the already present worst memory there.

6. Checking the stop criteria

   The harmony improvisation process terminates at the maximum number of iterations (*NI*), i.e., stopping criteria, as shown in Equation (5).

The nature-inspired meta-heuristic algorithm HSA offers perfect stability between the search process's exploration and exploitation stages [61]. Moreover, HSA has been effectively used in a variety of application areas, including image processing, wireless sensor networks, text clustering, and fuzzy clustering [62]. Therefore, it has high precision, faster convergence speed, and less complexity. Consequently, we selected HSA for weight optimization of the ANN edges in this study.

4.3. Proposed System Model

Recent research has focused on the integration of brown energy sources and RESs (solar and wind) because of low operational cost and carbon-free production of RESs. However, reliable forecasting of RESs is an important issue and needs keen attention. We have considered an ANN-based solar and wind energy forecasting model for efficient solar and wind energy production, thereby efficient supply and demand management, less energy cost, and less carbon emissions.

The literature review revealed that machine learning techniques are much better than time series techniques for solar and wind forecasting. The authors of [38] stated that machine learning techniques, such as, artificial neural network suffer from:

1. Loose precision due to high uncertainty of input data like solar and wind energy production;
2. Random determination of initial weights between different layers can affect the performance of an ANN.

In this study, we propose HSA-optimized ANN for solar and wind energy forecasting models, where initial weights between different layers of ANN are determined by a meta-heuristic algorithm named the harmony search algorithm. Our proposed forecasting model has strengths of machine learning (ANN) and meta-heuristic algorithm (HSA). Consequently, our proposed forecasting algorithm has high precision, faster convergence speed, and less complexity. Our proposed forecasting model for reliable solar and wind energy prediction is shown in Figure 5.

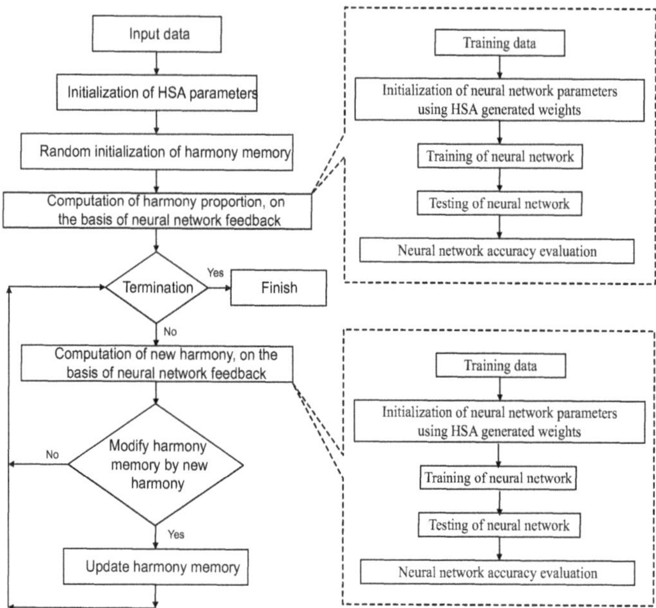

Figure 5. Proposed solar and wind energy forecasting model.

In Figure 5, it is evident that weights at the edges of ANN are being adjusted by the meta-heuristic algorithm HSA which is very famous for optimization problems. Here, the optimal weights assignment to the ANN's edges has a favourable impact on the forecasting of solar and wind energy. The findings mentioned in Section 6 contain supporting data.

5. Simulation Setup and Methodology

5.1. Simulation Setup

In this section, implementation specifications of our suggested model are described in terms of their performance indicators. Our model is tested on a system with a core i7, 16 GB of RAM, and a 4.8 GHz processor. Python and the Anaconda IDE environment are employed. Table 5 describes the simulation settings of our proposed system model.

Overfitting occurs when the model curve becomes too complex and performs too well on training data but fails or degrades performance on test data. The main cause of overfitting is that the model has not learned well from the training data. When underfitting occurs, the model does not perform well even on the training data because the model is too simple and/or the input features are not very expressive. If the number of epochs is too high, the model may overfit, and if the number of epochs is very low, the model may underfit. To avoid overfitting, we used an early stopping criterion in our model. If the model does not perform better after a certain number of epochs, e.g., between 50 and 60 epochs, it is automatically stopped even if the fixed number of epochs is 100. We tested our model for different numbers of epochs, i.e., from 100 to 300, and we found 200 to be the optimal number of epochs where we obtained good results for training and testing of the data.

Table 5. Simulation parameters.

Parameter(s)	Setting
Neural network	Artificial neural network
Number of hidden layers	2
Heuristic algorithm	Harmony search algorithm
Optimizer	Adam
Loss functions	MSE, MAE, MAPE, RMSE
Batch size	4
Number of epoch	200

5.2. Methodology

First, solar and wind energy datasets were downloaded from [63] for the time period of 1 January 2015 to 1 March 2018. A total of 70% of the total data was utilized for training purpose whereas, 30% of the data was used for testing the accuracy of the proposed forecasting model. Pre-processing of solar and wind energy datasets was performed using standard scalar to improve the training of our proposed model by means of data standardization. The following four different forecasting models were developed.

1. ANN-based solar irradiance forecasting model without HSA (using random weights assignment at the edges of ANN layers)
2. ANN-based wind speed forecasting model without HSA (using random weights assignment at the edges of ANN layers)
3. ANN-based solar irradiance forecasting model with HSA (using HSA optimized weights assignment at the edges of ANN layers)
4. ANN-based wind speed forecasting model with HSA (using HSA optimized weights assignment at the edges of ANN layers)

A basic structure of the ANN model with 2 hidden layers was created, and solar and wind datasets downloaded from [63] were loaded. The model was trained with 70% of the data ,and performance was measured by means of error criteria, i.e., mean square error (MSE), mean absolute error (MAE), mean absolute percentage error (MAPE) and root mean square error (RMSE). Later on, we used the harmony search algorithm for assignment of optimized weights at the edges of the ANN instead of random weight assignment to the edges of the ANN. Different weights harmonies were created using the harmony search algorithm, and this process was repeated in a loop until the number of iterations, i.e., 5 in our case. During this process, each time a new harmony (weight) was generated it was fitted to the ANN to obtain a loss value. Loss values of all the harmonies (weights) were compared and finally, the best harmony (weight) among all was selected and applied to the edges of the ANN. Tables 6 and 7 show the simulation results.

Table 6. Performance evaluation of solar irradiance forecasting.

Error Criteria	ANN [38]	GA Optimized ANN [38]	ANN [41]	SVR [41]	ANN (Ours)	HSA Optimized ANN (Proposed)
MSE	0.53	0.29	—	—	0.06354	0.04754
MAE	0.53	0.29	1.146	1.367	0.18520	0.18546
MAPE	7.6%	4.5%	—	—	0.32430%	0.32475%
RMSE	0.62	0.37	1.613	1.994	0.25208	0.21805

Table 7. Performance evaluation of wind speed forecasting.

Error Criteria	ANN [49]	SVM [49]	GRNN [51]	EWT-Elman [51]	ANN (Ours)	HSA Optimized ANN (Proposed)
MSE	—	—	—	—	0.46358	0.30944
MAE	0.929	0.963	0.89	0.66	0.66419	0.47172
MAPE	—	—	6.88%	5.08%	0.13988%	0.12896%
RMSE	1.293	1.349	1.27	0.83	0.68087	0.55627

6. Results and Discussions

6.1. Solar Irradiance Forecasting

As mentioned above, we proposed and developed an HSA-optimized ANN model for solar irradiance forecasting, and we used well-known error value measurement methods for accurate and reliable analysis of the results. Solar irradiance prediction was carried out for one week of all seasons, i.e., autumn, spring, summer, and winter.

In Table 6, the solar irradiance forecasting accuracy of our proposed model is compared with solar irradiance forecasting accuracy of the study at [38]. The authors of [38] developed two models for solar irradiance forecasting: (1) an ANN model was used with random determination of weights at its edges (ANN [38]) and (2) a genetic algorithm (GA) was used for optimized weights assignment at the edges of ANN (GA optimized ANN [38]). The authors of [41] implemented ANN and SVR models for solar irradiance forecasting. Instead, we used; (1) an ANN model with random determination of weights at its edges (ANN (Ours)) and (2) an ANN with HSA-optimized weight assignment at its edges (HSA-optimized ANN (proposed)). Simulation results prove the supremacy of our proposed solar irradiance forecasting model.

The results reported in the Table 6 show that the solar irradiance prediction accuracy is higher with our proposed HSA-optimized ANN model, achieving MSE = 0.04754, MAE = 0.18546, MAPE = 0.32430%, and RMSE = 0.21805. On the other hand, the first competitor ANN at [38] achieved the result values of MSE = 0.53, MAE = 0.53, MAPE = 7.6%, and RMSE = 0.62. Its second competitor, GA-optimized ANN, at [38] achieved results of MSE = 0.29, MAE = 0.29, MAPE = 4.5%, and RMSE = 0.37. Its third competitor ANN at [41] achieved results of MAE = 1.146 and RMSE = 1.613, whereas it fourth competitor SVR at [41] achieved results of MAE = 1.367 and RMSE = 1.994. Results of all the competitors of our proposed HSA-optimized ANN model are far behind. The authors of [41] did not consider MSE and MAPE as evaluation criteria in their study. During solar irradiance forecasting simulations, the computational time of ANN (Ours) was recorded = 60 s, whereas the computational time of our proposed HSA-optimized ANN was recorded = 176 s. The computational time of the HSA-optimized ANN is higher because it involves another meta-heuristic algorithm (HSA) for optimal weight assignment.

Figure 6 shows the result graph of actual solar irradiance values, ANN-predicted solar irradiance values without HSA, and ANN-predicted solar irradiance values with HSA for the whole dataset. The results of the one-week solar irradiance forecast for autumn, spring, summer, and winter seasons are shown in Figure 7a–d, respectively. The green lines in these figures represent the actual solar irradiance values, and the blue lines show the predicted solar irradiance values using ANN without HSA. The red lines, on the other hand, show the predicted solar irradiance values of the ANN model optimized with HSA, i.e., our proposed model.

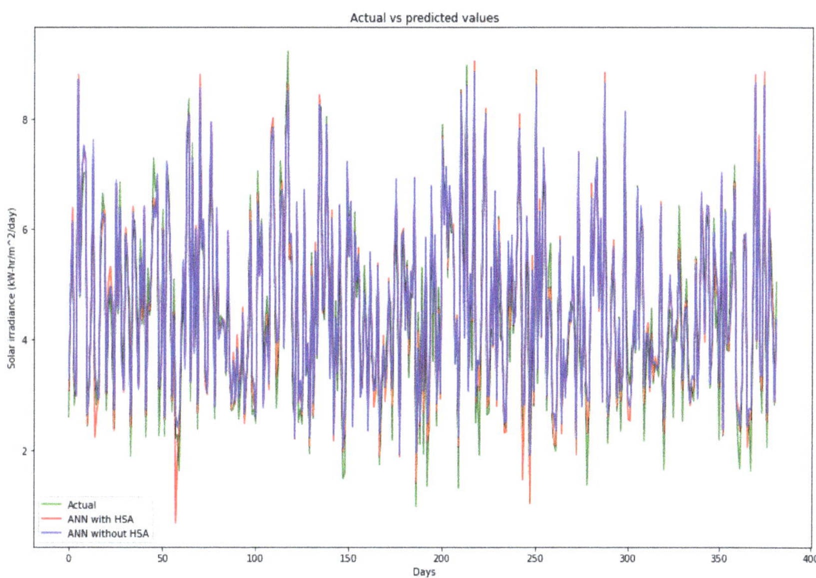

Figure 6. Actual vs. forecasted solar irradiation.

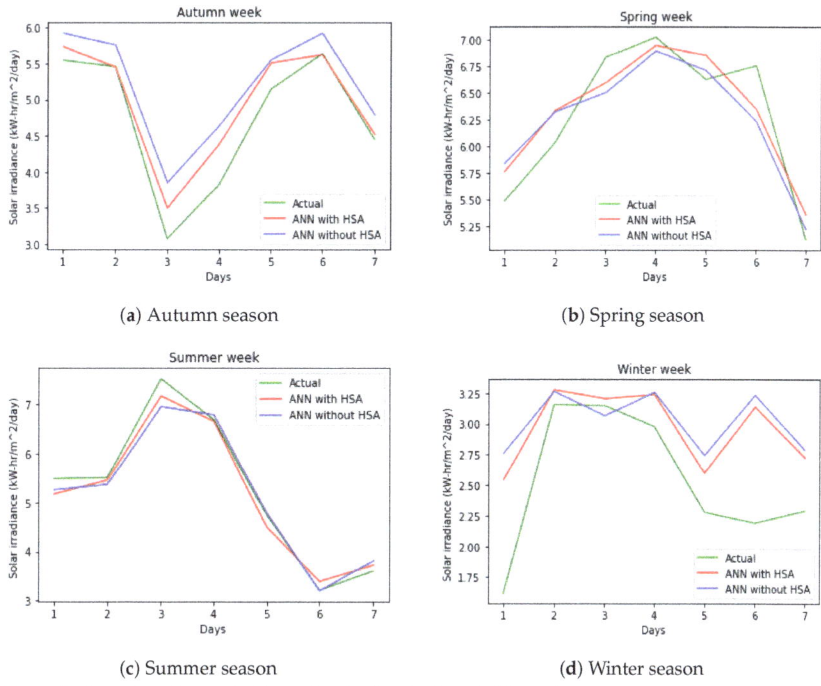

(**a**) Autumn season

(**b**) Spring season

(**c**) Summer season

(**d**) Winter season

Figure 7. Actual vs. forecasted 1-week solar irradiation of different seasons.

In Figure 7a–d it can be seen that sometimes the line of actual values (green line) and the line of predicted values (blue and red lines) cross each other. Ideally, this should not have happened. As mentioned earlier, whole dataset is split into 70% for training and 30% for accuracy testing purpose. We tested our models for a one-week solar irradiance forecast of four seasons. Each forecast model makes its predictions based on its learning process, where it learns about trends in changes. However, actual weather is 100% unpredictable, and we see sudden changes in actual weather parameters such as temperature, solar irradiance, etc. Therefore, the actual and predicted lines overlap in the results. We will deal with this issue in our future work.

6.2. Wind Speed Forecasting

Wind energy is less reliable compared to solar energy [22]. Therefore, results accuracy in the case of wind speed forecasting is less compared to solar energy forecasting. Wind speed prediction was carried out for one week of all seasons, i.e., autumn, spring, summer and winter. In Table 7, the wind speed forecasting accuracy of our proposed model is compared with wind speed forecasting accuracy of the studies at [40,51]. The authors of [49] discussed different wind speed forecasting models, and we have selected two representative models which are ANN and SVM. The authors of [51] discussed two models for wind speed forecasting: (1) a GRNN model and (2) an EWT-Elman model. We used: (1) an ANN model with random determination of weights at its edges (ANN (Ours)) and (2) an ANN with HSA-optimized weight assignment at the edges of the ANN (HSA-optimized ANN (proposed)). Simulation results prove the supremacy of our proposed wind speed forecasting model.

The results reported in the Table 7 show that the wind speed prediction accuracy is higher with our proposed HSA-optimized ANN model, achieving MSE = 0.30944, MAE = 0.47172, MAPE = 0.12896%, and RMSE = 0.55627. On the other hand, Its first competitor ANN at [49] achieved the values of MAE = 0.929 and RMSE = 1.29, whereas it second competitor SVM at [49] achieved the values of MAE = 0.963 and RMSE = 1.349. Its third competitor, GRNN, at [51] achieved the values MAE = 0.89, MAPE = 6.88%, and RMSE = 1.27, and the fourth competitor, EWT-Elman, at [51] achieved the values of MAE = 0.66, MAPE = 5.08%, and RMSE = 0.83. Results of all competitors of our proposed HSA-optimized ANN model were far behind. MSE was not considered by the authors of [49,51]. The authors pf [49] also did not consider MAPE in their study. However, we considered MSE as well in our proposed model, and its result values are shown in Table 7. During wind speed forecasting simulations, the computational time of ANN (Ours) was recorded = 60 s, whereas the computational time of our proposed HSA-optimized ANN was recorded = 323 s. Computational time of HSA-optimized ANN was higher because it involved another meta-heuristic algorithm (HSA) for optimal weight assignment.

Figure 8 shows the resuls graph of actual wind speed values, ANN predicted wind speed values without HSA, and ANN predicted wind speed values with HSA for the whole dataset. The results of the one-week wind speed forecast for autumn, spring, summer, and winter seasons are shown in Figure 9a–d, respectively. The green lines in these figures represent the actual wind speed values, and the blue lines show the predicted wind speed values using ANN without HSA. The red lines, on the other hand, show the predicted wind speed values of the ANN model optimized with HSA, i.e., our proposed model.

In Figure 9a–d it can be seen that sometimes the line of actual values (green line) and the line of predicted values (blue and red lines) cross each other. Ideally, this should not have happened. As mentioned earlier, whole dataset is split into 70% for training and 30% for accuracy testing purpose. We tested our models for a one-week wind speed forecast of four seasons. Each forecast model makes its predictions based on its learning process, where it learns about trends in changes. However, actual weather is 100% unpredictable and we see sudden changes in actual weather parameters such as wind speed, wind direction, temperature, wind pressure, etc. Therefore, the actual and predicted lines sometimes overlap each other in the results. However, we will try our best to deal with this problem in our future work.

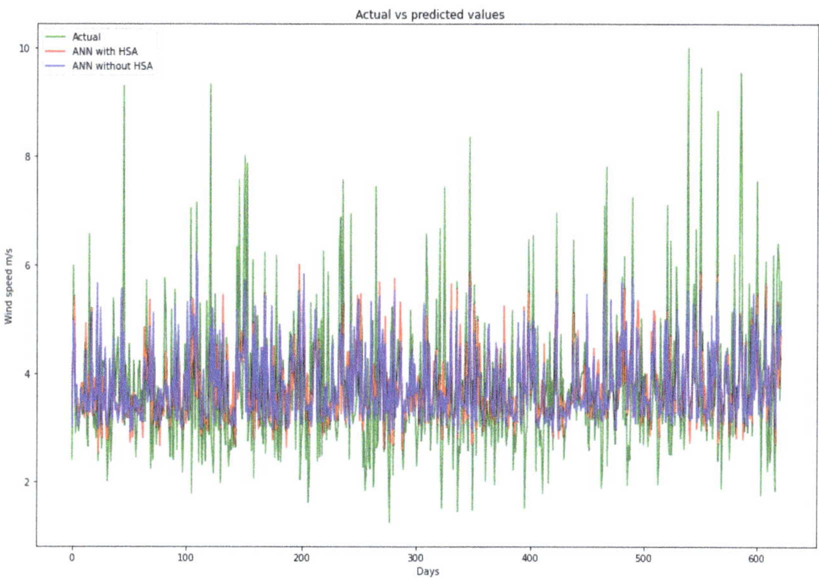

Figure 8. Actual vs. forecasted wind speed.

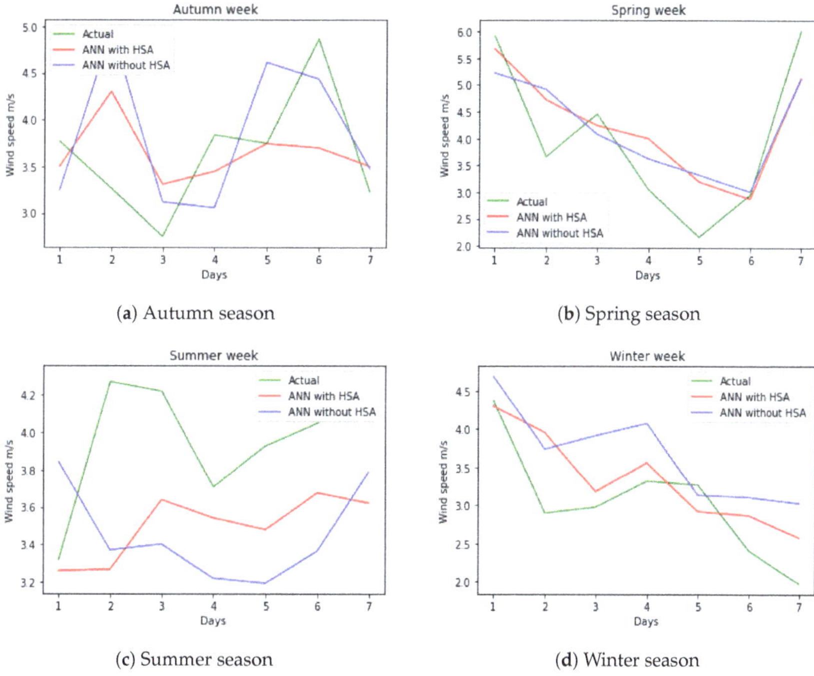

(**a**) Autumn season

(**b**) Spring season

(**c**) Summer season

(**d**) Winter season

Figure 9. Actual vs. forecasted 1-week wind speed of different seasons.

7. Conclusions and Future Work

Fossil fuel generated electric power leads to higher energy cost and environmental pollution. To deal with higher electricity costs and environmental implications, solar and wind energy are abundantly available renewable energy sources being used for green environment and low cost energy. As solar and wind energy are highly intermittent in nature, in this study we have proposed HSA-optimized ANN solar irradiance and wind speed forecasting models. We have assigned HSA-optimized weights at the edges of ANN layers, and simulation results prove the accuracy of our proposed forecasting models. Our proposed HSA-optimized ANN model for solar irradiation forecast achieved the values of MSE = 0.04754, MAE = 0.18546, MAPE = 0.32430(%), and RMSE = 0.21805, whereas our proposed HSA-optimized ANN model for wind speed prediction achieved the values of MSE = 0.30944, MAE = 0.47172, MAPE = 0.12896(%), and RMSE = 0.55627. Result accuracy of our proposed wind speed forecasting model is less compared to our proposed solar irradiance forecasting model. Accuracy enhancement of our proposed wind speed forecasting model is our future work. Furthermore, identifying the causes of the volatile character of wind speed and solar irradiance is also essential since doing so enables the adaptation or even mitigation of the intermittent nature of wind and solar energy.

Author Contributions: Methodology, S.M.M., T.M. and S.A.M.; Software, S.M.M.; Validation, T.M. and S.A.M.; Formal analysis, S.M.M. and T.M.; Investigation, T.M. and S.A.M.; Data curation, S.M.M.; Writing—original draft, S.M.M.; Writing—review & editing, T.M.; Visualization, S.M.M.; Supervision, T.M. and S.A.M.; Project administration, T.M. and S.A.M.; Funding acquisition, S.M.M. All authors have read and agreed to the published version of the manuscript.

Funding: This research received no external funding.

Data Availability Statement: Not applicable.

Conflicts of Interest: The authors declare no known conflict of interest.

References

1. Gao, P.X.; Curtis, A.R.; Wang, B.; Keshav, S. It's not easy being green. In *ACM SIGCOMM Computer Communication Review*; Association for Computing Machinery: New York, NY, USA, 2018; pp. 211–222.
2. Hepburn, C. Regulation by prices, quantities, or both: A review of instrument choice. *Oxf. Rev. Econ. Policy* **2006**, *22*, 226–247. [CrossRef]
3. Aslam, S.; Javaid, N.; Asif, M.; Iqbal, U.; Iqbal, Z.; Sarwar, M.A. A mixed integer linear programming based optimal home energy management scheme considering grid-connected microgrids. In Proceedings of the 2018 14th International Wireless Communications & Mobile Computing Conference (IWCMC), Limassol, Cyprus, 25–29 June 2018; pp. 993–998.
4. Aslam, S. An Optimal Home Energy Management Scheme Considering Grid Connected Microgrids with Day–Ahead Weather Forecasting Using Artificial Neural Network. Master's Thesis, COMSATS University Islamabad, Islamabad, Pakistan, 2018.
5. Chomać-Pierzecka, E.; Kokiel, A.; Rogozińska-Mitrut, J.; Sobczak, A.; Soboń, D.; Stasiak, J. Analysis and Evaluation of the Photovoltaic Market in Poland and the Baltic States. *Energies* **2022**, *15*, 669. [CrossRef]
6. Chomać–Pierzecka, E.; Sobczak, A.; Soboń, D. Wind Energy Market in Poland in the Back-ground of the Baltic Sea Bordering Countries in the Era of the COVID-19 Pandemic. *Energies* **2022**, *15*, 2470. [CrossRef]
7. Aurangzeb, K.; Aslam, S.; Mohsin, S.M.; Alhussein, M. A fair pricing mechanism in smart grids for low energy consumption users. *IEEE Access* **2021**, *9*, 22035–22044. [CrossRef]
8. Iqbal, Z.; Javaid, N.; Mohsin, S.M.; Akber, S.M.A.; Afzal, M.K.; Ishmanov, F. Performance analysis of hybridization of heuristic techniques for residential load scheduling. *Energies* **2018**, *11*, 2861. [CrossRef]
9. Bosker, B. Google Buys 20 Years Worth of Wind Energy to Power Data Centers. 2020. Available online: http://www.huffingtonpost.com (accessed on 5 September 2021).
10. Deng, W.; Liu, F.; Jin, H.; Li, B.; Li, D. Harnessing renewable energy in cloud datacenters: Opportunities and challenges. *IEEE Netw.* **2014**, *28*, 48–55. [CrossRef]
11. Gu, C.; Huang, H.; Jia, X. Green scheduling for cloud data centers using ESDs to store renewable energy. In Proceedings of the 2016 IEEE International Conference on Communications (ICC), Kuala Lumpur, Malaysia, 22–27 May 2016; pp. 1–7.
12. Grange, L.; Da Costa, G.; Stolf, P. Green IT scheduling for data center powered with renewable energy. *Future Gener. Comput. Syst.* **2018**, *86*, 99–120. [CrossRef]
13. Zhang, Y.; Wang, Y.; Wang, X. Greenware: Greening cloud-scale data centers to maximize the use of renewable energy. In *Middleware 2011, Proceedings of the ACM/IFIP/USENIX International Conference on Distributed Systems Platforms and Open Distributed Processing, Lisbon, Portugal, 12–16 December 2011*; Springer: Berlin/Heidelberg, Germany, 2011; pp. 143–164.
14. Gu, C.; Liu, C.; Zhang, J.; Huang, H.; Jia, X. Green scheduling for cloud data centers using renewable resources. In Proceedings of the 2015 IEEE Conference on Computer Communications Workshops (INFOCOM WKSHPS), Hong Kong, China, 26 April–1 May 2015; pp. 354–359.
15. Thi, M.T.; Pierson, J.M.; Da Costa, G.; Stolf, P.; Nicod, J.M.; Rostirolla, G.; Haddad, M. Negotiation game for joint IT and energy management in green datacenters. *Future Gener. Comput. Syst.* **2020**, *110*, 1116–1138. [CrossRef]

16. He, H.; Shen, H.; Liang, D. Cost minimizing online algorithm for internet green data centers on multi-source energy. *Concurr. Comput. Pract. Exp.* **2019**, *31*, e5044. [CrossRef]
17. Aurangzeb, K.; Aslam, S.; Haider, S.I.; Mohsin, S.M.; Islam, S.U.; Khattak, H.A.; Shah, S. Energy forecasting using multiheaded convolutional neural networks in efficient renewable energy resources equipped with energy storage system. *Trans. Emerg. Telecommun. Technol.* **2022**, *33*, e3837. [CrossRef]
18. Li, W.; Yang, T.; Delicato, F.C.; Pires, P.F.; Tari, Z.; Khan, S.U.; Zomaya, A.Y. On enabling sustainable edge computing with renewable energy resources. *IEEE Commun. Mag.* **2018**, *56*, 94–101. [CrossRef]
19. Aslam, S.; Javaid, N.; Khan, F.A.; Alamri, A.; Almogren, A.; Abdul, W. Towards efficient energy management and power trading in a residential area via integrating a grid-connected microgrid. *Sustainability* **2018**, *10*, 1245. [CrossRef]
20. Measurement and Instrumentation Data Center. Available online: http://www.nrel.gov/midc/ (accessed on 5 September 2022).
21. Li, C.; Zhang, W.; Cho, C.B.; Li, T. Solarcore: Solar energy driven multi-core architecture power management. In Proceedings of the 2011 IEEE 17th International Symposium on High Performance Computer Architecture, San Antonio, TX, USA, 12–16 February 2011.
22. Asiaban, S.; Kayedpour, N.; Samani, A.E.; Bozalakov, D.; De Kooning, J.D.; Crevecoeur, G.; Vandevelde, L. Wind and solar intermittency and the associated integration challenges: A comprehensive review including the status in the Belgian power system. *Energies* **2021**, *14*, 2630. [CrossRef]
23. Laganà, D.; Mastroianni, C.; Meo, M.; Renga, D. Reducing the operational cost of cloud data centers through renewable energy. *Algorithms* **2018**, *11*, 145. [CrossRef]
24. Agyekum, E.B.; PraveenKumar, S.; Alwan, N.T.; Velkin, V.I.; Shcheklein, S.E. Effect of dual surface cooling of solar photovoltaic panel on the efficiency of the module: Experimental investigation. *Heliyon* **2021**, *7*, e07920. [CrossRef] [PubMed]
25. Agyekum, E.B.; PraveenKumar, S.; Alwan, N.T.; Velkin, V.I.; Shcheklein, S.E.; Yaqoob, S.J. Experimental investigation of the effect of a combination of active and passive cooling mechanism on the thermal characteristics and efficiency of solar PV module. *Inventions* **2021**, *6*, 63. [CrossRef]
26. PraveenKumar, S.; Agyekum, E.B.; Velkin, V.I.; Yaqoob, S.J.; Adebayo, T.S. Thermal management of solar photovoltaic module to enhance output performance: An experimental passive cooling approach using discontinuous aluminum heat sink. *Int. J. Renew. Energy Res. (IJRER)* **2021**, *11*, 1700–1712.
27. PraveenKumar, S.; Agyekum, E.B.; Qasim, M.A.; Alwan, N.T.; Velkin, V.I.; Shcheklein, S.E. Experimental assessment of thermo-electric cooling on the efficiency of PV module. *Int. J. Renew. Energy Res. (IJRER)* **2022**, *12*, 1670–1681.
28. Heyd, E. America's Data Centers Consuming Massive and Growing Amounts of Electricity. 2014. Available online: https://www.nrdc.org/media/2014/140826 (accessed on 9 May 2022).
29. Google, DeepMind Target Machine Learning at Boosting Wind Energy Value. Available online: https://www.evwind.es/2019/02/28/google-deepmind-target-machine-learning-at-boosting-wind-energy-value/66291 (accessed on 7 November 2022).
30. Jebli, I.; Belouadha, F.Z.; Kabbaj, M.I.; Tilioua, A. Prediction of solar energy guided by pearson correlation using machine learning. *Energy* **2021**, *224*, 120109. [CrossRef]
31. Narvaez, G.; Giraldo, L.F.; Bressan, M.; Pantoja, A. Machine learning for site–adaptation and solar radiation forecasting. *Renew. Energy* **2021**, *167*, 333–342. [CrossRef]
32. Using Machine Learning to Make Wind Energy More Predictable. Available online: https://www.azocleantech.com/article.aspx?ArticleID=1518 (accessed on 7 November 2022).
33. Li, J.; Herdem, M.S.; Nathwani, J.; Wen, J.Z. Methods and Applications for Artificial Intelligence, Big Data, Internet-of-Things, and Blockchain in Smart Energy Management. *Energy AI* **2022**, *11*, 100208. [CrossRef]
34. Guha, S.; Hamann, H.F.; Klein, L.I.; Rodriguez, S.A.B. System and Method for Managing and Forecasting Power from Renewable Energy Sources. U.S. Patent Application No. 14/141,711, 2 July 2015.
35. Bing, J.M. Method and System for Predicting Solar Energy Production. U.S. Patent No. 2005/0039787 A1, 24 February 2005.
36. Westergaard, C.H. Method and System for Forecasting Wind Energy. U.S. Patent No. WO 2011/124226, 13 October 2011.
37. Aslam, S.; Herodotou, H.; Mohsin, S.M.; Javaid, N.; Ashraf, N.; Aslam, S. A survey on deep learning methods for power load and renewable energy forecasting in smart microgrids. *Renew. Sustain. Energy Rev.* **2021**, *144*, 110992. [CrossRef]
38. Hamidreza, J.; Ahmadian, A.; Golkar, M.A.; Elkamel, A.; Almansoori, A. Solar Irradiance Forecasting Based on the Combination of Radial Basis Function Artificial Neural Network and Genetic Algorithm. In Proceedings of the 6th European Conference on Renewable Energy Systems, Istanbul, Turkey, 25–27 June 2018.
39. Mohsin, S.M.; Javaid, N.; Madani, S.A.; Abbas, S.K.; Akber, S.M.A.; Khan, Z.A. Appliance scheduling in smart homes with harmony search algorithm for different operation time intervals. In Proceedings of the 2018 32nd International Conference on Advanced Information Networking and Applications Workshops (WAINA), Krakow, Poland, 16–18 May 2018; pp. 51–60.
40. El–Baz, W.; Tzscheutschler, P.; Wagner, U. Day-ahead probabilistic PV generation forecast for buildings energy management systems. *Sol. Energy* **2018**, *171*, 478–490. [CrossRef]
41. Ghimire, S.; Deo, R.C.; Downs, N.J.; Raj, N. Global solar radiation prediction by ANN integrated with European Centre for medium range weather forecast fields in solar rich cities of Queensland Australia. *J. Clean. Prod.* **2019**, *216*, 288–310. [CrossRef]
42. Alfadda, A.; Rahman, S.; Pipattanasomporn, M. Solar irradiance forecast using aerosols measurements: A data driven approach. *Sol. Energy* **2018**, *170*, 924–939. [CrossRef]

43. Raza, M.Q.; Mithulananthan, N.; Summerfield, A. Solar output power forecast using an ensemble framework with neural predictors and Bayesian adaptive combination. *Sol. Energy* **2018**, *166*, 226–241. [CrossRef]
44. Jeon, B.K.; Kim, E.J. Next-day prediction of hourly solar irradiance using local weather forecasts and LSTM trained with non-local data. *Energies* **2020**, *13*, 5258. [CrossRef]
45. Obiora, C.N.; Hasan, A.N.; Ali, A.; Alajarmeh, N. Forecasting hourly solar radiation using artificial intelligence techniques. *IEEE Can. J. Electr. Comput. Eng.* **2021**, *44*, 497–508. [CrossRef]
46. Guariso, G.; Nunnari, G.; Sangiorgio, M. Multi-step solar irradiance forecasting and domain adaptation of deep neural networks. *Energies* **2020**, *13*, 3987. [CrossRef]
47. Chu, T.P.; Jhou, J.H.; Leu, G.Y. Image-based solar irradiance forecasting using recurrent neural networks. In Proceedings of the 2020 International Conference on System Science and Engineering (ICSSE), Kagawa, Japan, 31 August–3 September 2020; pp. 1–4.
48. Aslam, M.; Lee, J.M.; Kim, H.S.; Lee, S.J.; Hong, S. Deep learning models for long-term solar radiation forecasting considering microgrid installation: A comparative study. *Energies* **2020**, *13*, 147. [CrossRef]
49. Cheng, L.; Zang, H.; Ding, T.; Sun, R.; Wang, M.; Wei, Z.; Sun, G. Ensemble recurrent neural network based probabilistic wind speed forecasting approach. *Energies* **2018**, *11*, 1958. [CrossRef]
50. Filik, U.B.; Filik, T. Wind speed prediction using artificial neural networks based on multi–ple local measurements in Eskisehir. *Energy Procedia* **2017**, *107*, 264–269. [CrossRef]
51. Liu, H.; Mi, X.W.; Li, Y.F. Wind speed forecasting method based on deep learning strategy using empirical wavelet transform, long short term memory neural network and Elman neural network. *Energy Convers. Manag.* **2018**, *156*, 498–514. [CrossRef]
52. Chen, J.; Zeng, G.Q.; Zhou, W.; Du, W.; Lu, K.D. Wind speed forecasting using nonline-ar-learning ensemble of deep learning time series prediction and extremal optimization. *Energy Convers. Manag.* **2018**, *165*, 681–695. [CrossRef]
53. Wang, H.Z.; Li, G.Q.; Wang, G.B.; Peng, J.C.; Jiang, H.; Liu, Y.T. Deep learning based ensemble approach for probabilistic wind power forecasting. *Appl. Energy* **2017**, *188*, 56–70. [CrossRef]
54. Yu, R.; Gao, J.; Yu, M.; Lu, W.; Xu, T.; Zhao, M.; Zhang, J.; Zhang, R.; Zhang, Z. LSTM-EFG for wind power forecasting based on sequential correlation features. *Future Gener. Comput. Syst.* **2019**, *93*, 33–42. [CrossRef]
55. Mi, X.; Liu, H.; Li, Y. Wind speed prediction model using singular spectrum analysis, empirical mode decomposition and convolutional support vector machine. *Energy Convers. Manag.* **2019**, *180*, 196–205. [CrossRef]
56. Mohandes, M.A.; Halawani, T.O.; Rehman, S.; Hussain, A.A. Support vector machines for wind speed prediction. *Renew. Energy* **2004**, *29*, 939–947. [CrossRef]
57. Zhang, C.; Wei, H.; Zhao, J.; Liu, T.; Zhu, T.; Zhang, K. Short-term wind speed forecasting using empirical mode decomposition and feature selection. *Renew. Energy* **2016**, *96*, 727–737. [CrossRef]
58. Catalão, J.D.S.; Pousinho, H.M.I.; Mendes, V.M.F. Short-term wind power forecasting in Portugal by neural networks and wavelet transform. *Renew. Energy* **2011**, *36*, 1245–1251. [CrossRef]
59. Kabir, E.; Kumar, P.; Kumar, S.; Adelodun, A.A.; Kim, K.H. Solar energy: Potential and future prospects. *Renew. Sustain. Energy Rev.* **2018**, *82*, 894–900. [CrossRef]
60. Geem, Z.W.; Kim, J.H.; Loganathan, G.V. A new heuristic optimization algorithm: Harmony search. *Simulation* **2001**, *76*, 60–68. [CrossRef]
61. Dubey, M.; Kumar, V.; Kaur, M.; Dao, T.P. A systematic review on harmony search algorithm: Theory, literature, and applications. *Math. Probl. Eng.* **2021**, *2021*, 5594267. [CrossRef]
62. Abualigah, L.; Diabat, A.; Geem, Z.W. A comprehensive survey of the harmony search algorithm in clustering applications. *Appl. Sci.* **2020**, *10*, 3827. [CrossRef]
63. Power | Data Access Viewer. Available online: https://power.larc.nasa.gov/data-access-viewer/ (accessed on 5 September 2022).

Article

Towards Electric Price and Load Forecasting Using CNN-Based Ensembler in Smart Grid

Shahzad Aslam [1,*], Nasir Ayub [2], Umer Farooq [3], Muhammad Junaid Alvi [4], Fahad R. Albogamy [5], Gul Rukh [6,*], Syed Irtaza Haider [7], Ahmad Taher Azar [8,9] and Rasool Bukhsh [10]

[1] Department of Statistics and Mathematics, Institute of Southern Punjab, Multan 66000, Pakistan
[2] School of Electrical Engineering and Computer Science, National University of Science and Technology, Islamabad 44000, Pakistan; nasir.ayubse@gmail.com
[3] Department of Computer Science, Islamia University Bahawalpur, Bahawalpur 63100, Pakistan; umer.dgk.se@gmail.com
[4] Electrical Engineering Department, NFC Institute of Engineering and Fertilizer Research, Faisalabad 38000, Pakistan; junaidalvi@iefr.edu.pk
[5] Computer Sciences Program, Turabah University College, Taif University, P.O. Box 11099, Taif 21944, Saudi Arabia; f.alhammdani@tu.edu.sa
[6] Department of Electrical Engineering, University of Engineering and Technology, Mardan 23200, Pakistan
[7] College of Computer and Information Sciences, King Saud University, Riyadh 11543, Saudi Arabia; sirtaza@ksu.edu.sa
[8] College of Computer and Information Sciences, Prince Sultan University, Riyadh 11586, Saudi Arabia; aazar@psu.edu.sa
[9] Faculty of Computers and Artificial Intelligence, Benha University, Benha 13518, Egypt
[10] Department of Computer Science, COMSATS University Islamabad (CUI), Islamabad 44000, Pakistan; rasoolbax.rb@gmail.com
* Correspondence: Shahzad.ddp@gmail.com (S.A.); gr@uetmardan.edu.pk (G.R.)

Abstract: Medium-term electricity consumption and load forecasting in smart grids is an attractive topic of study, especially using innovative data analysis approaches for future energy consumption trends. Loss of electricity during generation and use is also a problem to be addressed. Both consumers and utilities can benefit from a predictive study of electricity demand and pricing. In this study, we used a new machine learning approach called AdaBoost to identify key features from an ISO-NE dataset that includes daily consumption data over eight years. Moreover, the DT classifier and RF are widely used to extract the best features from the dataset. Moreover, we predicted the electricity load and price using machine learning techniques including support vector machine (SVM) and deep learning techniques such as a convolutional neural network (CNN). Coronavirus herd immunity optimization (CHIO), a novel optimization approach, was used to modify the hyperparameters to increase efficiency, and it used classifiers to improve the performance of our classifier. By adding additional layers to the CNN and fine-tuning its parameters, the probability of overfitting the classifier was reduced. For method validation, we compared our proposed models with several benchmarks. MAE, MAPE, MSE, RMSE, the f1 score, recall, precision, and accuracy were the measures used for performance evaluation. Moreover, seven different forms of statistical analysis were given to show why our proposed approaches are preferable. The proposed CNN-CHIO and SVM techniques had the lowest MAPE error rates of 6% and 8%, respectively, and the highest accuracy rates of 95% and 92%, respectively.

Keywords: smart grid; electricity price forecasting; energy management; electricity load forecasting; convolutional neural network; corona virus herd immunity optimization

Citation: Aslam, S.; Ayub, N.; Farooq, U.; Alvi, M.J.; Albogamy, F.R.; Rukh, G.; Haider, S.I.; Azar, A.T.; Bukhsh, R. Towards Electric Price and Load Forecasting Using CNN-Based Ensembler in Smart Grid. *Sustainability* **2021**, *13*, 12653. https://doi.org/10.3390/su132212653

Academic Editors: Marc A. Rosen and Alberto-Jesus Perea-Moreno

Received: 24 August 2021
Accepted: 5 November 2021
Published: 16 November 2021

Publisher's Note: MDPI stays neutral with regard to jurisdictional claims in published maps and institutional affiliations.

1. Introduction

Electricity is now a critical component of economic and social growth. It revolves around electricity. Our lives are thought to be stuck if we do not have electricity. Industrial, commercial, and residential electricity use are classified into three groups. According to [1], residential areas consume nearly 65% of total generated electricity. The majority of energy is lost in the conventional grid during the production, delivery, and supply of electricity. To resolve the aforementioned issues, the smart grid (SG) was developed. By incorporating information and communications technology (ICT) into a traditional grid, it can be transformed into an SG as shown in Figure 1.

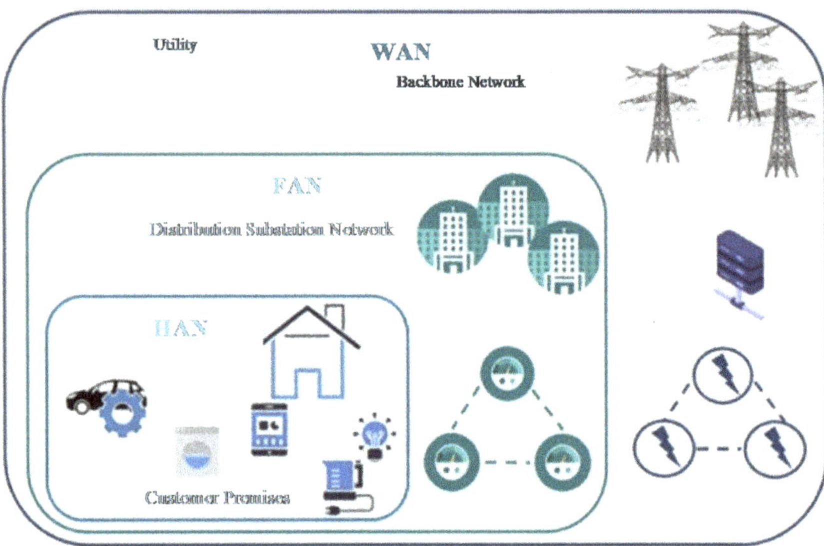

Figure 1. Hierarchical network of smart grid.

1.1. Smart Grid

SG is a smart power system that handles energy generation efficiently. Transmission incorporates emerging technology into a system of energy, allowing users and utilities to communicate in both directions. Power is also a necessity and a valuable asset. Because of the severe energy shortages in the summer, the youth of today are drawn to Singapore. Gadgets in the home are planned with DSM implementing meta-heuristic methods to minimize energy costs and highest point ratios and to find a satisfactory balance between energy costs and customer convenience [2]. By offering effective energy storage, SG assists consumers in achieving efficiency and sustainability. By encouraging customers and providers to exchange information in real-time, the smart meter made it possible to gather sufficient information about future power production. It will ensure that energy output and use are in order. The consumer engages in SG services by shifting demand from maximum to off-hours and conserving resources and saving money on energy [3]. DSM allows customers to monitor their energy usage patterns based on the price set by the utility. The load forecasting benefits market rivals more. Growth, distribution management energy, production planning, performance analysis, and quality control are all things that need to be taken into consideration that depend on upcoming load predictions. Another problem in the energy sector is efficient energy production and use. The primary objective of the consumer and the utility is utility maximization. Energy producers will increase their costs with the aid of reliable load forecasts, while consumers will profit from the low cost of buying electricity. In Singapore, there is no proper energy generation policy. A perfect balance between generated and consumed energy is needed to avoid extra generation. As a result, accurate load forecasting is more critical for market setup management. ISO-NE is also a local distribution system operated by an independent power system, charging the wholesale energy market's activities. Vermont, Massachusetts, Connecticut, and Rhode Island in New England are served by it. The analysis in the study was based on a large collection of ISO NE results. Price is not the only factor that influences the load; temperature, weather conditions, and other factors all affect the electrical load. There is a significant amount of real information [4].

The SG data were carefully scrutinized. The utility takes instructions from the huge quantities of data, which allow it to conduct research and enhance business activity planning and management. To enhance the supply side of SG, a decision-making model was developed. A method for producing is required. The successful choice process leads to a reduction in loss of power, lower energy costs, and lower PAR in the end consumer [5]. Researchers are concentrating on the power scheduling problem in light of these issues. Specific optimization approaches were utilized to address the energy issue [6].

1.2. Problem Statement and Motivation

Each technique in machine learning has advantages and disadvantages. In forecasting the electricity load, however, better performance and accuracy are the primary issues. A large volume of data, on the other hand, makes forecasting more difficult to achieve accuracy. As a result, several strategies have been developed and adapted to fix these problems within the time constraints; however, some challenges remain, such as varying power production and usage to monitor the varying behavior between the power consumption and production patterns [7]. Technique precision and adjusting the hyperparameters for the estimation of electricity demand data [8] and computational difficulty during fuzzy details, such as unnecessary and duplicate features in the data, which increases the learning process calculation time and decreases the reliability of energy load forecasting. A machine learning and deep learning-based model was proposed to solve these challenges. Furthermore, to achieve optimum precision, the hyperparameter values were fine-tuned to use an optimization algorithm. In the function engineering phase, RFE, X-G Boost, and RF were used to remove duplication and clean the files. Finally, the CHIO optimization algorithm was used to determine the optimal hyperparameter values for the convolutional neural network (CNN).

2. Background and Related Work

The term "smart grid" refers to the next generation of power grids, which are power systems in which integrated two-way communication is used to improve energy generation and management. They have the ability in interactions and pervasive computing for stronger control reliability, durability, and protection. Electricity is delivered between producers and customers through a SG. Digital technologies form two-way communication. It is in charge of intelligent appliances. For buyers' homes or buildings, it is used to save electricity and money and to improve trustworthiness, performance, and accountability [9]. The legacy power system is required to be updated by a SG. It automatically regulates, preserves, and maximizes the operation of the interconnected components. It includes everything from conventional main utilities to emerging regeneration distributed generators, as well as the transmission and distribution networks and systems that link them to industrial consumers and/or home users with heating systems, electric cars, and smart appliances [10]. The bidirectional link of energy and knowledge flows in a SG and creates an integrated, globally dispersed transmission network. It combines the advantages of digital communications with the legacy electricity grid to provide real-time information and to allow near-instantaneous supply and demand management [11]. Many of the SG systems are now in use in other industrial applications, such as sensor networks in manufacturing and wireless networks in telecommunications, and are being developed for use in this modern intelligent and integrated model. Advanced materials; sensing and measuring; enhanced interfaces and decision support, protocols, and classes; and integrated communications are the five main fields in which SG networking systems can be classified. Home area networks (HANs), business area networks (BANs), community area networks (NANs), data centers, and substation automation convergence schemes serve as general frameworks for SG networking infrastructures [12]. Using bi-directional information flow to monitor intelligent equipment at the consumer's side, SGs deliver electricity between generators (both conventional power generation and distributed generation sources) and end-users (industrial, private, and residential consumers), saving energy and lowering costs while improving system efficiency and service. Smart metering/monitoring strategies can include real-time energy usage as a review and can correlate to demand to/from utilities with the help of a network infrastructure. Customer power demand data and online market prices can be retrieved from data centers from network service centers in order to optimize electricity supply and delivery based on energy consumption. In a dynamic SG architecture, both utilities and consumers will benefit from the widespread rollout of modern SG elements and the integration of current information and control systems used in the legacy power grid [13]. Through incorporating digital connectivity technologies into SGs, it will also improve the reliability of legacy power generation, transmission, and distribution systems, as well as increase the use of sustainable renewable energy. The capacity for various organizations (e.g., intelligent instruments, dedicated software, systems, control center, etc.) to communicate with a network infrastructure is the foundation of a SG. As a result, the construction of a dependable and widespread network infrastructure is critical to the structure and service of SG communication networks [14]. In this regard, the construction of a secure connectivity system for developing robust real-time data transportation across wide area networks (WANs) to the delivery feeder and consumer level is a strategic necessity in supporting this mechanism [15]. Existing electrical utility WANs are built on a mix of networking technologies, including wired technologies like fiber optics, power line communication (PLC) systems, copper-wire lines, and wireless technologies like GSM/GPRS/WiMAX/WLAN and cognitive radio [16]. They are intended to enable monitoring/controlling technologies such as supervisory control and data aAcquisition (SCADA)/energy management systems (EMS), distribution management systems (DMS), enterprise resource planning (ERP) systems, generation plant automation, distribution feeder automation, and physical protection for facilities in a variety of locations with insufficient bandwidth. Many technologies, such as SG energy metering, have resulted from a decade of wireless sensor network research. However, sensor networks were unable to communicate with the Internet due to a lack of an IP-based network infrastructure, reducing their real-world influence. The LoWPAN and roll working groups were established by the Internet Engineering Task Force (IETF) to define specifications at different layers of the protocol stack to link low-power devices to the Internet. The authors of [17] explain how the scientific community effectively engages in this process by shaping the creation of these working groups' specifications and offering open-source implementations. The new transmission infrastructures can expand into virtually universal data transport networks capable of handling both power distribution applications and large amounts of new data generated by SG applications. These networks should be flexible to meet the current and future collection of functions that define the emerging SG networking technical platform, as well as being highly ubiquitous to support the deployment of last-mile communications (i.e., from a backbone to the terminal customers' locations) [18]. The remainder of this segment covers power line connections, distributed energy storage, smart metering, and tracking and regulation, as well as other important aspects of SG systems.

There are two pieces of the associated work. The literature on energy usage prediction is examined in the first sub-section. A systematic analysis of the literature on power price forecasts is presented in the second sub-section.

2.1. Forecasting Electricity Load

Many techniques for load forecasting have been used in the literature. The training data are difficult to work with as the data are so large and complex. The computing ability of a deep neural network (DNN) allows it to manage big data training. DNN has the capability of accurately forecasting and handling large amounts of data. A broad variety of estimation strategies are covered in the literature. Random forest, naive Bayes, and ARIMA, among other classifier-based techniques, are used for forecasting. Particle swarm optimization (PSO), shallow neural networks (SNN), artificial neural networks (ANN), deep neural networks (DNN), and other artificial intelligence techniques, including shallow neural networks (SNN), artificial neural networks (ANN), deep neural networks (DNN), and others are used. Forecasting the load or price employs a variety of methods. Neural networks have an advantage over other approaches due to automated feature extraction and training methods. In [19], SNN had poor outcomes and an overfitting issue. Regarding price and load forecasting, DNN outperforms SNN. The rectified linear unit (ReLU) and the restricted Boltzmann machine (RBM) were used by the author for forecasting [20]. Data processing and training are handled by RBM, while load forecasting is handled by ReLU. In [21], features were extracted using KPCA, and price forecasting was done using DE-based SVM. Deep auto encoders (DAE) [10] are used to forecast electricity load. DAE is superior in terms of data learning and accuracy. DAE is an unsupervised learning approach that outperforms other methods in terms of achieving high accuracy. In [22], the price was forecasted using the gated recurrent units (GRU) technique. To detect irregular load activity, the Parameter Estimation Method (PEM) was used in [23].

DAE is an unsupervised learning system that achieves high precision while outperforming other methods. It predicts the price using the gated recurrent units (GRU) approach [24]. In [25], the authors used the parameter estimation model (PEM) to identify irregular load activity. For DNN models, the predictability for outcomes is higher. Big data from SG will assist in determining the load and cost trends. It helps utilities create a market, distribution, and inspection routine, which is needed to achieve demand–supply stability. DNN models have a higher degree of predictability. The use of SG's big data would aid in the analysis of load and cost patterns. It assists utilities in developing a production, distribution, and maintenance strategy, both of which are essential for maintaining production equilibrium. Feature engineering is one of the applications of the classifier. The authors of [26] used a multi-layer neural network (MLNN) model to predict energy costs. However, the computational time and rate of neuron failure in this model are extremely high. Price prediction using hybrid structured deep neural networks was addressed by the authors in [27]. The HSDNN, LSTM, and CNN algorithms were merged. The accuracy of this framework was calculated for different benchmark schemes using performance evaluators like MAE and RMSE. The authors established the predictive performance with the suggested RNN and LSTM named GRU in [28]. Benchmark models such as SARIMA, Markov chain, and naive Bayes were also used in the comparison. To limit forecasting flaws in forecasting, [29] introduced a new framework for STLF named "back neural networks" (BPNN). SSA was used to pre-process the information. This model forecasts using CS and SVM. STLF accuracy was improved in this study. The envelope and embedded strategies were used in [30], and the training data were validated using extra tree regression (ETR) and recursive feature elimination (RFE). LSTM-RNN was used to forecast outcomes after splitting results into preparation and trial sets. It addressed the topic of load demand on the service side [31]. It also encouraged customers to shift their loads from maximum to off moments, saving them money. For DR, two types of architectures have been proposed: user- and utility-centric. The data pre-processing steps were addressed by the authors in [32]. The authors proposed a method for selecting and extracting features. Feature identification and filtration are essential in information pre-processing, and they play a crucial role in reliable forecasting. Forecasting accuracy is improved using normalized data. The actual data are inadequate to estimate demand correctly. A meta training methodology was employed to achieve better results, with the post approach being recommended. A battery was used to store energy in this model. It enabled facility users to discharge excess energy during peak hours when prices were high [33]. Additionally, the authors suggested the battery energy storage system (BESS) as a method for achieving effective cost forecasts. It describes the intra-hour term, which is used to evaluate if the cost is rising or decreasing at the time of publishing. The authors suggested a method for displaying the pattern of electricity use in [34]. With Apache Spark's library, the k-means method and the cluster validity indices (CVI) were proposed. The RF algorithm was used for prediction by the authors in [35]. Using hourly data from two separate University of North Florida buildings also determined the function value. This model can also predict monthly and yearly consumption patterns, and RF with support vector regression (SVR) were used to analyze the reliability of multiple aspects such as air, heat, moisture, and time type. SVM and artificial neural networks (ANN) were considered classifiers in this analysis. The authors considered improving forecasting accuracy in their proposed model. To extract and pick features from large datasets, RF and regression tree (CART) were employed. The collection of input determines the efficiency and accuracy of resources. The authors of this study concentrated on input selection and accuracy behavior when the training and testing sets were changed. Deep learning (DL) is a particular sub of machine learning that has advanced significantly in recent decades. The increased computing cost of training large models is one of the significant concerns of artificial neural networks. However, when a deep belief network is effectively formed utilizing a method known as greedy layer-wise pre-training, this problem is solved. The experts began to effectively train complex neural networks with more than one layer not visible. The precision and efficiency of these new designs have increased models that have been used to apply generalization technologies in software engineering applications such as object processing, voice identification, and other relevant disciplines. There are a few functions that are focused on deep processing. Estimating performance improved by 30% due to the sorting method in the literature, and some authors utilized CNN to obtain more precise modeling that outperformed the competition in energy-related areas, such as load and price forecasting, in terms of accuracy. The authors in [36–38] forecasted the short-term electricity load using the feature extraction methods and also the improved version of a general regression neural network and deep learning methods. The authors achieved accuracy in forecasting the electricity load. The authors of [39] recommended and described how to use DL time-series forecasting techniques for predicting electricity consumption. DL includes models such as the restricted Boltzmann machine (DBM), deep recurrent neural networks (RNN), the stack auto-encoder (SAE), CNN, and others. It is a subset of SAE in which the auto encoder is used as a foundation framework. It involves operational inference [40] to prevent overfitting. SAE aims to reduce the complexity of the data set. DBM is made up of layers, each of which contains hidden Boolean units that allow different layers to communicate with one another. This relation, however, does not exist between each layer [41]. The authors of [40–42] increased the accuracy of pricing and load predictions, but they did not account for processing time. Similarly, it solves the problem of load predictions, however, it does not address the issue of overfitting. Additionally, the authors presented the BPNN model for forecasting day-ahead electricity usage in 10; nevertheless, the recommended model's complexity has risen. Furthermore, we also discuss the literature on electricity price forecasting in the next section.

2.2. Forecasting Electricity Price

In [43], the authors proposed a cost forecast approach based on deep learning methods, which included DNN as an evolution of the DNN framework, the CNN framework, hybrid GRU, traditional MLP, and the hybrid LSTM-DNN framework. The suggested structure was then put up against 27 other schemes as a comparison. The suggested deep learning framework was discovered to improve prediction consistency. A single dataset was used to equate the proposed model to all other schemes. For all real-time experiments, a single dataset is insufficient. GCA, KPCA, and SVM were used to create a dual process for the choice of features, filtration of features, and a massive drop in measurements. However, since the authors used a broad dataset that included prices for wood, steam, gas, wind, and oil, the model's computation overhead increased. Furthermore, collecting all of these costs in a single real-time database is complex; these resources' prices cannot be collected in advance. The authors of [44] used DNN templates and the stacked DE noising auto-encoder (SDA). To improve market predictive accuracy, the authors compared various models such as multivariate regression DNN, classical neural networks, and SVM [45]. Additionally, the authors selected features using functional analysis and Bayesian optimization of variance. The creators have suggested the prototype for simultaneously predicting the prices of two markets. Furthermore, prediction can be improved by employing aspects of elimination methods. They reduce the chance of overfitting. The authors, on the other hand, compared the model that had been proposed. This study performed both price and load forecasting. The price prediction accuracy, on the other hand, is insufficient. The authors of [46] looked at a probabilistic model for predicting hourly prices. The generalized extreme learning machine (GELM) was used to make predictions. To speed up the model by reducing computational time, the authors used bootstrapping techniques. On the other hand, it does not work best for massive, large datasets, and the volume of information increased linearly. Oveis Abedinia et al. focused on attribute choice to improve prediction. For feature selection, these proposed models use information-theoretic parameters such as information gain (IG) and mutual information (MI). The hybrid filter-wrapper solution is another contribution of this study. In [47,48], the authors suggested a mixed algorithm for cost and demand modeling. Quasi-oppositional artificial bee colony (QOABC) and artificial bee colony optimization (ABCO) algorithms were updated by the researchers. Dogan Keles et al. are a group of researchers who came up with a novel solution. ANN [49] was used to propose a system. To find the best ANN parameters, the authors used a variety of clustering algorithms. The dynamic choice algorithm neural network (DCANN) was introduced by [50]. This design is used to predict rates for the next day. To unplug poor results and recognize acceptable inputs for a teaching method, this method integrates supervised and unsupervised training. The researchers of [51] developed a mixed design built on the neural network of Guo-Feng Fan et al. By combining the bi-square kernel regression model with the phase space reconstruction algorithm, the PSR-BSK model [52], a new model for predicting energy load, has been suggested. To validate the model's performance, the authors used an hourly dataset from NYISO in the New South Wales and the United States market. To extend the community in CS and to maximize the search space in [53], the authors suggested a dual SVR-chaotic cuckoo search (SVRCCS) framework. The authors recommended SSVRCCS, a seasonal CCS with SVR, to work with the load's periodic linear development. Owing to a large number of iterations, the computing period, on the other hand, was raised. It is impossible to overstate the value of contact between SG and its users. In the sense of creation in smart cities, the authors reduced energy usage and increased the traffic speed of device-to-device (D2D) interaction, also known as smart interaction, in [54]; smart communication is a serious issue that needs to be tackled. The authors broke the problem down into two parts to solving it: uplink subcarrier assignment (SA) and power allocation (PA) with joint optimization. The transmission power was distributed to all sub-carriers using a heuristic algorithm for SA. After that, an optimal PA algorithm was implemented to tackle the sub problem of convex similarity. The contact problems between SG and consumers were addressed by the authors in [55]. The authors provided a brief overview of the wireless and wired communication systems, as well as the various communication protocols. According to the cyber and physical frameworks, security concerns of hardware and software were also addressed. The authors discussed advanced metering infrastructure as well as automatic meter readings for customer information gathered via cables and portable links. The authors of [56] showed how SG and customers communicate digitally and with advanced control technologies. For connected and portable communication, the functionality, safety, robustness, efficiency, range, speed, power consumption, and protocols of wireless and other innovations were contrasted. API, HEMS, DA, DER, and EVS were some of the contact applications used by SG and consumers. To evaluate the efficiency of D2D delivery, in their study [57], the authors proposed an energy efficient delivery system (ECDS). D2D is a dependable and effective method of communication. It is not necessary to have any prior knowledge of content delivery, mobile mobility, or user demand. The energy conservation design system (ECDS) is used in smart cities to reduce their energy consumption. For a random and complex world, this method achieves the optimal solution.

Only a few logical operations were performed by ECDS, which made decisions based on local data from each system. The authors in [58,59] forecast the electricity price using the multi-step methods and dual decomposition methods. Furthermore, they tuned the parameters of the model using multi-objective optimization methods, which led to better forecasting.

SG can quickly predict the demand for consumers' energy usage after receiving information about the use of various devices through a green communication system. To maintain production and need equilibrium, load and price prediction are critical. Load balancing is necessary to prevent power shortages and over-production. When resources are abundant, storing them is very costly. A blackout could occur if a generation falls short of demand. The utility can generate electricity with the help of generators and other expensive tools. The cost of energy rises in all situations. A summary of related work is shown in Table 1.

Table 1. Summarized related work.

Methodology (s)	Aims and Objectives	Source(s) of Information/Achievement(s)	Drawback(s)
[10] NN with several layers.	Forecasting prices	Price forecasting with reasonable accuracy.	The loss rate is high, as is the computational time.
[13] HSDNN (LSTM and CNN combined).	Forecasting electricity costs	PJM (half-hour).	The amount of time required for computation is considerable.
[14] Recurrent units with gates (GRU).	Estimating prices	Turkish power sector forecast for the day.	The problem of over-fitting has gotten worse.
[15] Neural networks with back propagation (BPNN).	Load forecasting for the short term	Texas Electric Reliability Council, USA, a day ahead.	The level of complexity has risen.
[16] CS-SSA-SVM is a combination of CS-SSA and SVM.	Forecasting Loads	New South Wales: half-hourly, hourly, regular day, and non-working day results (ten weeks).	The calculation takes a very long time.
[18] LSTM and RNN.	Predictions Loads	Hourly and monthly payments are accepted. France Metropolitan.	Over-fitting is a risk that cannot be avoided.
[20] DNN, CNN, and LSTM are all examples of deep neural networks.	Forecasting the price of electricity	Estimated price.	The effect of dataset size is not measured, and redundancy is not eliminated.
[22] The UC-DADR and CC-DADR algorithms.	Decrease high growth and increase consumer benefits by reducing generation capacity.	Interconnection of the states of New Jersey, Maryland (PJM), and Pennsylvania,	The degree of defect detection is smaller.
[23] Storage device with batteries.	Estimating prices	Ontario's power market data is updated hourly.	The model isn't stable or accurate.
[24] Clustering validity indices (CVIs) are a form of validity index that is used to group together similar items.	The use of electricity	The upcoming day. The University of Seoul has eight buildings.	The amount of time it takes to compute something is enormous.

Table 1. *Cont.*

Methodology (s)	Aims and Objectives	Source(s) of Information/Achievement(s)	Drawback(s)
[26] SVM and ANN (are two different types of artificial neural networks).	Forecasting loads	The day ahead. PJM and Tunisian electricity industry.	The calculation takes a very long time.
[27] DCA, KPCA, SVM (are all types of simulation models).	Forecasting prices	Estimated cost and hybrid feature range.	Irrelevant features in the dataset add to the processing time.
[28] SVM/DNN	Forecasting short-term power costs	Different models are compared, and short-term price predictions are made.	Only for a particular situation.
[29] DNN	Forecasting prices	By using Bayesian optimization, you can improve accuracy and finish feature selection.	There was no thought given to redundancy or dimensionality reduction.
[30] Multivariate model	Price forecast on an hourly basis	Reduced the probability of overfitting by using a multivariate model instead of a univariate model.	Except for the unit-variate approach, the model's output is not comparable to that of other techniques.
[31] LSTM, DNN	Predictions of cost and load	Predict both the price and the volume of a product.	Price forecasting is unreliable.
[32] GELM	Price prediction on an hourly basis	Using bootstrapping techniques, predict hourly price and increase model speed.	For large datasets, this method does not function well.
[33] IG/MI	Hybrid algorithm for feature selection	Accuracy has improved as a result of a better choice of features.	The classifier's optimization was not taken into consideration.
[34,35] LSSVM, QOABC	Forecasting prices with loads	Artificial bee colony forecasting of price and load, as well as conditional feature selection and modification.	Their established scenario was the only one that works for them.
[36] ANN	ANN parameters: finding the best	Parameters for ANN and price estimation have been optimized.	Problems of overfitting were not taken into account.
[37] DCANN	Price prediction for the next day	Price prediction and development architecture that uses a neural network to automate scenarios.	The computational time is extremely long.

3. System Models

This stud proposes two methods for forecasting energy load and price. Since they use similar strategies, these two models are related.

The last design, on the other hand, is utilized to predict electricity demand, while the second framework is utilized to predict energy costs. The models that were suggested are the electricity price forecasting model and the electricity load forecasting model.

3.1. Model for Predicting Electricity Load and Price

Figure 2 depicts the load forecasting model. To predict the electricity load and price, take the following steps:

1. Data input (i.e., dataset).
2. Feature extraction using RFE.
3. Feature selection using RF and XG-Boost.
4. Splitting of data into training and testing.
5. Load the CNN layers and parameters.
6. Tuning the CNN parameters using CHIO and then model compiling.
7. Predicted price and load.
8. Performance evaluation.
9. Statistical analysis.

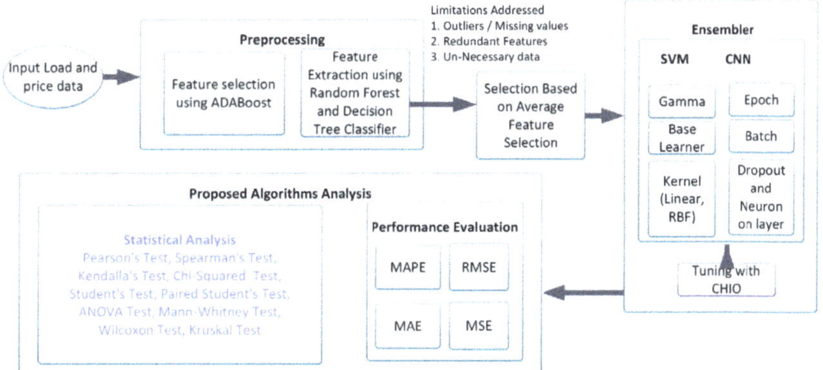

Figure 2. System model for electricity price and load forecasting.

3.2. Data Collection

ISO-NE is the name given to the electricity energy sector in New England [60]. It is in charge of producing, processing, and delivering electric energy to end-users in the processing, retail, and industrial areas. ISO-NE provides a large amount of data about, among other things, load, cost, production, and supply. The load data for 2018 come from ISO-NE, and they were used to incorporate the proposed models. This study used daily energy load data from Independent System Operator New England (ISO NE) (https://www.iso-ne.com, accessed on: 29 June 2021) for three years, from January 2017 to December 2019. It provides power to a number of English towns. Weather, temperature, humidity, and other dependent and independent data were included in the dataset. Our goal data are in a column called "electricity load". The target data were affected by all functionality other than the target features. The energy load demand pattern of a similar month in each year is roughly the same. As a result, we took three years' worth of results, or 36 months. To that end, the dataset was split into two sets: preparation and research. As a result, 90% of the data was used for teaching, and 10% was used for research, since the more data generated for training, the higher the model's learning rate would be. Furthermore, data from previous years' equivalent months, such as January 2017, January 2018, and

January 2019, were combined to provide a short-term load forecast for December 2019. The data in the dataset were organized by month, which aids in the better training of our model to determine the load pattern of months. All data from the first week of December 2019, i.e., 1 December 2019 to 7 December 2019, were used as preparation for weekly forecasting. In the first week of December, the teaching model was put to the test. Furthermore, the first five months of 2019 were also taken into account for preparation and research. Similarly, except for January 2019, all data were used for preparation and monitoring. In addition, the same situation was pursued in February, March, April, and May 2019. The suggested model's effectiveness is shown by the simulation and the results. The data description and function names are shown in Figure 3.

Date	DA_Demand	RT_Demand	DA_LMP	DA_EC	DA_CC	DA_MLC	RT_LMP	RT_EC	RT_CC	RT_MLC	Dry_Bulb	Dew_Point	System_Load	Reg_Capacity_Price	Reg_Service_Price
1/1/2017	16679	16391	63	62	0	1	45	45	0	0	30	9	16688	6	0
1/2/2017	16924	16649	57	56	0	1	22	22	0	0	31	15	16934	7	0
1/3/2017	15999	16624	53	52	0	0	58	58	0	0	27	25	16958	12	1
1/4/2017	14404	15765	41	40	0	0	77	76	0	1	41	40	16058	21	1
1/5/2017	18036	18557	83	82	0	1	63	62	0	1	19	-2	18867	8	3
1/6/2017	18504	19259	82	82	0	1	130	130	0	1	15	10	19554	28	0
1/7/2017	19409	19883	115	114	0	1	127	126	0	1	10	-11	20159	24	0
1/8/2017	19192	19813	107	106	0	1	90	89	0	1	19	0	20143	11	0
1/9/2017	17996	18155	85	85	0	1	60	60	0	1	23	8	18454	5	0
1/10/2017	17734	17620	102	101	0	1	81	80	0	1	17	-3	17899	8	0
1/11/2017	17029	17601	113	113	0	1	87	86	0	1	27	12	17874	11	2
1/12/2017	17489	17694	97	96	0	1	85	84	0	1	34	32	17993	17	0
1/13/2017	19463	19034	116	115	0	1	104	101	3	1	15	-2	19307	11	1
1/14/2017	18440	18901	97	96	0	1	93	93	0	1	21	15	19213	9	1
1/15/2017	17852	18212	107	106	0	1	99	98	0	1	26	17	18513	19	1
1/16/2017	18037	17913	114	112	0	1	88	87	0	1	20	-1	18219	23	1
1/17/2017	17640	17321	104	103	0	1	60	60	0	0	20	0	17574	9	0
1/18/2017	15825	15683	53	53	0	0	7	7	0	0	44	41	15929	13	0
1/19/2017	17233	17185	73	72	0	0	49	49	0	0	35	23	17443	17	0
1/20/2017	17716	17683	72	71	0	1	56	56	0	0	30	9	17917	17	0

Figure 3. Dataset overview.

3.3. Feature Extraction Using (RFE)

Recursive feature elimination (RFE) is a tool for obtaining a set of attributes from a database [61–63]. It replaces the lowest feature recursively before the required set of attributes is achieved. RFE involves the selection of many features; however, determining how many features are most important is difficult in advance as in Figure 4. To solve this dilemma, cross-validation was combined with RFE. Cross-validation tests the reliability of various categories and picks the most reliable.

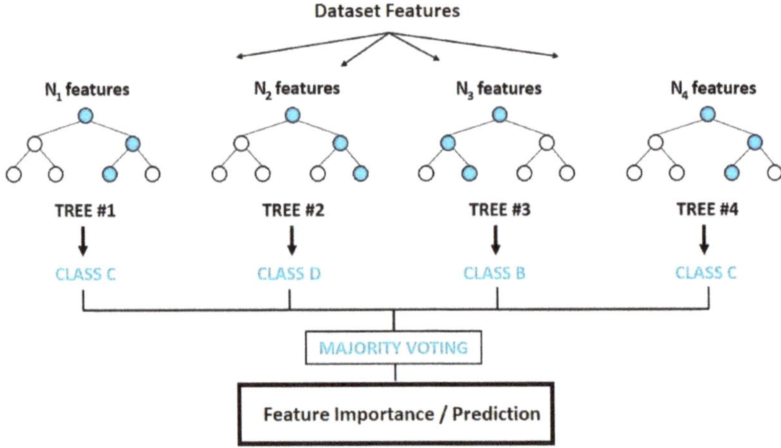

Figure 4. Random forest classifier.

3.4. Feature Selection

The method of selecting more important features is known as feature selection [64–66]. The number of features in the data set was reduced. Every feature's importance was calculated using RF. It was done to exclude the less relevant functions, and a hybrid solution was proposed for the final selection, which was a mixture of XG-boost and RF as shown in Figure 5.

XG-Boost

XG-boost gradient boosting (Extreme) is an optimized gradient boosting library [67]. It is made to be extremely compact, adaptable, and efficient. It uses the gradient method boosting and tree boosting in tandem to effectively and reliably produce accurate classification issues. It can be used to address estimation, grouping, and rating concerns. It is a library that is free to use. It comes in a variety of languages, including C++ and Python, for a variety of platforms of activity. The abstract diagram of XG-boost is shown in Figure 6.

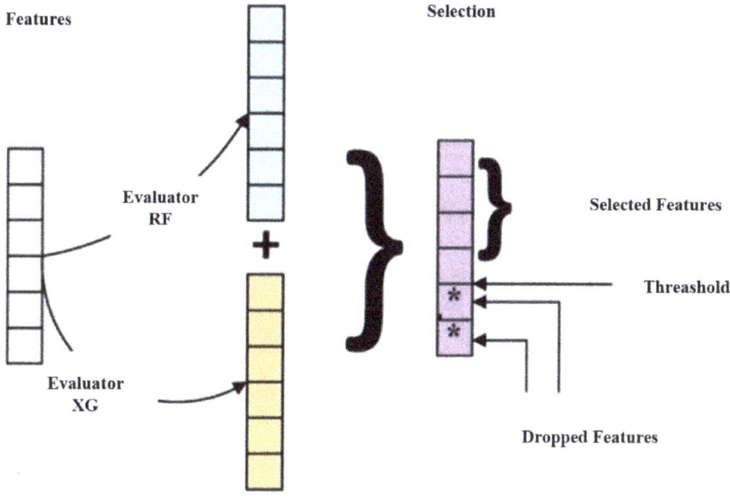

Figure 5. Feature selection.

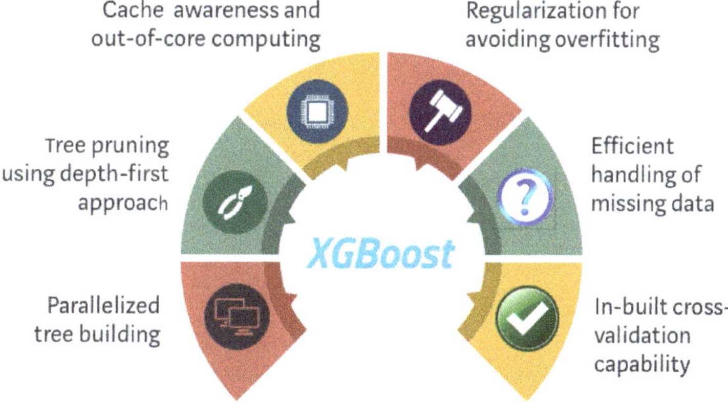

Figure 6. XG-boost abstract model.

3.5. Convolutional Neural Network

CNN is a type of neural network that belongs to the category of supervised deep learning prototypes [68]. In CNN implementation, firstly, a sequential model is implemented. It builds model layers upon layers. A prediction framework is built using four distinct levels in this design. A second surface, the convolution layer, is added to verify the neurons with outcomes that are related to the input layer. The convolutional layer receives m*r as its input. The dimensions of the height and width of the matrix are denoted by m and r, respectively. In cases where the matrix's dimension is less than the query, the kernel size will be used as a filter. The network's linked structure is determined by the filter's height. The equation will be used to calculate Relu, which will be used as an activation function. If the input value is negative, Relu returns 0; otherwise, it produces the same result, where x is the inputs:

$$\max(0, x) = \text{Relu}(x) \tag{1}$$

Following it, as a network's third tier, max-pooling is used to provide a matrix with small numbers. Max pooling, for example, chooses the most significant value from the various matrices. Then, using these values, it makes a small matrix.

For example, where p stands for padding and f stands for the range of filters, and n is the length of content: $32 \times 32 \times 1$. To prevent the issue of over-fitting, flatten layer was used as the fourth layer to turn all of the neurons into a single associated layer using a dropout layer. Each entity in the system is attached to the others. Early on in the process, the importance of the neuron failure rate was revealed. If the value of a network's failure rate in a stable state cannot be found by soon stopping the process it can be tested again. Then, to prevent overfitting, one switches to the dropout layer and applies the dense layer once more. The prediction result is finally shown in the output layer. The optimizer in this model is called "Adam". CNN forecasted energy demand and price under various scenarios in this study. Algorithm 1 illustrates the proposed model step by step. The architecture of CNN is shown in Figure 7.

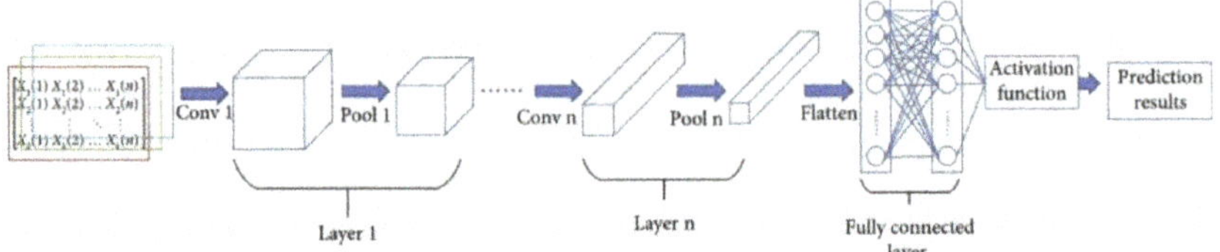

Figure 7. CNN architecture.

3.6. Coronavirus Herd Immunity Optimization

In this study, we utilized the CHIO algorithm [68] to tune the parameters of Adaboost. CHIO is used to minimize time complexity and increase precision in AdaBoost performance measurement. The concept of coronavirus herd immunity optimization (CHIO) was inspired by preventing the COVID-19 disease outbreak. The rate at which coronavirus infection spreads is regulated by how affected people interact with others in society. To protect all members of the community from the condition, health authorities advise social distancing. Herd immunity is a state attained by a species when the majority of its population is immune, inhibiting disease transmission. These concepts are represented by optimization principles. CHIO is a combination of herd immunity and social distancing strategies. Human cases are classified into three types for herd immunity: vulnerable, immuned, and contaminated. This is to determine whether the newly developed method employs social distancing strategies to update the genes. Figure 8 depicts the flow of the CHIO algorithm.

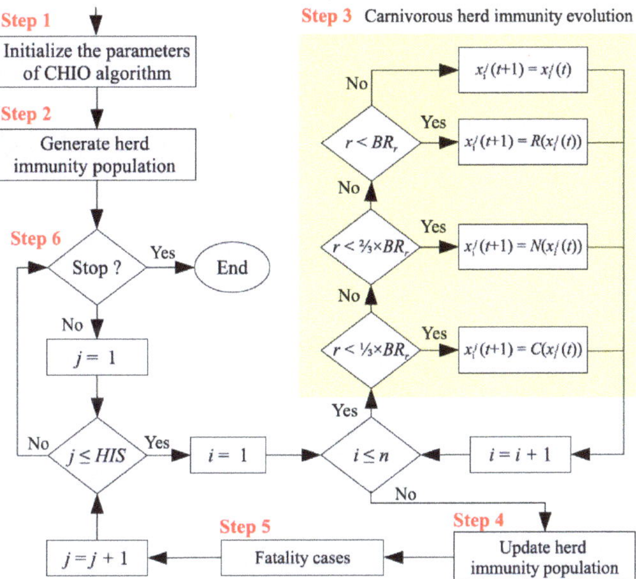

Figure 8. CHIO algorithm flow chart.

Algorithm 1 illustrates the proposed model step by step. The proposed algorithm of our work is:

Algorithm 1: Proposed Work Algorithm

Result: Electricity price and load forecasting
X: data features;
Y: data with a purpose;
 /* *Separate the data into two categories: preparation and testing.* */ ;
split (x, y) = x train, x test, y train, y test;
RFE (5, x train, y train); Selected_ function;
/* *Selection of hybrid features* */ ;
Incorporate$_{imp}$ = RF$_{imp}$ + XG$_{imp}$;
/* *Using RF and XG-boost, measure value* */ ;
RF imp = RF calculates importance;
/* *RFE is a technique for extracting features.* */ ;
if *Incorporate imp* $\geq$ *RFE and the threshold* == *right* **then**
  | Select the feature;
else
  | decline feature;
end
CNN-CHIO predicting the future with fine-tuned;
Performance evaluation test, compare predictions;

3.7. Performance Evaluation

Based on efficiency metrics, the suggested models were evaluated: MSE, MAPE, MAE, and RMSE. Equations (2)–(5) [22] provide the MSE, MAE, RMSE, and MAPE formulas. On the data collection of ISO-NE, Tables 2 and 3 displays the measurement of output measures of various methods. The MAPE is calculated using the formula:

$$\text{MAPE} = \frac{1}{y} \sum_{yn=1}^{yn} 100 \left| \frac{S_b - G_b}{A_b} \right| \tag{2}$$

The RMSE is calculated using the formula:

$$\text{RMSE} = \sqrt{\frac{1}{Y} \sum_{yn=1}^{YM} (S_b - G_b)^2} \tag{3}$$

The MAE and MSE are calculated using the formula:

$$\text{MSE} = \frac{1}{Y} \sum_{yn=1}^{YN} (S_b - G_b)^2 \tag{4}$$

$$\text{MAE} = \frac{\sum_{yn=1}^{YN} |(G_b - S_b)|}{Y} \tag{5}$$

Table 2. Performance evaluation values of electricity load forecasting.

Techniques	Accuracy (%)	F1-Score (%)	Recall (%)	Precision (%)	RMSE (%)	MAPE (%)	MSE (%)	MAE (%)
SVM	90.89	90.32	94.456	88.21	8.43	7.23	12.34	10.77
RF	84.54	72.98	89.33	82.22	24.27	24.56	27.65	25.78
LR	81.22	75	71.555	84.94	24	22.78	27	21
LDA	76.21	74.12	82.22	65.22	29	28.78	35.22	31.56
CNN-CHIO	95.789	96.22	98.55	94.639	6.23	5.67	10.82	7.22

Table 3. Electricity price forecasting performance evaluation values.

Techniques	Accuracy (%)	Precision (%)	F1-Score (%)	Recall (%)	RMSE (%)	MSE (%)	MAPE (%)	MAE (%)
LR	75.22	78.94	69	65.545	24	27	22.78	21
RF	79.54	77.22	67.98	84.33	24.27	27.65	24.56	25.78
SVM	88.89	85.21	87.32	91.466	8.34	11.34	7.23	10.77
LDA	71.21	60.22	69.12	77.22	29	35.22	28.78	31.56
CNN-CHIO	90.789	89.639	91.22	93.55	6.23	9.82	5.67	7.22

4. Simulation Results and Discussions

The implementation effects of our proposed model are explained in terms of their performance metrics in this section. We simulated our model on the following system specifications: 16 GB RAM and a 4.8 GHZ Core i7 processor. The IDE environment Anaconda (Spyder) and the Python language were used.

4.1. Electricity Load Forecasting

Figures 9 and 10 show the feature importance calculated by machine learning techniques, i.e., AdaBoost and RF. The feature importance means how much a feature impacts the target feature, i.e., electricity load. The high importance value of the feature means an important influence on the targeted function. The high impact of the feature shows the high relevancy towards the target. Changes in these relevant features can cause a huge impact on the target. Features with a low importance value were considered as low-impact features. If these features are removed, they had no impact or low impact on the target. Getting rid of the features that are not needed improves the simulation time and reduces computational complexity. Figure 9 shows the feature score/importance calculated by the AdaBoost technique, and Figure 10 displays the importance of features calculated by RF.

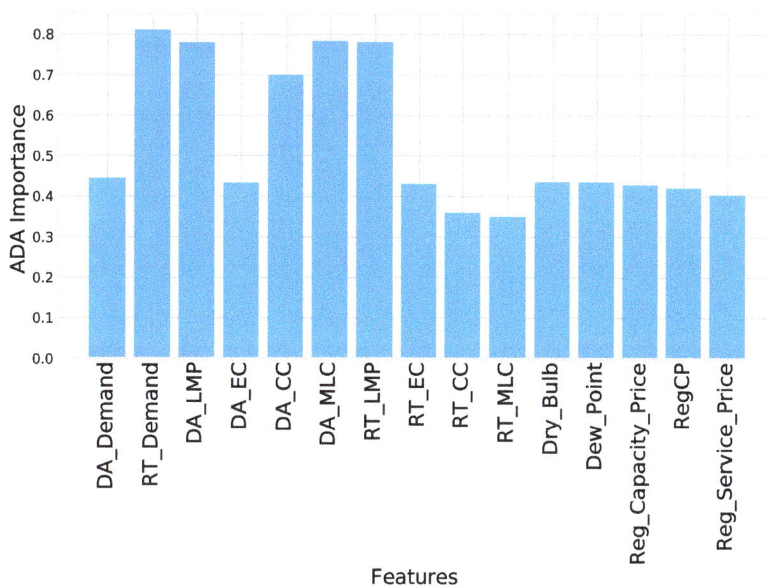

Figure 9. ADABoost-computed feature importance.

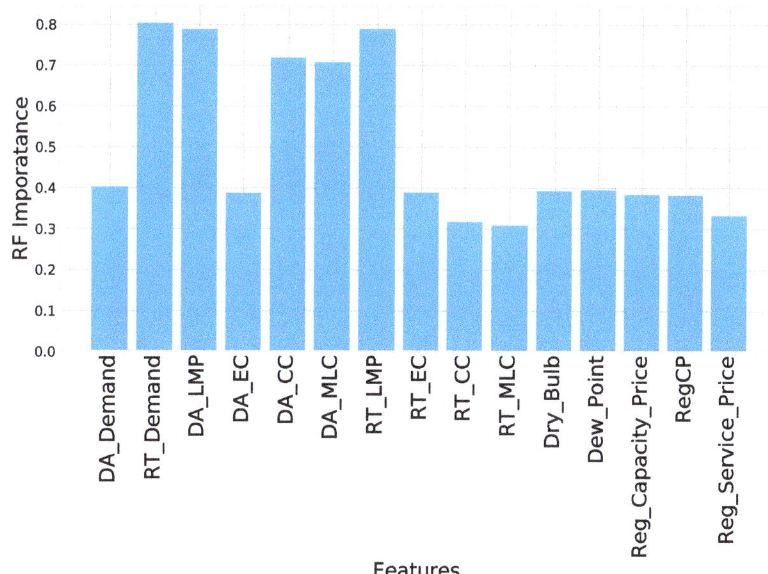

Figure 10. Random-forest-computed feature importance.

Figure 11 shows the daily normal load electricity of the years 2012–2020. We can see that the normal load had some different patterns with respect to time. Figure 11 also comprises the historical consumption pattern of consumers.

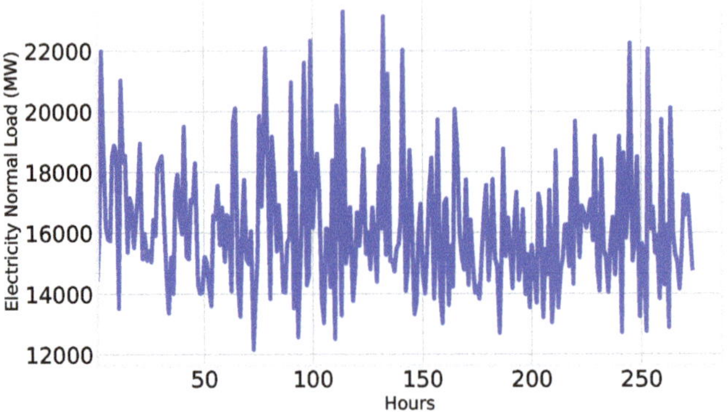

Figure 11. Normal electricity load. of ISO-NE 2012–2020.

Using the modified machine learning algorithm SVM and the deep learning algorithm CNN embedded with a GRU layer, we forecast the electricity load of one day as shown in Figure 12.

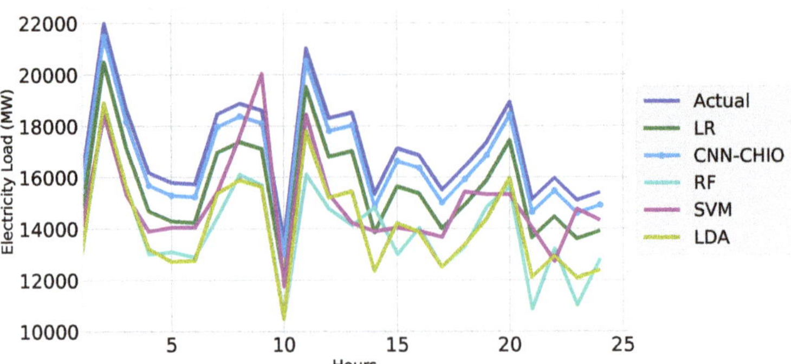

Figure 12. One-day electricity. load forecast.

Furthermore, with the same methodology, we forecasted two-day, three-day, and one-week upcoming electricity loads with a high accuracy of 96%.

In Figures 13–15, we can see that our proposed algorithm forecasts better than the other benchmark algorithms. The proposed algorithm CNN-CHIO performed better than the other proposed algorithms, and SVM performed better than the most up-to-date algorithms.

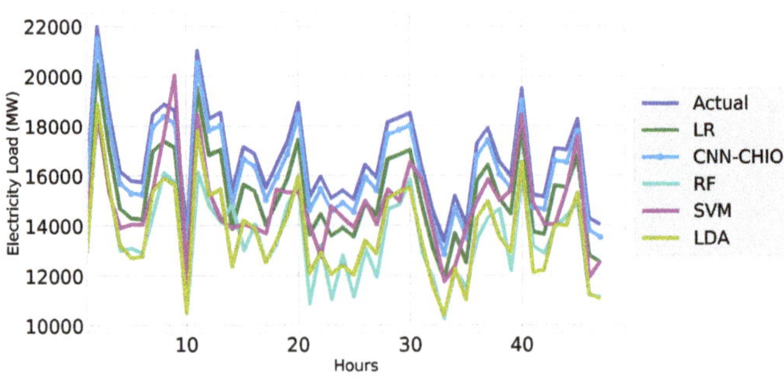

Figure 13. Two-day load forecast.

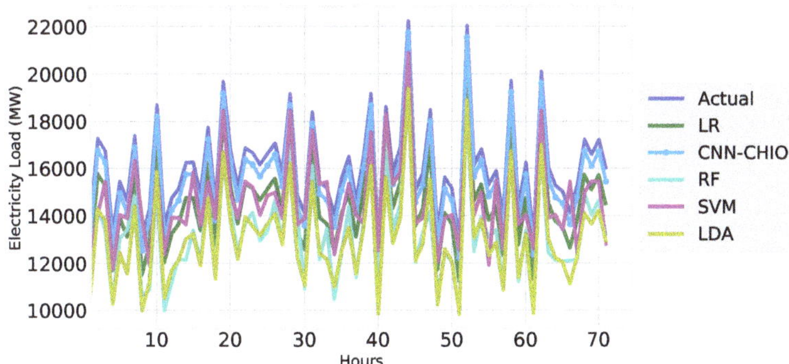

Figure 14. Three-day load forecast.

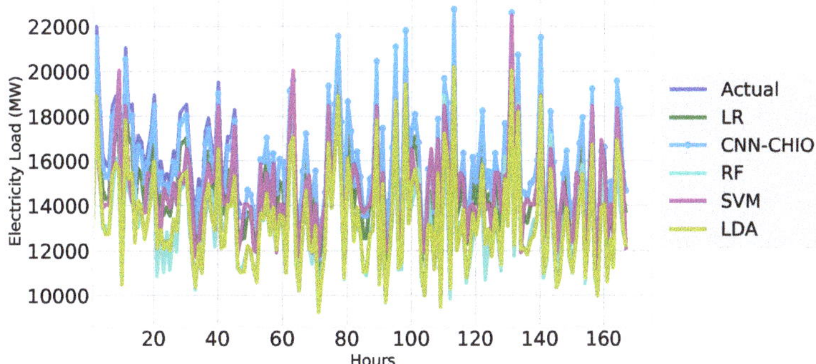

Figure 15. One-week load forecast.

Figures 16 and 17 shows the accuracy and loss curve of our proposed model. In Figure 16, we can see that the curve of training and the testing accuracy was increasing, while Figure 17 shows the decrease in the model loss value. The increase in accuracy and the decrease in the loss curve shows the superiority of the model that we proposed, which means our proposed model performed better in achieving the accuracy.

4.2. Electricity Price Forecasting

Figure 18 shows the normal electricity price from 2012–2020. The price of electricity varied with time. It also shows the seasonal change in the electricity price.

Figures 19–22 shows the electricity price forecasting of 24 h, two days, three days, and one week. From Figures 19–22, it was determined that the proposed algorithm worked well in terms of predicting the electricity. In comparison with the actual electricity price, we can see that the curve of the proposed algorithm is near to the actual. In forecasting the short-term electricity price, our proposed model outperformed benchmark algorithms.

Figure 23 describes the proposed model's loss and accuracy. The proposed model's accuracy was increasing, and the loss value was decreasing with the number of iterations. Our proposed methodology performed better in achieving the accuracy of 92% and 90%, respectively.

4.3. Performance Evaluation of Electricity Price and Load Forecasting

This section evaluates the proposed model and benchmark schemes using performance evaluation techniques, performance error metrics, and statistical analysis. Figure 24 shows the performance evaluation using the error metrics MAPE, MSE, RMSE, and MAE. We can determine in Figure 24 that the proposed models SVM and CNN-CHIO had the lowest error rate compared with the RF, LDA, and RF techniques. The LDA technique had the highest error rate in forecasting the electricity price and load. The lowest error showed the superiority of the proposed techniques.

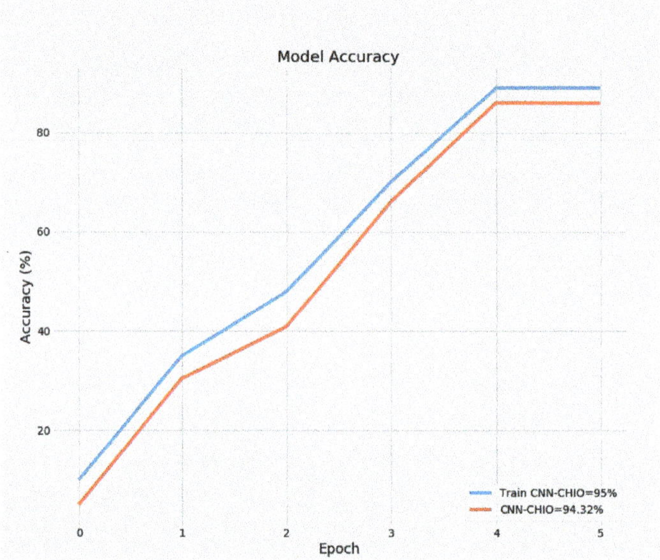

Figure 16. Accuracy curve of electricity load model.

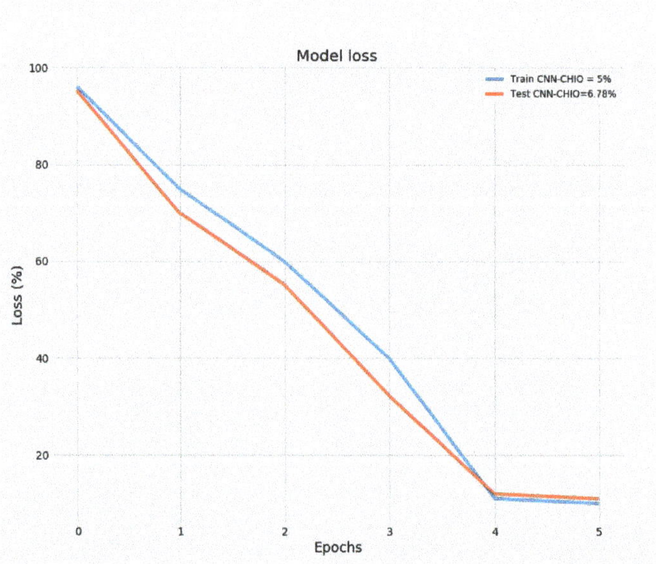

Figure 17. Loss curve of electricity load model.

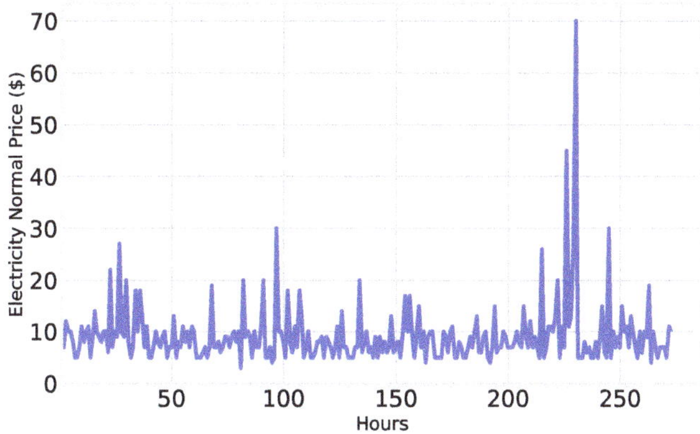

Figure 18. Normal electricity price of ISO-NE 2012–2020.

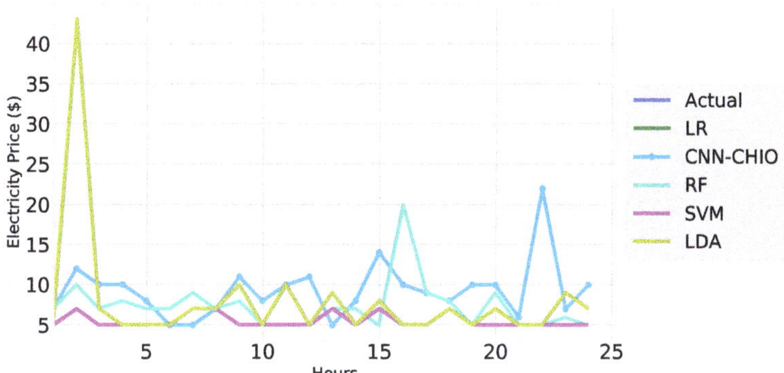

Figure 19. 24-h electricity price forecast.

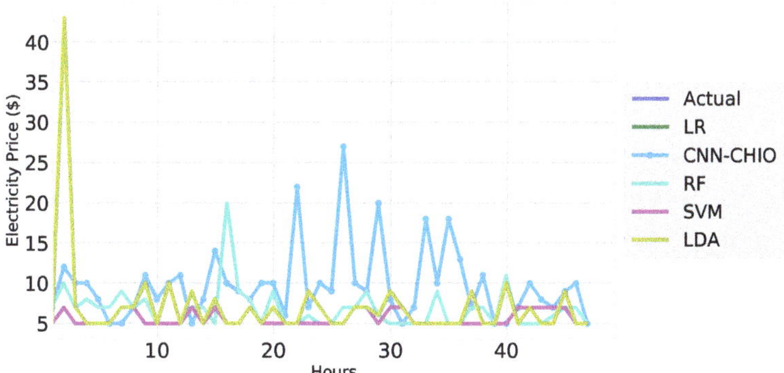

Figure 20. Two-day electricity price forecast.

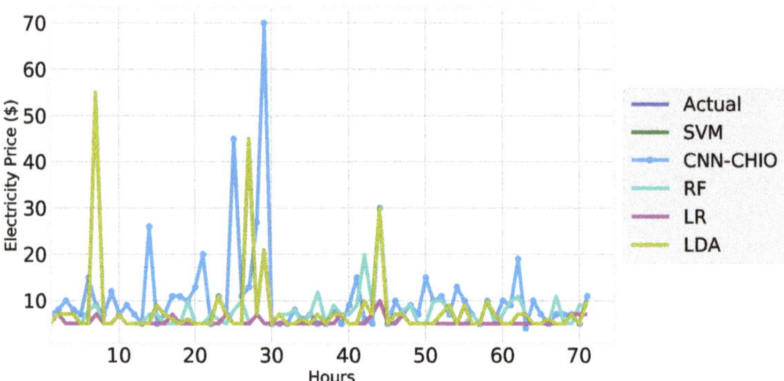

Figure 21. Three-day electricity price forecast.

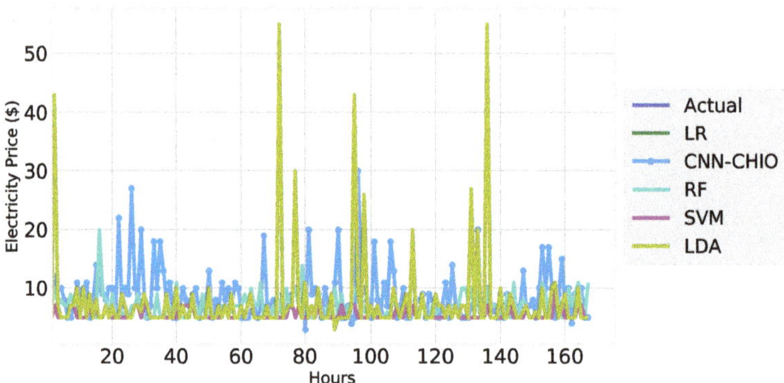

Figure 22. One-week electricity price forecast.

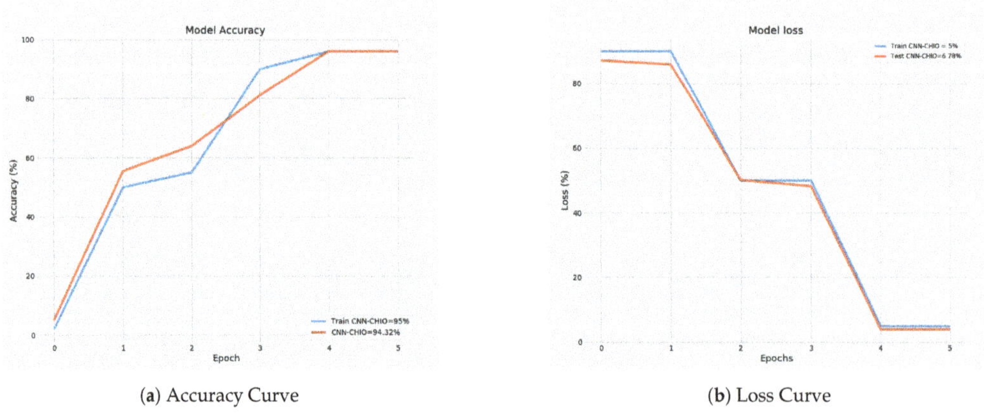

(a) Accuracy Curve　　　　　　　　　　(b) Loss Curve

Figure 23. Electricity price forecasting model accuracy and loss.

The performance evaluation metrics, i.e. precision, F-score, accuracy, and recall, were also used to assess the proposed model and to compare with the benchmark algorithm.

In Figure 25, the performance evaluation of the electricity price and the electricity load forecasting model is shown. Figure 25 clearly shows that the accuracy of CNN-CHIO and SVM was higher than the other benchmark algorithm. The optimization part of the proposed model provided the exact values to the models, which increased the accuracy of our proposed model.

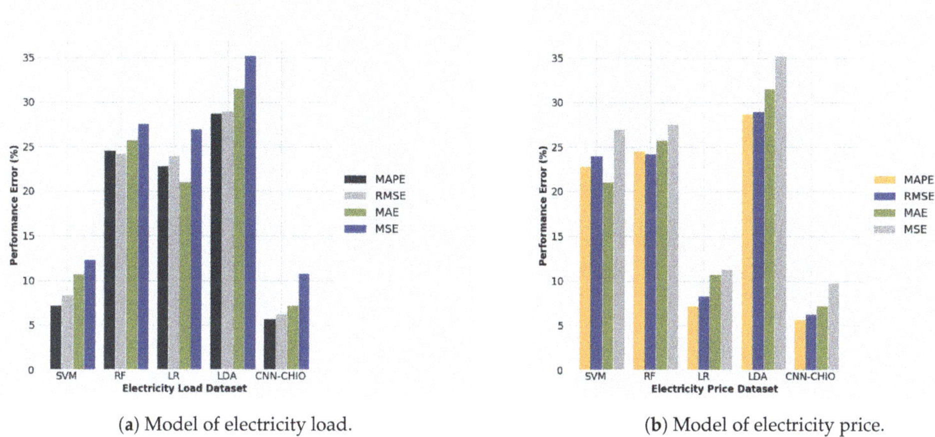

(**a**) Model of electricity load.　　　　(**b**) Model of electricity price.

Figure 24. Performance error metrics of proposed and benchmark techniques.

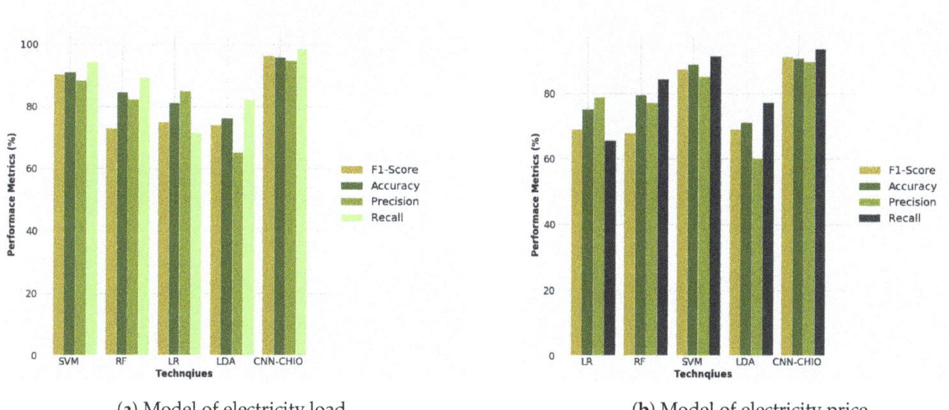

(**a**) Model of electricity load.　　　　(**b**) Model of electricity price.

Figure 25. Evaluation metrics performance of proposed and benchmark techniques.

Our proposed model's, i.e., SVM's and CNN-CHIO's, accuracy in electricity price forecasting, was 92% and 90%, respectively. Furthermore, SVM achieved 95% accuracy, while CNN-CHIO achieved 92% accuracy in terms of the electricity load forecasting model.

Tables 2 and 3 shows the performance evaluation of electricity load and price forecasting values in tabular form. Our proposed technique CNN-CHIO achieved 95% accuracy, and SVM achieved 90.89% accuracy in load forecasting with 90% and 87.32% accuracy in price forecasting, respectively, as shown in Figures 26 and 27. Our proposed technique outperformed the state of the art.

Table 4 shows the statistical analysis of the proposed algorithm. We applied ten statistical techniques to analyze our proposed model. The supremacy of the proposed model can also be identified in the analysis table.

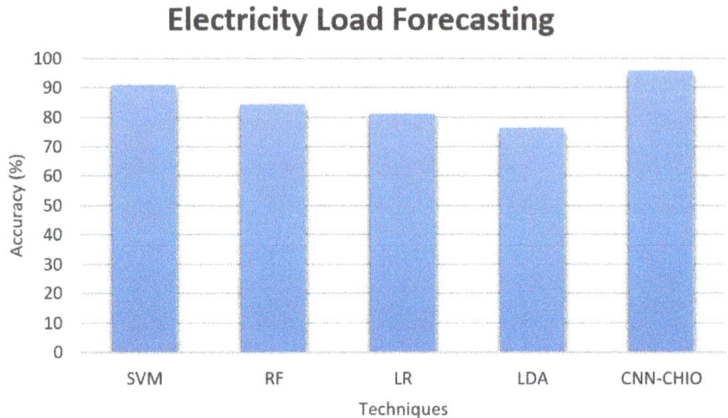

Figure 26. Electricity load forecasting accuracy proposed vs. benchmark techniques.

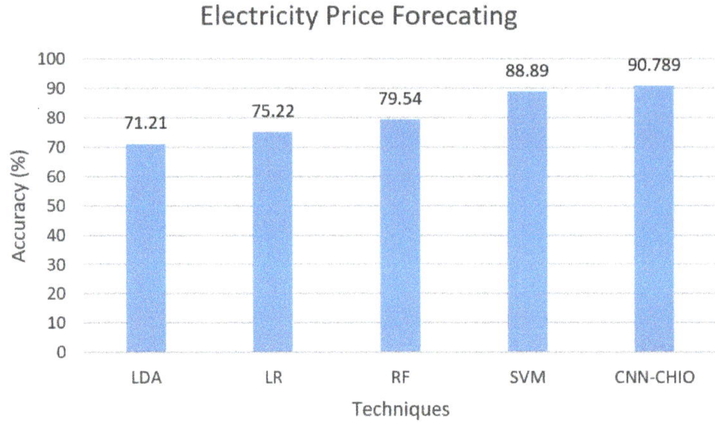

Figure 27. Electricity price forecasting accuracy proposed vs. benchmark techniques.

Table 4. Statistical analysis of proposed techniques vs. benchmark algo.

Techniques	Kendallas	Spearmans	ANOVA	Mann-Whitney	Kruskal	Chi-Squared
SVM F-statistic	−0.128	−0.149	99.775	13,344.500	194.502	168.491
SVM p-Value	0.014	0.014	0.000	0.000	0.000	0.000
RF F-statistic	0.785	0.856	28.779	39,227.000	35.686	107.540
RF p-Value	0.911	0.995	0.000	0.785	0.000	0.042
CNN-CHIO F-stat	1.000	1.000	0.000	37,538.000	0.000	6028.000
CNN-CHIO p-Val	1.000	0.000	1.000	0.500	1.000	0.000
LDA F-statistic	0.801	0.867	3.100	41,232.000	70.847	109.440
LDA p-Value	0.849	0.839	0.079	0.811	0.000	0.053
LG F-statistic	1.000	1.000	0.000	37,538.000	0.000	6028.000
LG p-Value	0.000	0.000	1.000	0.500	1.000	0.000

5. Conclusions

We proposed a CNN-GRU hybrid model tuned with a novel optimization technique CHIO was used to simulate energy use and energy price in residential buildings in this study. The proposed model was validated using a publicly accessible dataset from ISONE. Since the input data were non-linear, we first normalized them using a regular min–max scalar, then we fed the normalized data into the feature selection method using AdaBoost and extracted the feature importance and selected the features with high importance. We applied RF and RFE to remove the redundant features and selected the optimum and most relevant features. The preprocessing process was performed to improve the training of our model and to decrease the computational complexity. Following that, we looked at various machine learning and deep learning approaches before settling on a mixed model that merged CNN and GRU. We first used feature engineering to extract spatial features. We then fed them into our tuned CNN-CHIO and SVM to simulate temporal characteristics corresponding to the time series data entry. As opposed to other baseline models, the proposed model performed well, suggesting that our presented, existing buildings model must be able to be found in actual life. Furthermore, our proposed model of CNN-CHIO and SVM achieved 95% and 92% accuracy in load forecasting and 92% and 89% accuracy in price forecasting, respectively. In future work, we intend to validate the proposed CNN-GRU and SVM model on various datasets and enhance the model's accuracy by incorporating fuzzy logic concepts. The model is currently being based on residential building results, but it will also be tested on commercial loads and price datasets. We predicted short-term electricity consumption and electricity prices in this study; however, our long-term aim is to assess the model's efficiency in predicting medium- and long-term electricity consumption and electricity prices.

Author Contributions: Conceptualization, S.A. and N.A.; methodology, S.A., N.A., U.F. and M.J.A.; software, S.A. and M.J.A.; validation, F.R.A. and G.R.; formal analysis, U.F. and A.T.A.; investigation, A.T.A.; resources, A.T.A.; data curation, S.A. and U.F.; writing—original draft preparation, S.A., N.A. and U.F.; writing—review and editing, F.R.A., G.R., S.I.H. and R.B.; visualization, A.T.A.; supervision, A.T.A. and R.B.; project administration, A.T.A.; funding acquisition, F.R.A. All authors have read and agreed to the published version of the manuscript.

Funding: The APC is funded by Taif University Researchers Supporting Project Number (TURSP-2020/331), Taif University, Taif, Saudi Arabia.

Institutional Review Board Statement: Not applicable.

Informed Consent Statement: Not applicable.

Data Availability Statement: The dataset used in this study can be found here: https://www.iso-ne.com.

Acknowledgments: The authors would like to acknowledge the support from Taif University Researchers Supporting Project Number (TURSP-2020/331), Taif University, Taif, Saudi Arabia.

Conflicts of Interest: The authors declare no conflict of interrest.

References

1. Aslam, S.; Herodotou, H.; Mohsin, S.M.; Javaid, N.; Ashraf, N.; Aslam, S. A survey on deep learning methods for power load and renewable energy forecasting in smart microgrids. *Renew. Sustain. Energy Rev.* **2021**, *144*, 110992. [CrossRef]
2. Liu, Y.; Yuen, C.; Huang, S.; Hassan, N.; Wang, X.; Xie, S. Peak-to-average ratio constrained demand-side management with consumer's preference in residential smart grid. *IEEE J. Sel. Top. Signal Process.* **2014**, *8*, 1084–1097. [CrossRef]
3. Aurangzeb, K.; Aslam, S.; Mohsin, S.M.; Alhussein, M. A fair pricing mechanism in smart grids for low energy consumption users. *IEEE Access* **2021**, *9*, 22035–22044. [CrossRef]
4. Hor, C.L.; Watson, S.; Majithia, S. Analyzing the impact of weather variables on monthly electricity demand. *IEEE Trans. Power Syst.* **2005**, *20*, 2078–2085. [CrossRef]
5. Siano, P. Demand response and smart grids—A survey. *Renew. Sustain. Energy Rev.* **2014**, *30*, 461–478. [CrossRef]
6. Aslam, S.; Iqbal, Z.; Javaid, N.; Khan, Z.A.; Aurangzeb, K.; Haider, S.I. Towards efficient energy management of smart buildings exploiting heuristic optimization with real time and critical peak pricing schemes. *Energies* **2017**, *10*, 2065. [CrossRef]

7. Liu, Y.; Wang, W.; Ghadimi, N. Electricity load forecasting by an improved forecast engine for building level consumers. *Energy* **2017**, *139*, 18–30. [CrossRef]

8. Jin, X.B.; Zheng, W.Z.; Kong, J.L.; Wang, X.Y.; Bai, Y.T.; Su, T.L.; Lin, S. Deep-Learning Forecasting Method for Electric Power Load via Attention-Based Encoder-Decoder with Bayesian Optimization. *Energies* **2021**, *14*, 1596. [CrossRef]

9. Li, Y.; Kubicki, S.; Guerriero, A.; Rezgui, Y. Review of building energy performance certification schemes towards future improvement. *Renew. Sustain. Energy Rev.* **2019**, *113*, 109244. [CrossRef]

10. Carmichael, R.; Gross, R.; Hanna, R.; Rhodes, A.; Green, T. The Demand Response Technology Cluster: Accelerating UK residential consumer engagement with time-of-use tariffs, electric vehicles and smart meters via digital comparison tools. *Renew. Sustain. Energy Rev.* **2021**, *139*, 110701. [CrossRef]

11. Ghosal, A.; Conti, M. Key management systems for smart grid advanced metering infrastructure: A survey. *IEEE Commun. Surv. Tutor.* **2019**, *21*, 2831–2848. [CrossRef]

12. Dileep, G. A survey on smart grid technologies and applications. *Renew. Energy* **2020**, *146*, 2589–2625. [CrossRef]

13. Kuo, P.H.; Huang, C.J. An electricity price forecasting model by hybrid structured deep neural networks. *Sustainability* **2018**, *10*, 1280. [CrossRef]

14. Ugurlu, U.; Oksuz, I.; Tas, O. Electricity price forecasting using recurrent neural networks. *Energies* **2018**, *11*, 1255. [CrossRef]

15. Eapen, R.; Simon, S. Performance analysis of combined similar day and day ahead short term electrical load forecasting using sequential hybrid neural networks. *IETE J. Res.* **2019**, *65*, 216–226. [CrossRef]

16. Zhang, X.; Wang, J.; Zhang, K. Short-term electric load forecasting based on singular spectrum analysis and support vector machine optimized by Cuckoo search algorithm. *Electr. Power Syst. Res.* **2017**, *146*, 270–285. [CrossRef]

17. Patil, M.; Deshmukh, S.; Agrawal, R. Electric power price forecasting using data mining techniques. In Proceedings of the 2017 International Conference on Data Management, Analytics and Innovation (ICDMAI), Pune, India, 24–26 February 2017; pp. 217–223.

18. Bouktif, S.; Fiaz, A.; Ouni, A.; Serhani, M. Optimal deep learning lstm model for electric load forecasting using feature selection and genetic algorithm: Comparison with machine learning approaches. *Energies* **2018**, *11*, 1636. [CrossRef]

19. Keles, D.; Scelle, J.; Paraschiv, F.; Fichtner, W. Extended forecast methods for day-ahead electricity spot prices applying artificial neural networks. *Appl. Energy* **2016**, *162*, 218–230. [CrossRef]

20. Ma, Z.; Zhong, H.; Xie, L.; Xia, Q.; Kang, C. Month ahead average daily electricity price profile forecasting based on a hybrid nonlinear regression and SVM model: An ERCOT case study. *J. Mod. Power Syst. Clean Energy* **2018**, *6*, 281–291. [CrossRef]

21. Lago, J.; De Ridder, F.; De Schutter, B. Forecasting spot electricity prices: Deep learning approaches and empirical comparison of traditional algorithms. *Appl. Energy* **2018**, *221*, 386–405. [CrossRef]

22. Ghadimi, N.; Akbarimajd, A.; Shayeghi, H.; Abedinia, O. Two stage forecast engine with feature selection technique and improved meta-heuristic algorithm for electricity load forecasting. *Energy* **2018**, *161*, 130–142. [CrossRef]

23. Jindal, A.; Singh, M.; Kumar, N.; Response, C.A. Scheme for Peak Load Reduction in Smart Grid. *IEEE Trans. Ind. Electron.* **2018**, *65*, 8993–9004. [CrossRef]

24. Chitsaz, H.; Zamani-Dehkordi, P.; Zareipour, H.; Parikh, P. Electricity price forecasting for operational scheduling of behind-the-meter storage systems. *IEEE Trans. Smart Grid* **2017**, *9*, 6612–6622. [CrossRef]

25. Pérez-Chacón, R.; Luna-Romera, J.M.; Troncoso, A.; Martínez-Álvarez, F.; Riquelme, J.C. Big data analytics for discovering electricity consumption patterns in smart cities. *Energies* **2018**, *11*, 683. [CrossRef]

26. Wang, Z.; Wang, Y.; Zeng, R.; Srinivasan, R.; Ahrentzen, S. Random Forest based hourly building energy prediction. *Energy Build.* **2018**, *171*, 11–25. [CrossRef]

27. Lahouar, A.; Slama, J. Day-ahead load forecast using random forest and expert input selection. *Energy Convers. Manag.* **2015**, *103*, 1040–1051. [CrossRef]

28. Wang, K.; Xu, C.; Zhang, Y.; Guo, S.; Zomaya, A. Robust big data analytics for electricity price forecasting in the smart grid. *IEEE Trans. Big Data* **2017**, *5*, 34–45. [CrossRef]

29. Wang, L.; Zhang, Z.; Chen, J. Short-term electricity price forecasting with stacked denoising autoencoders. *IEEE Trans. Power Syst.* **2016**, *32*, 2673–2681. [CrossRef]

30. Lago, J.; De Ridder, F.; Vrancx, P.; De Schutter, B. Forecasting day-ahead electricity prices in Europe: The importance of considering market integration. *Appl. Energy* **2018**, *211*, 890–903. [CrossRef]

31. Raviv, E.; Bouwman, K.; Van Dijk, D. Forecasting day-ahead electricity prices: Utilizing hourly prices. *Energy Econ.* **2015**, *50*, 227–239. [CrossRef]

32. Mujeeb, S.; Javaid, N.; Akbar, M.; Khalid, R.; Nazeer, O.; Khan, M. Big data analytics for price and load forecasting in smart grids. In Proceedings of the International Conference on Broadband and Wireless Computing, Communication and Applications, Taichung, Taiwan, 27–29 October 2018; pp. 77–87.

33. Rafiei, M.; Niknam, T.; Khooban, M.H. Probabilistic forecasting of hourly electricity price by generalization of ELM for usage in improved wavelet neural network. *IEEE Trans. Ind. Inform.* **2016**, *13*, 71–79. [CrossRef]

34. Abedinia, O.; Amjady, N.; Zareipour, H. A new feature selection technique for load and price forecast of electrical power systems. *IEEE Trans. Power Syst.* **2016**, *32*, 62–74. [CrossRef]

35. Ghasemi, A.; Shayeghi, H.; Moradzadeh, M.; Nooshyar, M. A novel hybrid algorithm for electricity price and load forecasting in smart grids with demand-side management. *Appl. Energy* **2016**, *177*, 40–59. [CrossRef]

36. Liang, Y.; Niu, D.; Hong, W.C. Short term load forecasting based on feature extraction and improved general regression neural network model. *Energy* **2019**, *166*, 653–663. [CrossRef]
37. Chen, K.; Chen, K.; Wang, Q.; He, Z.; Hu, J.; He, J. Short-term load forecasting with deep residual networks. *IEEE Trans. Smart Grid* **2018**, *10*, 3943–3952. [CrossRef]
38. Deng, Z.; Wang, B.; Xu, Y.; Xu, T.; Liu, C.; Zhu, Z. Multi-scale convolutional neural network with time-cognition for multi-step short-term load forecasting. *IEEE Access* **2019**, *7*, 88058–88071. [CrossRef]
39. Shayeghi, H.; Ghasemi, A.; Moradzadeh, M.; Nooshyar, M. Simultaneous day-ahead forecasting of electricity price and load in smart grids. *Energy Convers. Manag.* **2015**, *95*, 371–384. [CrossRef]
40. Wang, J.; Liu, F.; Song, Y.; Zhao, J. A novel model: Dynamic choice artificial neural network (DCANN) for an electricity price forecasting system. *Appl. Soft Comput.* **2016**, *48*, 281–297. [CrossRef]
41. Varshney, H.; Sharma, A.; Kumar, R. A hybrid approach to price forecasting incorporating exogenous variables for a day ahead electricity market. In Proceedings of the 2016 IEEE 1st International Conference on Power Electronics, Intelligent Control and Energy Systems (ICPEICES), Delhi, India, 4–6 July 2016; pp. 1–6.
42. Fan, G.F.; Peng, L.L.; Hong, W.C. Short term load forecasting based on phase space reconstruction algorithm and bi-square kernel regression model. *Appl. Energy* **2018**, *224*, 13–33. [CrossRef]
43. Dong, Y.; Zhang, Z.; Hong, W.C. A hybrid seasonal mechanism with a chaotic cuckoo search algorithm with a support vector regression model for electric load forecasting. *Energies* **2018**, *11*, 1009. [CrossRef]
44. Kai, C.; Li, H.; Xu, L.; Li, Y.; Jiang, T. Energy-efficient device-to-device communications for green smart cities. *IEEE Trans. Ind. Inform.* **2018**, *14*, 1542–1551. [CrossRef]
45. Kabalci, Y. A survey on smart metering and smart grid communication. *Renew. Sustain. Energy Rev.* **2016**, *57*, 302–318. [CrossRef]
46. Mahmood, A.; Javaid, N.; Razzaq, S. A review of wireless communications for smart grid. *Renew. Sustain. Energy Rev.* **2015**, *41*, 248–260. [CrossRef]
47. Zhou, L.; Wu, D.; Chen, J.; Dong, Z. Greening the smart cities: Energy-efficient massive content delivery via D2D communications. *IEEE Trans. Ind. Inform.* **2017**, *14*, 1626–1634. [CrossRef]
48. Abdullah, A.; Sopian, W.; Arasid, W.; Nandiyanto, A.; Danuwijaya, A.; Abdullah, C. Short-term peak load forecasting using PSO-ANN methods: The case of Indonesia. *J. Eng. Sci. Technol.* **2018**, *13*, 2395–2404.
49. Fallah, S.; Deo, R.; Shojafar, M.; Conti, M.; Shamshirband, S. Computational intelligence approaches for energy load forecasting in smart energy management grids: State of the art, future challenges, and research directions. *Energies* **2018**, *11*, 596. [CrossRef]
50. Huang, G.B.; Zhu, Q.Y.; Siew, C.K. Extreme learning machine: Theory and applications. *Neurocomputing* **2006**, *70*, 489–501. [CrossRef]
51. Rojas-Domínguez, A.; Padierna, L.C.; Valadez, J.M.C.; Puga-Soberanes, H.J.; Fraire, H.J. Optimal hyper-parameter tuning of SVM classifiers with application to medical diagnosis. *IEEE Access* **2017**, *6*, 7164–7176. [CrossRef]
52. Li, Z.L.; Xia, J.; Liu, A.; Li, P. States prediction for solar power and wind speed using BBA-SVM. *IET Renew. Power Gener.* **2019**, *13*, 1115–1122. [CrossRef]
53. Morley, S.; Brito, T.; Welling, D. Measures of model performance based on the log accuracy ratio. *Space Weather* **2018**, *16*, 69–88. [CrossRef]
54. Koohi-Fayegh, S.; Rosen, M. A review of energy storage types, applications and recent developments. *J. Energy Storage* **2020**, *27*, 101047. [CrossRef]
55. Rahimi, F.; Ipakchi, A. Demand response as a market resource under the smart grid paradigm. *IEEE Trans. Smart Grid* **2010**, *1*, 82–88. [CrossRef]
56. Vadari, S. *Electric System Operations: Evolving to the Modern Grid*; Artech House: Braga, Portugal, 2020.
57. Ayub, N.; Irfan, M.; Awais, M.; Ali, U.; Ali, T.; Hamdi, M.; Alghamdi, A.; Muhammad, F. Big Data Analytics for Short and Medium-Term Electricity Load Forecasting Using an AI Techniques Ensembler. *Energies* **2020**, *13*, 5193. [CrossRef]
58. Yang, W.; Wang, J.; Niu, T.; Du, P. A novel system for multi-step electricity price forecasting for electricity market management. *Appl. Soft Comput.* **2020**, *88*, 106029. [CrossRef]
59. Yang, W.; Wang, J.; Niu, T.; Du, P. A hybrid forecasting system based on a dual decomposition strategy and multi-objective optimization for electricity price forecasting. *Appl. Energy* **2019**, *235*, 1205–1225. [CrossRef]
60. Ahmad, W.; Ayub, N.; Ali, T.; Irfan, M.; Awais, M.; Shiraz, M.; Glowacz, A. Towards short term electricity load forecasting using improved support vector machine and extreme learning machine. *Energies* **2020**, *13*, 2907. [CrossRef]
61. Amin, S.; Wollenberg, B. Toward a smart grid: Power delivery for the 21st century. *IEEE Power Energy Mag.* **2005**, *3*, 34–41. [CrossRef]
62. Vaccaro, A.; Villacci, D. Performance analysis of low earth orbit satellites for power system communication. *Electric Power Syst. Res.* **2005**, *73*, 287–294. [CrossRef]
63. Albahli, S.; Shiraz, M.; Ayub, N. Electricity Price Forecasting for Cloud Computing Using an Enhanced Machine Learning Model. *IEEE Access* **2020**, *8*, 200971–200981. [CrossRef]
64. Cupp, J.; Beehler, M. Implementing smart grid communications. *TECHBriefs* **2008**, *4*, 5–8.
65. Ghassemi, A.; Bavarian, S.; Lampe, L. Cognitive radio for smart grid communications. In Proceedings of the 2010 First IEEE International Conference on Smart Grid Communications, Gaithersburg, MD, USA, 4–6 October 2010; pp. 297–302.

66. Ko, J.; Terzis, A.; Dawson-Haggerty, S.; Culler, D.; Hui, J.; Levis, P. Connecting low-power and lossy networks to the internet. *IEEE Commun. Mag.* **2011**, *49*, 96–101.
67. Aimal, S.; Javaid, N.; Rehman, A.; Ayub, N.; Sultana, T.; Tahir, A. Data analytics for electricity load and price forecasting in the smart grid. In Proceedings of the Workshops of the International Conference on Advanced Information Networking and Applications, Matsue, Japan, 27–29 March 2019; pp. 582–591.
68. Al-Betar, M.A.; Alyasseri, Z.A.A.; Awadallah, M.A.; Doush, I.A. Coronavirus herd immunity optimizer (CHIO). *Neural Comput. Appl.* **2021**, *33*, 5011–5042. [CrossRef] [PubMed]

 *sustainability*

Article

Efficient Energy Optimization Day-Ahead Energy Forecasting in Smart Grid Considering Demand Response and Microgrids

Fahad R. Albogamy [1], Ghulam Hafeez [2,3,*], Imran Khan [4], Sheraz Khan [4], Hend I. Alkhammash [5], Faheem Ali [6] and Gul Rukh [4]

[1] Computer Sciences Program, Turabah University College, Taif University, P.O. Box 11099, Taif 26571, Saudi Arabia; f.alhammdani@tu.edu.sa

[2] Centre of Renewable Energy, Government Advance Technical Training Centre, Hayatabad, Peshawar 25100, Pakistan

[3] Department of Electrical and Computer Engineering, Islamabad Campus, COMSATS University Islamabad, Islamabad 44000, Pakistan

[4] Department of Electrical Engineering, University of Engineering and Technology, Mardan 23200, Pakistan; imran@uetmardan.edu.pk (I.K.); sheraz@uetmardan.edu.pk (S.K.); gr@uetmardan.edu.pk (G.R.)

[5] Department of Electrical Engineering, College of Engineering, Taif University, P.O. Box 11099, Taif 21944, Saudi Arabia; Khamash.h@tu.edu.sa

[6] Department of Electrical Engineering, University of Engineering and Technology, Peshawar 25000, Pakistan; faheem@uetpeshawar.edu.pk

* Correspondence: ghulamhafeez393@gmail.com, Tel.: +92-300-5003574 or +92-348-8818497

check for updates

Citation: Albogamy, F.R.; Hafeez, G.; Khan, I.; Khan, S.; Alkhammash, H.I.; Ali, F.; Rukh, G. Efficient Energy Optimization Day-Ahead Energy Forecasting in Smart Grid Considering Demand Response and Microgrids. *Sustainability* **2021**, *13*, 11429. https://doi.org/10.3390/su132011429

Academic Editors: Sheraz Aslam, Herodotos Herodotou and Nouman Ashraf

Received: 4 October 2021
Accepted: 11 October 2021
Published: 16 October 2021

Publisher's Note: MDPI stays neutral with regard to jurisdictional claims in published maps and institutional affiliations.

Abstract: In smart grid, energy management is an indispensable for reducing energy cost of consumers while maximizing user comfort and alleviating the peak to average ratio and carbon emission under real time pricing approach. In contrast, the emergence of bidirectional communication and power transfer technology enables electric vehicles (EVs) charging/discharging scheduling, load shifting/scheduling, and optimal energy sharing, making the power grid smart. With this motivation, efficient energy management model for a microgrid with ant colony optimization algorithm to systematically schedule load and EVs charging/discharging of is introduced. The smart microgrid is equipped with controllable appliances, photovoltaic panels, wind turbines, electrolyzer, hydrogen tank, and energy storage system. Peak load, peak to average ratio, cost, energy cost, and carbon emission operation of appliances are reduced by the charging/discharging of electric vehicles, and energy storage systems are scheduled using real time pricing tariffs. This work also predicts wind speed and solar irradiation to ensure efficient energy optimization. Simulations are carried out to validate our developed ant colony optimization algorithm-based energy management scheme. The obtained results demonstrate that the developed efficient energy management model can reduce energy cost, alleviate peak to average ratio, and carbon emission.

Keywords: energy optimization; day ahead energy prediction; artificial neural network; renewable energy sources; demand response; microgrid; smart grid

1. Introduction

The traditional power system is inefficient because it entirely depends on fossil fuels, and having centralized generation that is far away from consumers. In these circumstances, the generated electricity needs to be transmitted and distributed to consumers via transmission and distribution lines over long distances, spending many resources on construction, maintenance of all systems involved and high levels of technical losses [1]. According to [2], centralized power systems suffer from severe transmission and distribution energy losses because of long distances between consumers and generating stations. Furthermore, centralized generation usually causes more environmental pollution than distributed generation technologies. From the consumer's point of view, practically, electricity is used in unintelligent manner without control. Keeping in view both perspectives, new paradigms of distributed generation and smart grid have emerged for electricity generation and consumption. These paradigms are prominent in the electricity market around the world because they have low transmission losses and provide electricity intelligently with control to consumers. According to [1], the distributed generation definitely refers to microgrids, which is small-sized, near to consumers, and directly connected to the distribution system. Smart grids are power grids with advanced communication and control technologies between consumers and generating stations, delivering optimized power usage, clean energy at reduced cost, and improving the quality of energy and efficiency of the power grid. They also provide reductions in technical losses and greenhouse gas emissions, and solve the high carbon emissions problem, where 23% and 41% pollution emission is caused by transport sector, and energy sector, respectively, around the globe [3]. Thus, the conventional power gird is not able to meet future electricity demand due to dependence on limited and environment foe fossil fuels.

With microgrids, evolution dependence on fossil fuels is reduced, and high carbon emission problem is resolved. Furthermore, the microgrid can be connected with smart homes in modes like grid-connected and islanded. In islanded mode, microgrids and commercial grids couldnot initiate the purchase/sell mechanism of energy. On the other hand, in grid-connected mode, the microgrid purchases and sells electricity from/to the external power grid. In the recent past, a significant variety of forecasts have been employed. The selection of a forecasting model is typically dependent on the available data, the model network mechanism's aims, and the energy planning operation. In this paper, we examine renewable energy and power forecast models used as an energy planning tool in a critical and precise way, the application of these techniques for forecasting, their accuracy for geographical and temporal prediction, and their significance to policy and planning purposes are all explored. Machine learning models handling enormous data while also providing precise predictive analyses. By integrating various models and using ensemble methods, we may improve forecasting accuracy. Artificial neural networks, when utilised correctly, can help one make better decisions because they can extract and model previously unknown correlations and characteristics [4]. The microgrid includes renewable energy sources (RES) such as solar, wind etc., to generate electricity, contributing to pollution emission minimization. Furthermore, RES are intermittent in nature; thus, one cannot rely on them. Therefore, ESSs and EVs are used with RES to solve this problem.

One of the problems to be catered within smart grid/smart home is energy management, whose purpose is to give greater control to the user's over their power usage to promote the efficient use of electricity, which is possible with the implementation of demand response (DR). The DR is classified in two classes, namely: direct DR program, in which the electric utility company (EUC) operator disconnects or interrupts the load as per an interruption contract signed with the consumer; indirect DR program, where the consumer changes/adapts their demand in response to the offered pricing signaled by the EUC operator [5]. The latter DR program is the focus of this work.

The indirect DR program plays a vital role in cost efficient and reliable power system operations [6]. In [7], users reduced cost and improve comfort level via scheduling smart home appliances using indirect DR program. Similarly, the DR program is used to schedule multiple homes' loads with the same living patterns in [8]. However, the assumptions made by the authors seem impractical because, usually, multiple homes do not have the same devices with the same operation time and power rating. A novel energy management scheme can create optimal schedule of appliances power usage, which illustrates the same profile as the microgrid generated power. The users reduce electricity cost and import a minute amount of energy from the commercial power grid by adopting the optimal schedule. In this manner, efficient energy management ensures stable and reliable microgrid operation. In [9], EMC is introduced, which schedules the power usage pattern of a single home to alleviate peak demand. This work considered RES and day-ahead weather information to ensure energy cost minimization and an energy efficient operation. The paper [10] proposed a strategy for cost minimization considering weather parameters.

The authors of [11] introduced the deterministic method to schedule the operation of appliances and charging/discharging of EVs. Furthermore, EVs are considered mobile storage. In [12], authors developed a building energy management strategy for peak energy consumption mitigation. They considered ESSs and EVs as storage devices to mitigate fluctuations accompanied by RES and improve building energy efficiency regarding cost minimization. A hybrid energy system of PV, WT, and ESSs is developed with day-ahead energy forecasting in [13]. Furthermore, they considered diesel generator as a backup for power generation; however, power is produced at high cost and pollution emission.

The valuable research reviewed above either focused on load scheduling or charging/discharging scheduling of ESSs and EVs or failed to fully utilize beneficial aspects of smart grid technologies and DR program implementation. Some studies used a diesel generator as a backup, which is not appropriate for RES-based microgrids. Some solutions such as load shedding may cause user frustration, and minimize welfare and system efficiency. With this motivation, this work proposes an efficient management strategy to schedule user activities, ESSs and EVs are connected to a smart microgrid. In addition, we assess a smart home and microgrid equipped with PV panels, WT, electrolyzer, hydrogen tank, ESSs, and EVs connected to the external power grid. As a backup resource, a MGT is employed in place of diesel drive subject to cost and carbon emission minimization concerns. We use mobile storage (EVs) and static storage (ESSs) simultaneously to cater for uncertainties in EVs (parked in home or goes out on driving), and ensure the reliable provision of electricity. Moreover, a prediction model based on modified Enhanced Differential Evolution (mEDE) and Artificial Neural Network (ANN) is developed for microgrid generation capacity accurate prediction instead of assuming. The main contribution and novelty of this paper is outlined below.

- An efficient energy management scheme is proposed, which considers the RTP curve with variations that systematically schedule appliance operation and charging/discharging of EVs to maintain a balance between energy supply and demand.
- Ant colony optimization (ACO) algorithm is adapted, which takes into account constraints, occupant energy consumption pattern, users priorities, and uncertainties in the presence of RTP to schedule load and EVs charging/discharging for efficient energy management.
- Adapted ACO algorithm successfully solves the presented problem, allowing a high monetary reduction in the energy cost paid by consumers, alleviating the peak formation in electricity demand, minimizing carbon emission, and improving the comfort of the users.
- For efficient energy management, an accurate forecast model ANN based on mEDE (ANN-mEDE) is developed to forecast a generation profile of microgrid using weather information and mathematical models of the WT and PV.
- Simulation results demonstrate that the newly devised scheme based on the ACO technique is effective, which considerably reduces the consumer's cost, PAR, and peak electricity demand reduction in the commercial grid.

The rest of the paper is structured as follows: Section 2 presents related work. In Section 3, a proposed efficient system model is discussed. Section 4 presents mathematical modeling of the proposed system model. The problem formulation is presented in Section 5. The simulation results are presented in Section 6 and finally in Section 7, the paper is concluded.

2. Related Work

Several research works have been conducted over the last few decades in the literature to address energy management problems in a smart gird. To solve energy management problems, many heuristic, mathematical, and controller base methods are developed in the literature, which are discussed in detail as follows.

2.1. Mathematical Techniques

In [14], the authors power usage scheduling framework based on Mixed Integer Linear Programming (MILP) to reduce energy cost and alleviate PAR. The authors in [15] proposes a game theory-based appliance scheduling framework for electricity cost and peak energy demand reduction. However, the user frustration level is not taken into consideration. The authors of [16] proposes a scheme for residential sector based on MILP technique to analyze EVs bi-directional flow. In [17], authors mitigate the cost of electricity in a smart house and determine the modes of operation of various loads; an exact solution technique is used to retain the surplus electricity from RESs in batteries. To minimize overall energy bill in the microgrid, an optimal energy management model based on MILP is presented [18]. The authors use the Power Grand Composite Curves (PGCC) technique for adjusting system functioning in response to short-term energy requirement, and their study provides a unique way for identifying appropriate Power Management Strategies (PMS) in RES-based smart grids. The authors of [19] evaluate prosumers-based Energy Management and Sharing (PEMS) as well as the issues that come with it. It will assist in the interpretation and analysis of the prosumers effect on future smart grids. Their study gives a detailed evaluation of these goals, PEMS in the smart grid environment, and its impacts on power system reliability and energy sustainability are studied. In [20], authors developed a technique based on the MILP paradigm, employed in this project to provide an optimal solution in terms of tasks such as energy usage and renewable resource management. The suggested technique achieves an optimal schedule under dynamic electrical limitations while maintaining thermal comfort based on user requirements. The mathematical and deterministic methods suffer from system and computational complexity.

2.2. Controller-Based Methods

Researchers adopted controller-based methods to resolve problems accompanied with deterministic methods. For example, in [21], the authors presented a Distributed Model Predictive Controller (DMPC) for consumers and EUCs DSM. The smart grid have generating units such as RESs, ESSs, and smart load. The DMPC schedules the smart load for efficient DSM. The authors of [22] proposed an integration and control automation of RESs such as a PV plant, a solid oxide fuel cell with battery, and load in smart grid. The Energy Management System (EMS) is based on Proportional Integral (PI) and Adaptive Neuro-Fuzzy Inference System (ANFIS) techniques to effectively balance supply and demand. A DMPC method is developed for grid-connected cooperative energy management in [23]. The proposed DMPC's optimization results are similar to Centralised MPC (CMPC) Pareto solutions according to a real-time hardware-in-the-loop, while the computation speed is significantly faster than CMPC. The authors in [24] proposed an energy management scheme to forecast interrupted data for a microgrid with a centralized dispatching mechanism. The proposed energy management scheme is based on MPC and ensemble learning network approaches. A novel coordinated MPC method is developed that schedules the operation of the microgrid while considering the variations of stochastic RESs as well as meteorological circumstances [25]. A dynamic energy management system based on MPC is developed for a power grid-connected microgrid linked to and serving a residential area [26]. The dynamic energy management system collects data from various components of the electrical system via a smart metering system. However, the controller-based methods are computationally intensive and become too slow for energy optimization.

2.3. Stochastic Techniques

To resolve the problems accompanied by deterministic and controller-based methods, stochastic techniques have emerged as a promising solution. For instance, in [27], the Genetic Algorithm (GA) based non-sorting scheme was adopted for scheduling household with primary objectives such as carbon emissions, power usage, and energy bill minimization. A new algorithm for programming the EMC to schedule household load is developed in [28]. The EMC schedules home appliances in such a manner that the load is shifted to the battery during peak hours. The work [29] proposed a hybrid optimization technique of GA and MILP for economic dispatch. The authors of [30] incorporate a mutation operator with ACO algorithm to resolve the problem trapping at local optima with primary DSM objectives such as PAR, and cost minimization. The authors in [31] scheduled the home appliances to mitigate PAR and minimize the delay time of appliances. However, the Carbon emission is not taken into account in their study. The authors of [32] implemented the Firefly Algorithm (FA) for energy management under the pricing schemes such as Real-Time Price (RTP) and Critical Peak Price (CPP). However, user comfort is compromised. The Grey Wolf Accretive Satisfaction Algorithm (GWASA) to solve the residential demand-side management problem with the lowest cost and highest ratio of satisfaction in [33]. The [34] used the Candidate Solution Updation Algorithm (CSUA) for reducing PAR and consumer delay time by increasing user comfort level. The authors of this paper considered uncertainties in loads and ensured optimal scheduling to facilitate residents. The authors presented an intelligent energy management principles and technology issues for smart grid applications to help the Distributed Electric System (DES). The authors in [35] conducted a comprehensive analysis of IoT-based energy management in smart communities. Following that, the foundation and software model for an IoT-based system at the network's edge is presented. The authors developed optimum power scheduling technique using RTP and Inclining Block Rate (IBR) tariff to minimize electricity costs, reduce the PAR, and minimize user discomfort. The authors in [36] proposed hierarchical architecture using cloud computing and edge computing to a distributed architecture which provides autonomous strategic decisions with agent-based intelligence for massive information. In households and grids, large-scale information gathering, communications, processing, and control are performed through agents for cooperative energy management. The results of the experiments show that the agent-based solution is promising in cooperative energy management. In [37], the authors developed a game-theoretical model to schedule entire electricity consumption scheduling for efficient power consumption planning and DSM. The [38] proposed load schedling and distributed storage approach to improve user satisfaction and minimize customer energy costs. In [39], an optimization-based energy management structure is developed to schedule consumer energy consumption pattern using RTP signals under utility, PV, and ESSs. The purpose is to reduce energy bills, carbon emissions, and peak power consumption while mitigating pricing rebound peak generation. The authors proposed a multi-domain communication network with federation concept in [40]. This model introduced how IEC 61850 and the extensible Message Presence Protocol (XMPP) may be used to provide a common communication framework for Virtual Power Plants (VPPs) management in smart grids. A privacy-preserving technique is introduced. The nodes use accurate data for estimation and broadcasting the noisy version in [41]. The authors show that the proposed algorithm can protect privacy and retain the final solution's convergence and optimality. Extensive simulations indicate that their proposed strategies are effective. The authors used a GA-, game theory-, and fuzzy logic-based framework that seeks to maximize profit by choosing the optimum alternative and forecasting future energy demands [42]. The authors implemented a Cuckoo Search Algorithm (CSA) to study the impact Battery Energy Storage (BES) on power system operation [43]. A game-theoretic (Stackelberg) model is introduced in [44] to examine coordination of generators with microgrids. A novel technique with Artificial Neural Networks (ANN) is developed for expressing and conveying the energy flexibility of distributed energy resources for efficient energy management [45]. In [46], the authors proposed a scheduling technique based on the markov decision process for energy management in a smart grid . This work aims to lower a customer's energy expenditure. The authors in [47] presented a method for regulating the active and reactive power flow in an islanded renewable generating system's Point of Common Connection (PCC). In [48], the authors employed optimum DR technique to maintain balance between supply and demand. The DR program engaged the Plug in Hybrid EVs (PHEVs) at parking stations as distributed energy storage and source to participate in DSM.

The related works discussed above are summarized in Table 1. All the methods discussed above are capable and effective in energy management. However, the researchers did not all use key features such as advanced metering infrastructure, forecasting, and bidirectional communication of the smart grid. Furthermore, mostly researchers assumed consumers behavior and RESs profile. Some authors focused on electricity cost minimization, others catered for PAR alleviation, a few handled both electricity cost and PAR minimization, while some authors considered user comfort maximization. However, the objectives such as electricity cost, carbon emission, PAR, and user comfort are not handled simultaneously by any authors in the literature. Thus, it is concluded from the above literature that electricity cost, carbon emission, PAR, and user comfort are energy management objectives, where carbon emission and PAR are utility objectives which are directly linked to customer objectives such as energy cost and user comfort. The utility and customer objectives are contradictory in nature and challenging to handle simultaneously.

Table 1. A detailed analysis of the most relevant study in terms of sources, storage, objectives, and proposed algorithms. Abbreviations used in the Table are; RESs = Renewable Energy Sources, CG = Commercial Grid, BES = Battery Energy Storage, ESS = Energy Storage System, and MGT = Micro Gas Turbine.

References	Sources	Storage	Objective(s)	Proposed Algorithm
[14]	RES + CG	ESS + MGT	Reducing PAR, minimizing cost, maximizing user comfort	MILP
[15]	RES + CG	ESS	Minimizing cost and PAR	Game-theory framework
[16]	RES + CG	ESS	Reducing electricity cost	MILP
[17]	RES + CG	ESS	Reducing electricity cost	MILP
[18]	RES + CG	ESS	Short term energy demands	PGCC and MILP
[19]	RES + CG	ESS	Issues faced by prosumer	PEMS and MILP
[20]	RES + CG	ESS	Saving consumer cost	MILP
[21]	RES + CG	ESS	Electricity price varying	MPC
[22]	RES + CG	ESS	Minimizing electricity cost	PI and ANFIS
[23]	RES + CG	ESS	Minimizing electricity cost	DMPC
[24]	RES + CG	ESS	Reducing electricity cost	MPC
[25]	RES + CG	ESS	Reducing green house gases emissions	MPC
[26]	RES + CG	ESS	Energy-saving and gain	MPC
[27]	RES + CG	ESS	Carbon emission, energy consumption and reducing electricity cost	GA
[28]	RES + CG	ESS	Cost minimization, maximize comfort level	Control algorithm
[29]	RES + CG	ESS	Minimizing cost and PAR and economic dispatch	GA and MILP
[30]	RES + CG	ESS	Minimizing PAR and cost	ACO algorithm
[31]			Reducing PAR cost, and consumer delay time	GmEDE
[32]	RES + CG	ESS	Minimizing electricity cost	FA
[33]	RES + CG	ESS	Minimize cost and high level of user satisfaction	GWASA
[34]	RES + CG	ESS	Reducing PAR and increase user comfort level	CSUA
[49]	RES + CG	ESS	Minimizing cost and PAR	GA, GWO, mEDE and GmGWO
[50]	RES + CG	ESS	Issues of EVs integration smart grid	DES
[35]	RES + CG	ESS	Cost minimization	IoT-based system
[51]	RES + CG	ESS	Minimizing cost and reducing PAR	Aquifer Thermal Energy Storage (ATES)
[36]	RES + CG	ESS	Reducing electricity cost	Hierarchical architecture
[37]	RES + CG	ESS	TCLs electricity consumption scheduling and minimizing RES fluctuation	Game-Theoretic Demand Side Management
[52]	RES + CG	ESS + BES	Minimizing cost and PAR	GA and PSO and ACO
[38]	RES + CG	ESS	Minimizing cost and User comfort level	Distributed storage strategy
[39]	RES + CG	ESS	Minimizing cost and PAR	HGACO
[40]	RES + CG	ESS	VPPM	XMPP based IEC 61850 communication
[41]	RES + CG	ESS	Accurate data for state changing	Privacy-preserving technique
[42]	RES + CG	ESS	Predicting future energy demands using GA	Game-theory based fuzzy logic
[43]	RES + CG	ESS	BES on the functioning of power systems	CSA
[44]	RES + CG	ESS	Maximizing their payoffs	Stackelberg game theoretic framework
[45]	RES + CG	ESS	Energy flexibility of distributed energy resources	ANN
[46]	RES + CG	ESS	Lowering Consumer's electricity cost	MDP
[47]	RES + CG	ESS	Reactive islanded power flow	PCC
[48]	RES + CG	ESS	Energy Balance and flexible loads	Multilayer individual-based optimization algorithm

Thus, an optimization technique is needed that considers user priority, DR program, and comfort constraints to cater for uncertainties in load and renewable generation for efficient energy utilization, energy cost reduction, PAR alleviation, carbon emission mitigation, and end-user satisfaction, to satisfy both utility providers and consumers at the same time. In this regard, an innovative framework composed of residential smart homes and a grid-connected smart microgrid is proposed. Furthermore, an efficient energy management scheme based on the ACO algorithm is developed for the proposed framework to systematically schedule load and EV's charging/discharging connected in a smart microgrid.

3. Framework of Efficient Energy Management System

Efficient energy management framework is presented in this section, followed by a description of the day-ahead energy generation prediction model. The complete implementation diagram of the proposed model is depicted in Figure 1, which is discussed in detail in subsequent sections.

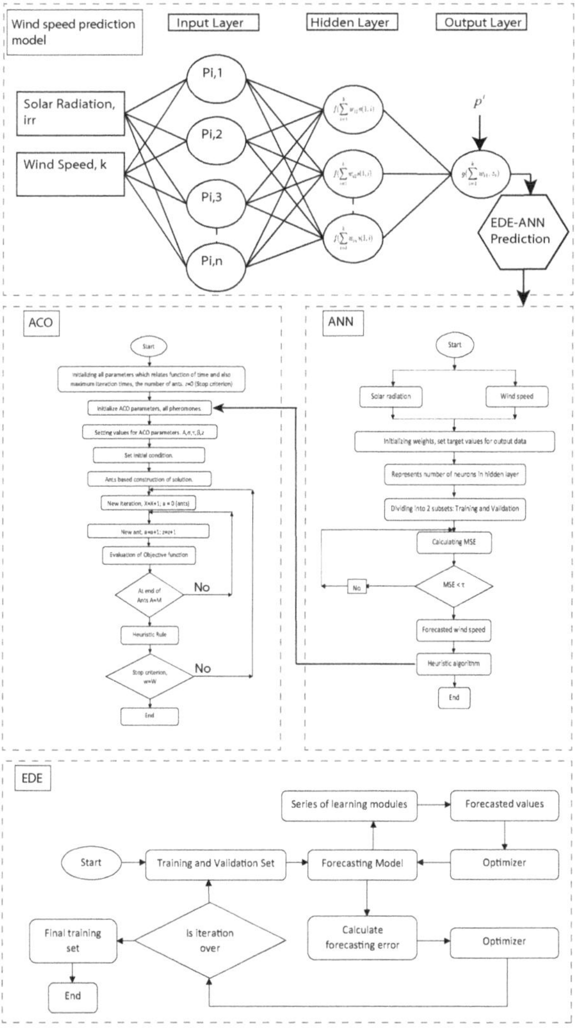

Figure 1. Proposed scheme implementation flow chart for efficient energy management

3.1. System Model

The proposed system model is composed of a residential smart home equipped with smart meter and smart appliances, and a smart microgrid comprising WT, PV panels, electrolyzer, and hydrogen tank. The proposed system model is depicted in Figure 2. However, residential consumers load and electricity generation of a microgrid is stochastic and uncertain. Thus, an ANN-mEDE-based prediction model is developed for efficient energy management. An efficient EMC based on ACO algorithm (whose control parameters are listed in Table 2) is installed to systematically schedule load and EVs charging/discharging connected to a smart microgrid according to RTP signal. In our proposed work, the simulation time T is assumed to be 1 day, which is split into sub-timeslots of equal duration denoted by $t(1h)$. The detailed description of each component of the proposed system model is as follows.

Table 2. ACO Parameters.

Technique	Parameters	Values
Ant Colony Optimization	Number of Ants	15
	Maximum number of Iterations	250
	Evaporation Rate	7
	Pheromone Factor	3
	Stopping criteria	When Maximum iteration reached

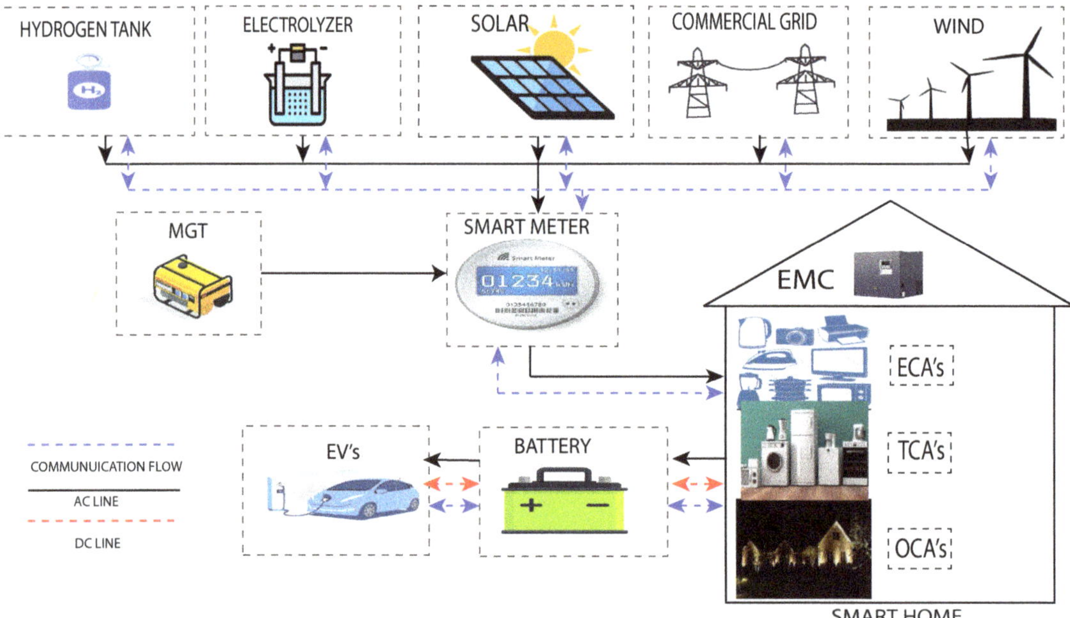

Figure 2. Proposed system model schematic diagram

3.2. Energy Generation Prediction Model

This section describes the prediction model to predict solar irradiation and wind speed for effective generation estimation. In this regard, a framework of ANN-mEDE is developed to predict solar irradiation and wind speed for effective estimation of generation. The proposed prediction model is composed of three parts: (i) feature selector, (ii) forecaster, and (iii) optimizer. Feature selector phase of the prediction model based on mutual information technique uses time-series solar irradiation and wind speed as input. The mutual information technique rank the inputs and passed the ranked inputs to redundancy and irrelevancy filters for the removal of irrelevant and redundant features. Then, the selected inputs are split into training and testing datasamples.

The forecaster phase is based on an artificial neural network (ANN), which uses training and testing sets to forecast solar irradiation and wind speed for a day-ahead time horizon. The ANN is a multilayer feed-forward network, where each succeeding layer gets input from the preceding layers. In other words, the outputs of one layer's nodes are used as inputs in the next layer. A weighted linear combination is used to combine the inputs, which are changed by a nonlinear function [53] to generate output.

The ANN has three layers layout having input, hidden, and output layers, where Artificial Neurons (ANs) in each layer are exploited by sigmoidal activation function [54], which is shown in Equation (1) as follows.

$$f(S, b) = \frac{1}{1 + e^{-\beta(S-b)}} \tag{1}$$

The input signal is S with attributes (as discussed in the first module). The parameter β is for steepness control of the activation function and b indicates the bias value. As discussed above, the developed prediction model is enabled via training to learn and accurately estimate future energy generation. In the literature, learning mechanisms such as unsupervised, supervised, and re-enforcement learning exist. The developed forecasting framework learns from time series analysis with the supervised learning approach, which employs multivariate autoregressive rules due to high convergence than benchmark learning rules [55]. Solar irradiation and wind speed prediction model is illustrated in Figure 3. The RESs three-year weather data is adopted from [56]. The acquired data is divided into training and testing sets of 80% and 20%, respectively, as per the mechanism available [57–59].

$$MAPE(i) = \frac{1}{n} \sum_{j=1}^{m} \frac{\left| p^{actual}(i,j) - p^{forecast}(i,j) \right|}{p^{actual}(i,j)} \tag{2}$$

The training set trains the forecasting framework to predict future values, and testing set validates the forecasting framework to show the accuracy of the obtained predicted results compared to the ground-truth observations. Mean absolute percentage error (MAPE) is a validation metric (to illustrate the relationship between ANN predicted output and observed values), which is formulated as follows in Equation (2) [54].

Where $p^{actual}(i,j)$ denotes the actual and $p^{forecast}(i,j)$ represents forecasted solar irradiation and wind speed. m denotes number of days under observation. The control parameters are tuned using Levenburg Marquart algorithm until the error is minimized. The forecasted values (solar irradiation and wind speed) by ANN forecaster is fed to optimizer phase of the prediction model for further error minimization.

In the optimizer phase, the error is calculated between estimated and observed value, and this error is further minimized with the use of the mEDE algorithm in the optimizer phase. The meta-heuristic algorithms such as GA, PSO, FA, DE and GWO are widely used in prediction problems. The DE among these algorithms is selected due to its better performance in the aspects of fast converging speed, computational efficiency, and avoidance of premature convergence while seeking a global optimal solution. This better performance of DE is due to the different vector-based mechanism, which expands its search space [60]. On this note, the authors of [61–64] used DE and GA techniques, respectively; with this as motivation, we developed modified enhanced DE to optimize control parameters of the prediction model for returning accurate estimation without trapping into local optima and avoiding premature convergence. The energy estimations returned from the optimizer phase are more accurate and utilized for efficient energy management.

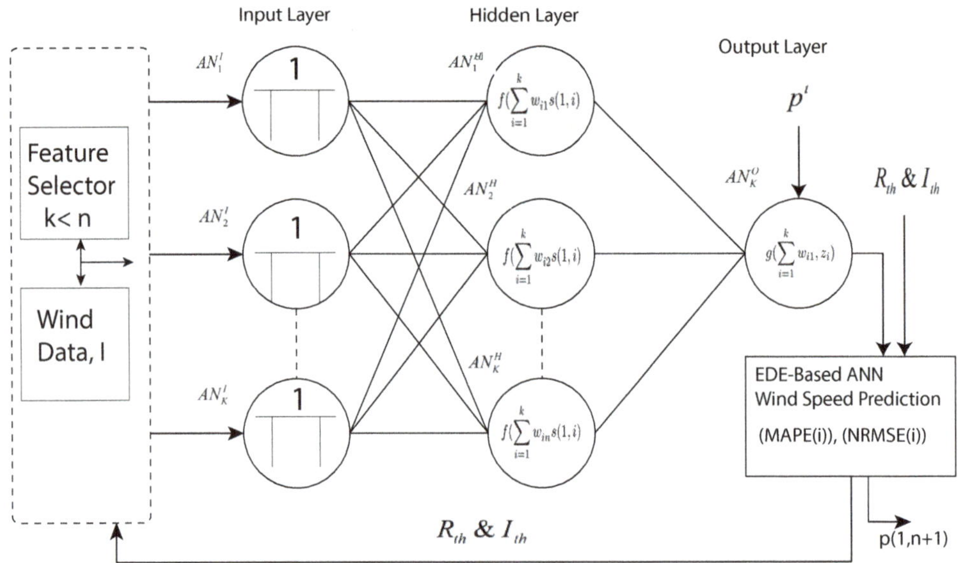

Figure 3. Solar irradiance and wind speed prediction model for estimation of electricity

4. Mathematical Modeling

This section presents the mathematical modeling of a smart microgrid equipped with controllable appliances, PV panels, WT, electrolyzer, hydrogen tank, EVs, and ESS.

4.1. Modeling of Appliances Operating within Smart Home

This section presents explanation and modeling modeling of appliances operating within smart home. Each consumer is in contact with smart appliances. Moreover, each appliance is scheduled according to consumer demand. These appliances are further categorized into three categories: (i) Electrically Controllable-Appliances (ECA), (ii) Thermostatically Controllable-Appliances (TCA), and (iii) Optically Controllable-Appliances (OCA). Each category will be discussed in detail as follows.

4.1.1. Scheduling of ECA

An adaptive approach based on RTP signal for ECA scheduling such as PHEVs is proposed. The EMC schedules ECA as soon as it receives RTP signal from the utility. Consumers typically schedule smart appliances within the assigned time interval to avoid high payment due to operation in peak price hours. Furthermore, the ECA parameters such as the time interval in between the Earliest Starting Time (EST) and Latest Finishing Time (LFT), and Length of Operation Time (LOT) are adjusted using In-Home Display (IHD) and the adjusted parameters will be communicated to EMC via Home Gate (HG). These parameters are adopted from [65] and listed in Table 3.

Table 3. Appliances Classification.

Category	Appliance	EST (h)	LFT (h)	LOT (h)	Power (kw)
Electrically Controllable Appliances	Dishwasher	09:00	17:00	7	2.4
	Dryer	13:00	18:00	5	2.5
	Washing Machine	09:00	17:00	8	2
	Pump	12:00	19:00	9	1
	EV	18:00	8:00	15	3.5
Thermostatically Controllable Appliances	Heater	09:00	19:00	6	2.4
	Fridge	00:00	24:00	24	0.5
	Freezer	09:00	20:00	10	0.3
	AC	08:00	16:00	3	0.7
Optically Controllable Appliances	Lighting	19:00	24:00	6	0.84

The appliance consumption cycle is started using the starting probability function, which is defined as P_{step} in Equation (3):

$$P_{start}(A, W, \delta, \sigma_{flat}, h, d) = P_{season}(A, W) P_{hour}(A, h, d) P_{step}(\delta) P_{social}(\sigma_{flat}) \qquad (3)$$

where A represents an ECA appliance, h represents the hour of the day, d represents the day of the week, W represents the week of the year, and δ represents the computing time step (second or minute), σ is the standard deviation for P_{social} which represents the social random factor. Moreover, seasonal changes are customized by P_{season}, P_{hour} the probability hourly factor, P_{step} stands for scaling factor, which scales the probabilities in consideration with δ. P_{start} is described for each time interval δ it takes a value between 0 and 1. Further detailed elaboration is given in [65].

4.1.2. Scheduling of TCA

TCA may be electrical or thermal equipment such as an air conditioner or a water heater, which can be scheduled using the desired temperature and RTP signal. The air conditioner, refrigerator, heater, and freezer temperatures are denoted by T_t^{ac}, T_t^{fridge}, T_t^{heater}, and $T_t^{freezer}$, respectively, presented in Equations (4)–(7) which should be adjusted by consumers as per their specified temperature to avoid user-frustration. These circumstances can be stated as follows:

$$T_{min}^{ac} \leq T_t^{ac} \leq T_{max}^{ac} \qquad (4)$$

$$T_{min}^{fridge} \leq T_t^{fridge} \leq T_{max}^{fridge} \qquad (5)$$

$$T_{min}^{heater} \leq T_t^{heater} \leq T_{max}^{heater} \qquad (6)$$

$$T_{min}^{freezer} \leq T_t^{freezer} \leq T_{max}^{freezer} \qquad (7)$$

where T_{min}^{ac}, T_{min}^{fridge}, T_{min}^{heater} and $T_{min}^{freezer}$ are air conditioner, fridge, heater and freezer lower bound, while T_{max}^{ac}, T_{max}^{fridge}, T_{max}^{heater} and $T_{max}^{freezer}$ are air conditioner, fridge, heater and freezer upper bound, respectively. These appropriate temperatures may differ from one household to the next, since they are determined by each individual consumer to ensure user comfort. Further detailed elaboration can be found in [65].

4.1.3. Scheduling of OCA

The OCA mostly includes lighting loads, which are scheduled according to illumination. The lighting load is modeled using illumination level index, which is based on the activity probability, which indicates the house occupancy in the lighting load calculation. The following is how the illumination of a room (a) in the house is expressed in Equation (8).

$$L_t^a + L_t^{OUT} \geq (1 + k_t) L_t^{a,min} \qquad (8)$$

A detailed elaboration is available in [65].

4.2. Microgrid

The microgrid considered in this work is equipped with controllable appliances, PV panels, WT, electrolyzer, hydrogen tank, EVs, and ESS for electricity generation. The net energy generated from the sources equipped with microgrid is $m \, \epsilon \, M$ in time interval $t \, \epsilon \, T$ is presented in Equation (9) as follows.

$$E(t) = \sum_{m \epsilon M} em(t) \tag{9}$$

where em represents microgrid generation. Equation (10) calculates microgrid total electricity generation.

$$E(t) = \sum_{t=1}^{T} \sum_{m \epsilon M} em(t) \tag{10}$$

As the microgrid is equipped with RESs which are intermittent in nature, thus, the ANN-mEDE prediction model is implemented to predict solar irradiation and wind speed for next-day timeslots T for accurate future electricity generation estimation. The detailed explanation of generating sources equipped with microgrid is presented as follows.

4.2.1. Wind Turbine

The WT is entirely dependent on the wind speed for electricity generation, which is mathematically modeled in Equation (11) as follows.

$$P^{wt}(t) = 1/2 \times C_p \times (\lambda) \times \rho \times A \times (V_t^{wt})^3 \tag{11}$$

where P^{wt} represents WT electricity generation at timeslot t, A is the area swept by turbine blades through which WT generates power, V_t^{wt} denotes wind speed and air density is ρ. The electricity generation from WT is directly proportinal to wind speed, i.e., the higher the wind speed, the higher the generation, and vice versa, as presented in Equation (11). The WT electricity generation is subjected to specified constraints [66], which are defined in Equations (12)–(14).

$$V^{cut-in} \leq V^{wt}(t) \leq V^{cut-out} \tag{12}$$

$$0 = V^{wt}(t) \geq V^{cut-out}, \; \forall \, t \, \varepsilon \, T \tag{13}$$

$$0 = V^{wt}(t) \leq V^{cut-in}, \; \forall \, t \, \varepsilon \, T \tag{14}$$

4.2.2. PV Panels

PV panels generates electricity from the sunlight, which is mathematically modeled in Equation (15) as follows [65].

$$P_t^{pv} = \eta^{pv} \times A^{pv} \times Irr(t) \times (1 - 0.005 \times (Temp(t) - 25)) \tag{15}$$

where P_t^{pv} shows amount of electricity produced per hour from PV panels. The symbols A^{pv} and η^{pv} show area and efficiency of PV panel, respectively. Moreover, the $Temp(t)$ and $Irr(t)$ indicate amount of solar irradiation and the outside temperature at timeslot t.

4.2.3. Electrolyzer

An enhanced electrolyzer has been developed in our proposed study. In the electrolyzer model, electrochemical processes are employed. Furthermore, the maximum temperature of 100 °C was taken into account. An electrolyzer cell's electrode dynamics can be approximated using an empirical voltage relationship. Several electrolyzer empirical models have been proposed [67–70].

$$U = U_{rev} + \frac{r}{A}I + s\log(\frac{t}{A} + 1) \tag{16}$$

In addition, in order to model the temperature dependency of an over voltages, Equation (16) can be altered.

4.2.4. Hydrogen Tank

As this Bennedict Webb Rubin (BWR) equation comprises of eight parameters, and is more exact than that of three-parameter equations, particularly at temperatures above critical and pressures not exceeding extremely high, as indicated in Equation (17) [71].

$$p = RT\rho + (B_0RT - A_0 - \frac{C_o}{T^2})\rho^2 + (bRT - a)\rho^3 + a\alpha\rho^6 + \frac{c\rho^3}{T^2} * [(1 + \gamma\rho^2)\exp(-\gamma\rho^2)] \quad (17)$$

where ρ represents gas density which relates to compressibility factor by (18):

$$z = 1 + (B_0 - \frac{A_0}{RT} - \frac{c_0}{RT^3})\rho + (b - \frac{a}{RT})\rho^2 + \frac{a\alpha}{RT}\rho^5 + \frac{c\rho^2}{RT^3}[(1 + \gamma\rho^2)\exp(-\gamma\rho^2)] \quad (18)$$

where as, p is represented in atm, T is in kelvin (k), ρ is in mol/L, $R = 0.08205$ atm L/mol K. Moreover, the remaining parameters are as follows:

$a = -9.2211 \times 10^{-3}$ (L/mol)^3 atm;
$A_0 = 9.7319 \times 10^{-2}$ (L/mol)^2 atm;
$b = 1.7976 \times 10^{-4}$ (L/mol)^2;
$B_0 = 1.8041 \times 10^{-2}$ (L/mol);
$c = -2.4613 \times 10^2$ (L/mol)^3 K^2 atm;
$C_0 = 3.8914 \times 10^2$ (L/mol)^2 K atm;
$\alpha = -3.4215 \times 10^{-6}$ [(L/mol)^3];
$\gamma = 1.89 \times 10^{-3}$ [(L/mol)^2]

4.3. Micro-Gas Turbine

MGT is a generator used for electricity generation with a generation capacity range of 15–300 (kW). It is usually preferred due to its ease of installation and maintenance [72]. In this work, the MGT is employed as a backup and shiftable generating source, and operates as per the direction and requirement of users. Its generation is cheaper and environment friendly compared to diesel generators, and significantly contributes to carbon emission minimization [29]. On/off are the two operation statuses; it produces constant energy of 2 kW in on status while in off status, its generation is 0 kW [13].

4.4. Energy Storage System

ESSs includes static storage and mobile storage systems that are equipped with RES in the microgrid to store energy during on-peak hours (act as a load) and discharge energy to load in low price hours (act as a source). Complete discussion is presented below.

4.4.1. Static Energy Storage System: BES

The smart home considered in this work is equipped with BES with a storage capacity of 3 kWh, the same as is discussed in [14,73]. The BES is subjected to various constraints such as ESS_{min} and ESS_{max}, which represent minimum and maximum charge limit, respectively. Every BES has a predefined discharging limit, namely lower discharging limit, which is set at 20%, the same as is used in [74,75]. The energy stored in BES is mathematically modeled in Equation (19).

$$SE(t) = SE(t-1) + k \times \eta^{ESS} \times ES^{ch}(t) - k \times ES^{dis}(t)/\eta^{ESS} \quad (19)$$

subjected to

$$ES(t)^{ch} \leq ES_{max} \quad (20)$$

$$ES(t)^{dis} \geq ES_{min} \quad (21)$$

$$ESS(t)^{ch} < ESS_{upl} \tag{22}$$

where SE shows stored energy at timeslot t in BES measured in Ah, η^{ESS} is the efficiency of BES, $ES^{ch}(t)$ and $ES^{dis}(t)$ represent charging and discharging status at timeslot t, respectively. Moreover, the charging/discharging and idle status of BES is mathematically modeled in Equation (23) as follows.

$$\alpha_e(t) = \begin{cases} 1 & if \ ESS \ is \ ch\arg ing, \\ -1 & if \ ESS \ is \ disch\arg ing, \\ 0 & if \ ESS \ is \ idle. \end{cases} \tag{23}$$

When the status of BES is equal to 1, the battery is charging from RES and acts as a load. On the other hand, when the status of BES is equal to -1, the battery is in discharging mode and acts as a source; otherwise, it will be in idle mode.

4.4.2. Mobile Energy Storage System: EVs

EVs are employed as mobile energy storage systems as well as being used for driving purposes. The main objective is to mitigate fluctuations accompanied by RES and minimize energy cost. The EVs energy utilization for arrival and departure usage mode is mathematically modeled in Equation (48) as follows [14].

$$EV_e = \sum_{t=t_a^e}^{t_d^e} (\psi_e(t) \times \alpha_e(t)) \tag{24}$$

where, in Equation (24), t_a^e represents EVs arrival time and t_d^e denotes EVs departure time. The symbols $\psi_e(t)$ and $\alpha_e(t)$ denote charging/discharging in Ah and status of EV at timeslot t, respectively. When $\alpha_e = 1$ EV is in charging mode, $\alpha_e = -1$ EV is in discharging mode, and $\alpha_e = 0$ EV is in idle mode.

5. Problem Formulation

In energy management, the main objectives are energy cost, PAR, carbon emission, and user discomfort minimization to facilitate both end-users and utility. The energy management objectives can be achieved by actively engaging consumers/end-users in distributed generation and DR programs. In this work, end-users are engaged in both distributed generation and DR programs to obtain the energy management objectives. On this note, the EMC based on the ACO algorithm is employed, which receives a DR signal (RTP) and broadcasts it to consumers ahead of time. In response, consumers send their electricity consumption pattern to the EMC based on the ACO algorithm, whose control parameters are listed in Table 2. The EMC schedules power usage of consumers and charging/discharging of EVs and BES to manage the electricity consumption pattern of consumers in such a manner that energy cost is reduced, PAR is alleviated, carbon emission is mitigated, and user discomfort is minimized. Thus, the proposed objective function is modeled as optimization function to minimize the above-discussed objectives. Each optimization objective will be individually mathematically formulated following the whole energy management problem formulation.

5.1. Energy Cost

Energy cost is a payment made by consumers to utility providers for the electricity consumed for a specific period of time. The energy cost is formulated with the RTP signal offered by the utility provider. In 2009, FERC reported that consumers who implemented DR programs for power usage scheduling achieved 65% monetary benefit in aspects of cost reduction. The energy cost payment to the utility providers for energy consumption under the RTP signal without considering the microgrid is formulated in Equation (25).

$$\zeta(t) = (\Gamma_{ECA}(t) + \Gamma_{TCA}(t) + \Gamma_{OCA}(t)) \times EP(t) \tag{25}$$

where $\Gamma_{ECA}(t)$, $\Gamma_{TCA}(t)$, and $\Gamma_{OCA}(t)$ are electricity demand of ECA, TCA and OCA, respectively. Moreover, the total energy cost charged per day is formulated in Equation (26) as follows.

$$\zeta(T) = \sum_{t=1}^{T} (\Gamma_{ECA}(t) + \Gamma_{TCA}(t) + \Gamma_{OCA}(t)) \times EP(t) \tag{26}$$

The energy imported per hour after considering the microgrid equipped with RES, BES, EV, and MGT is formulated in Equation (27) as follows.

$$\Phi(t) = ((\Gamma_{ECA}(t) + \Gamma_{TCA}(t) + \Gamma_{OCA}(t)) - (E(t) + EV.\alpha_{EV(t)} + ESS.\alpha_{ESS(t)} + MGT.\alpha_{MGT(t)})) \tag{27}$$

$$\Phi(t) = \begin{cases} \Phi(t) & if\ \Phi(t) > 0, \\ 0 & otherwise \end{cases} \tag{28}$$

Equation (28) shows that if $\Phi(t) > 0$ the electricity is imported and if $\Phi(t) = 0$, the energy is not imported. The total energy imported per day is formulated in Equation (29) as follows.

$$\Phi(T) = \sum_{t=1}^{T} \Phi(t) \tag{29}$$

The consumers hourly and daily energy cost after considering the microgrid is formulated in Equations (30) and (31), respectively, as follows.

$$\delta(t) = (\Phi(t)EP(t)) \tag{30}$$

$$\delta(T) = \sum_{t=1}^{T} (\Phi(t) \times EP(t)) \tag{31}$$

where $\delta(t)$ is the per hour energy cost and $\delta(T)$ is the total energy cost per day. The symbol $\Phi(t)$ represents imported electricity per hour and $EP(t)$ is RTP signal.

5.2. PAR

PAR is a ratio of peak electricity consumption to average electricity consumption. The PAR is important for EUCs and users because it smooths out the load curve for utility providers, which stops operation of peak power plants during peak hours and thus reduces consumers energy cost. In this work, utility providers stimulate users to participate in DR either by load and EVs charging/discharging scheduling, and to install the EMC, both of which significantly contribute to PAR alleviation. The PAR is mathematically formulated in Equation (32) as follows.

$$\mu = \frac{\max\left(\Gamma_{ECA}(t) + \Gamma_{TCA}(t) + \Gamma_{OCA}(t) + ES^{ch}(t)\right)}{\Gamma_{ECA}(t) + \Gamma_{TCA}(t) + \Gamma_{OCA}(t) + ES^{ch}(t)\Big/4} \tag{32}$$

where μ represents PAR.

5.3. User Comfort

User comfort is measured in various aspects such as waiting time, energy consumption, temperature, sound, illumination, and demographic energy demand patron of consumers. This study measures user discomfort in aspects of delay, i.e., how much of a delay a user confronts after scheduling. Furthermore, a tradeoff exists between energy bill and user discomfort; the users who tolerate more delay would be charged a lower energy cost and those who are not delay-tolerant would be charged a higher energy cost. The user comfort is measured in aspects of delay/waiting time, which is mathematically formulated in Equation (33).

$$w_a = \frac{\sum\limits_{t=1}^{T}\sum\limits_{a=1}^{n}\left|\left(T_{a,t}^{o,unsch} - T_{a,t}^{o,sch}\right)\right|}{T_a^{lo}} \tag{33}$$

where w_a is the waiting time that each appliance may face after scheduling, $T_{a,t}^{o,unsch}$ shows status of an appliance before scheduling, $T_{a,t}^{o,sch}$ denotes status of an appliances after scheduling, and T_a^{lo} represents the total LOT of an appliance. The EMC based on the ACO algorithm schedules appliances in response to the RTP signal and consumer priority. An appliance that can tolerate maximum delay/waiting time is formulated in Equation (34) as follows.

$$w_a^d = T_a^t - T_a^{lo} \tag{34}$$

where w_a^d represents maximum waiting time/delay an appliance faces after scheduling and T_a^t denotes the total time interval of appliances. User frustration increases with the increase in w_a^d and hence their comfort is compromised. The user discomfort is at its peak when $w_a = w_a^d$, and this represents the worst case, which never happens. Percentage user discomfort is mathematically modeled in Equation (35).

$$D = \frac{w_a}{w_a^d} \times 100 \tag{35}$$

5.4. Carbon Emission

Carbon emission is defined as the release of carbon dioxide in atmosphere while generating and using electricity in the energy sector. In this work, distributed generation and MGT are included in the microgrid instead of fossil fuels and diesel generators, while load and charging/discharging of EVs are scheduled, which reduces carbon emissions and ensures a cleaner and greener environment. The carbon emission is mathematically formulated in Equation (36) as follows.

$$Y = \frac{mean(EP(t))}{\varepsilon \times \varsigma \times \Im} \tag{36}$$

The term Y represents carbon emission which is measured in pounds, where $mean(EP(t))$ denotes mean electricity price, ε indicates price per kWh, ς is the electricity emission factor, and $\Im$ is the hour in the day.

5.5. Objective Function

Now, the overall problem is modeled as optimization problem to achieve our desired objectives: minimum energy cost, carbon emission, PAR, and user discomfort with load and charging/discharging scheduling, which is expressed is in Equation (37) as follows.

$$\text{minimize} \sum_{t=1}^{T} \left(\begin{array}{c} (\Gamma_{ECA}^t + \Gamma_{TCA}^t + \Gamma_{OCA}^t) - \\ (E^t + ESS^t + EV^t + MGT^t) \end{array} \right) \times EP^t \tag{37}$$

Subjected to:

$$E(t) = P^{pv}(t) + P^{wt}(t) + P^{hyd}(t) + P^{elec}(t), \tag{38}$$

$$\begin{array}{c} \Gamma_{ECA}^t + \Gamma_{TCA}^t + \Gamma_{OCA}^t = E^t + ESS^t \\ + EV^t + MGT^t + \Phi^t, \end{array} \tag{39}$$

$$\sum_{a=1}^{n} \eta = LOT(a), \tag{40}$$

$$\sum_{a=1}^{n} \alpha \leq \eta \leq \beta, \tag{41}$$

$$U_{rev} = \frac{\Delta G}{zF} \tag{42}$$

$$z = \frac{pV_m}{RT} \tag{43}$$

$$\Phi_t \leq KI, \tag{44}$$

$$V^{cut-in} \leq V(t)^{wt} \leq V^{cut-out}, \qquad \forall t \, \varepsilon \, T, \tag{45}$$

$$0 < Irr(t) < kc, \qquad \forall t \, \varepsilon \, T, \tag{46}$$

$$0 < ESS_{min} < ESS_{max}, \qquad \forall t \, \varepsilon \, T, \tag{47}$$

$$EV^e_{min} \leq \rho EV^e_a + EV_e \leq EV^e_{max} \tag{48}$$

The symbols used in the above equations are defined in Section 3 and presented in Nomenclature table.

6. Analysis of Simulation Results

Analysis of simulation results are discussed in this section. The extensive simulations are carried out in a MATLAB environment installed in the computer system with Intel Core i5 2.4 GHz processor, 8 GB RAM, and Windows 10 operating system, in order to show optimal operation and charging/discharging scheduling of EVs. This study uses four performance metrics including energy cost, PAR, carbon emission, and waiting time/delay for performance evaluation of the proposed model in comparison with existing models. The proposed system model was developed for a residential smart home with three types of load: ECA, TCA, and OCA, which communicate with an EMC based on a ACO algorithm via the Internet, and the EMC schedule operation of appliances and charging/discharging of EVs per the RTP tariff received from the utility provider. The appliance description is discussed in Section 3 and their parameters are listed in Table 3. Furthermore, we developed a forecasting framework for solar irradiation and wind speed to accurately estimate electricity generation for efficient energy management. Simulations are conducted for cases: (I) energy management without a microgrid, (II) energy management with a microgrid for a 24 h time horizon. The detailed discussion is as follows.

6.1. Energy Management without Microgrid

This section presents the results and discussions of scenario I, where microgrid is not considered.

An RTP from Figure 4, it is obvious that the proposed scheme schedules the load during off-peak hours. When the proposed technique is not implemented, meanwhile, the maximum amount of energy is purchased during peak hours. In this figure, the hourly cost is also shown.

The electricity consumption profile of appliances like ECA, TCA, and OCA within home in presence of RTP tariff received from utility with and without proposed approach is illustrated in Figure 4 and their 24 h status is listed in Table 4. It is obvious that the proposed scheme schedules operation of most appliances during low and mid price hours considering the priority of users to ensure both objectives: energy cost and user discomfort minimization. In contrast, the MILP schedules the operation of most appliances during low price hours cause user discomfort and rebound peak creation. However, the MILP is better than the case without scheduling. The energy cost hourly pattern of this scenario is depicted in Figure 5. It is observed in the energy cost profile that our proposed scheme optimally schedules the operation of smart home appliances keeping in view both user priority and energy cost constraint, and hence have little bit more energy cost during peak hours. On the other hand, with MILP-based schedules, consumers pay minimum cost over our proposed case without scheduling because user priority constraints are ignored.

Table 4. Operation schedule of smart home appliances with our proposed scheme.

Category	Appliances	1	2	3	4	5	6	7	8	9	10	11	12	13	14	15	16	17	18	19	20	21	22	23	24
	Dish Washer	0	1	1	0	1	0	0	0	0	0	0	0	0	0	1	1	0	1	1	1	0	1	1	1
	Dryer	0	1	1	0	0	0	0	0	0	0	0	0	0	0	0	1	1	0	1	1	0	1	1	1
ECAs	Cloth Washer	1	0	1	1	1	1	0	0	0	0	0	0	0	0	1	0	1	1	0	1	1	1	1	0
	Pump	1	0	1	1	1	1	0	0	0	0	0	0	0	0	1	0	1	1	0	1	1	1	1	0
	EV	1	0	1	1	1	1	0	0	0	0	0	0	0	0	1	0	1	1	0	1	1	1	1	0
	Heater	0	0	0	0	0	0	0	0	0	0	0	0	0	0	0	0	0	0	1	0	1	0	1	0
TCAs	Fridge	0	1	0	0	0	0	0	0	0	0	0	0	0	0	0	1	0	0	1	1	1	1	1	1
	Freezer	0	1	0	0	0	0	0	0	0	0	0	0	0	0	0	1	1	0	0	1	1	1	0	1
	AC	0	1	0	0	0	0	0	0	0	0	0	0	0	0	0	1	1	0	0	1	1	1	0	1
OCAs	Lighting	0	1	0	0	0	0	0	0	0	0	0	0	0	0	0	1	1	0	0	1	1	1	0	1

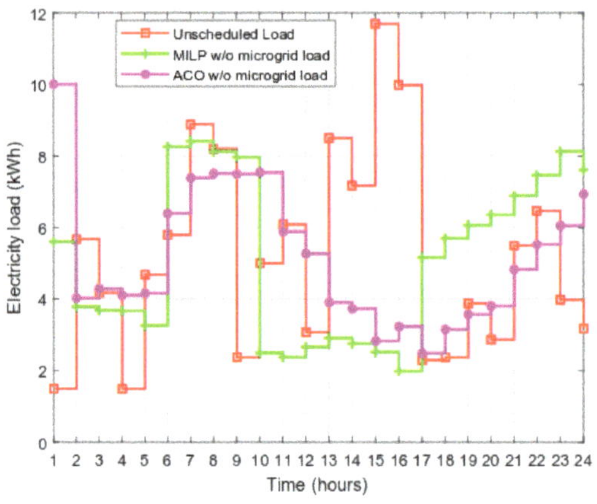

Figure 4. Energy consumption profile for energy management without microgrid scenario.

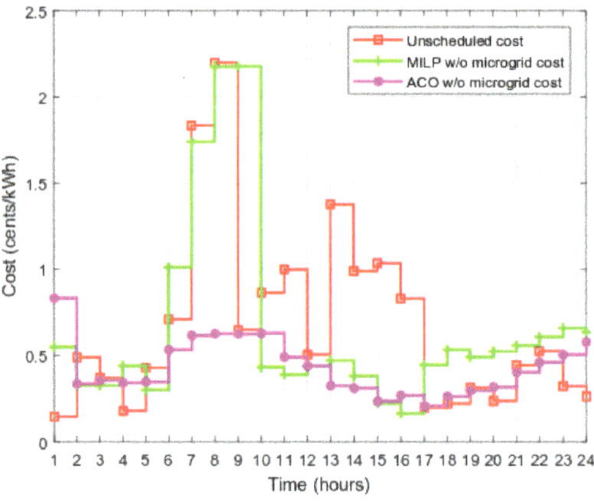

Figure 5. Energy cost profile without microgrid.

Figure 6 illustrates the consumer's daily energy cost with and without using the proposed scheme. It is clearly seen in the Figure that our proposed scheme reduces energy cost significantly compared to without scheduling scheme and MILP based scheme. The solid reason for this performance is that the proposed scheme considers users priority and cost constraints, and minimizes both energy cost and user discomfort. In addition, it also avoids rebound peaks to ensure stable and reliable power grid operation.

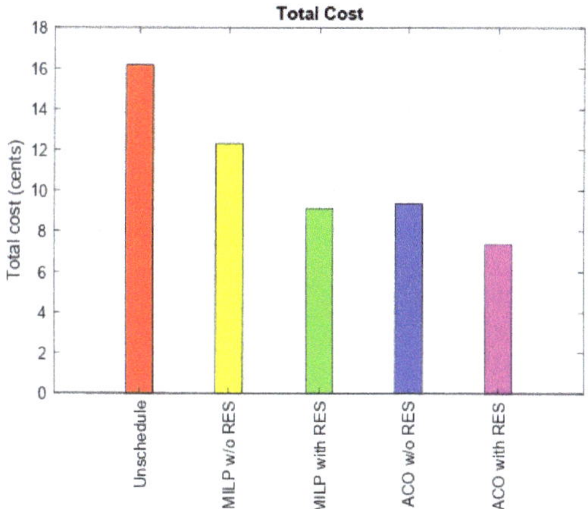

Figure 6. Net Energy cost evaluation without microgrid scenario.

The PAR is illustrated in Figure 7 for existing and proposed schemes. The proposed scheme minimized the PAR to 40%, which is lower as compared to MILP and the case without scheduling.

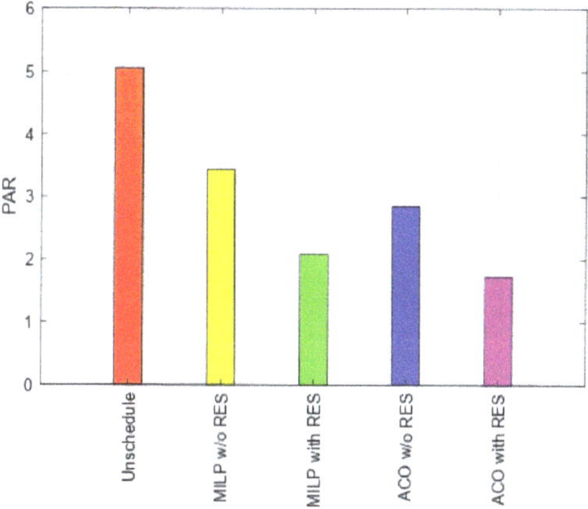

Figure 7. Net PAR evaluation without scheduling scenario.

Energy Management with a Microgrid

Simulations are conducted to evaluate the proposed scheme performance with and W/O microgrid in the proposed system model. microgrid is equipped with PV panels, WT, electrolyzer, hydrogen tank. The RES equipped with a microgrid are intermittent; therefore, the BES and EVs in the proposed microgrid are also included. The electricity is generated from sources such as PV panels, WT, electrolyzers, and hydrogen tanks depicted in Figure 8. It is visualized in figure that the PV generation is high during day time and lower at night time. This behavior is due to the well-known fact that the solar irradiation is high during daytime and lower during the night. The electricity generation from WT entirely depends on wind speed, i.e., electricity generation is maximum when wind speed is high and below the cut-out speed, and vice versa. Thus, we forecast wind speed to accurately estimate electricity generation from wind. Further, to ensure the accuracy of the prediction model, MAPE and NRMSE metrics are used, which are shown in Figure 9 and Figure 10, respectively. It is clear from the above figures that observed and forecasted values of wind are closely related ensures accurate estimation of electricity generation.

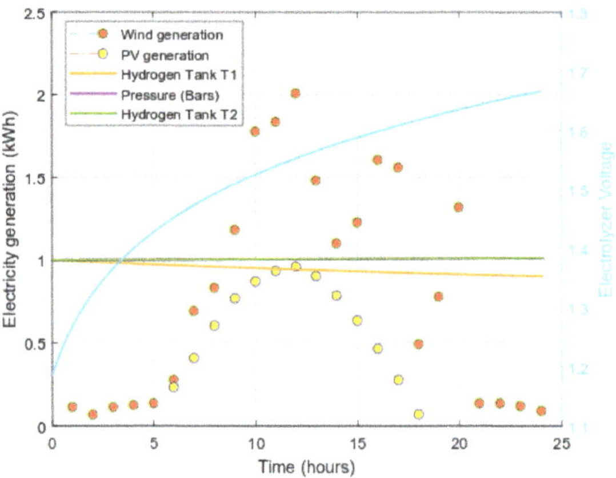

Figure 8. Microgrid electricity generation profile.

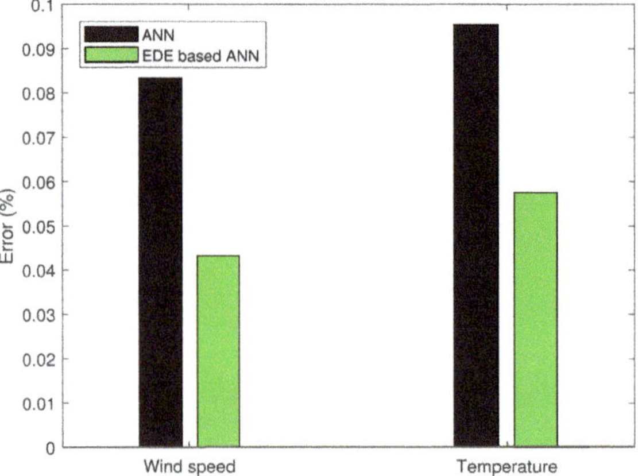

Figure 9. Prediction model evaluation using NRMSE.

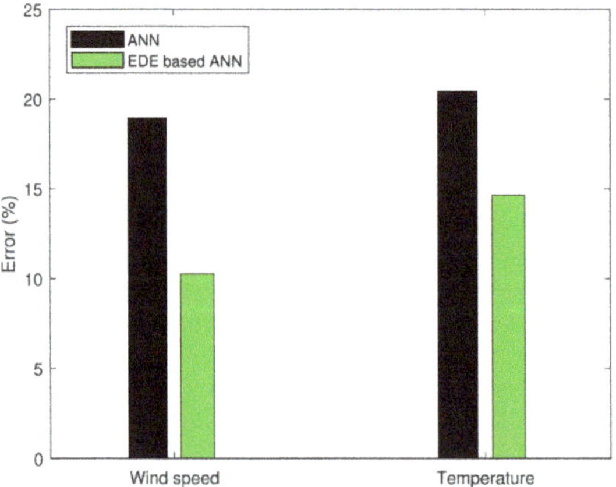

Figure 10. Prediction model evaluation using MAPE.

Solar energy is generated from sunlight, and is a sustainable and environmentally favorable energy source. Light from the Sun reaches the planet, which supplies energy.

However, we require power every hour in the generation process. Energy is generated for commercial, industrial, and residential customers to serve their load. As a result, it is more efficient. As seen from Figure 11 the solar generation is high during the daytime due to high solar irradiance.

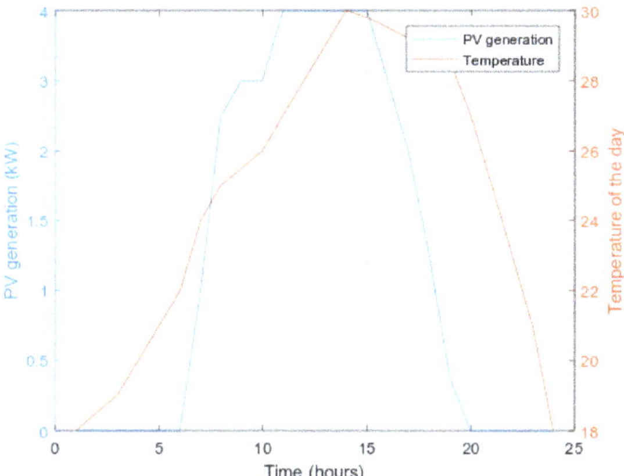

Figure 11. PV electricity generation profile.

From Figure 12, it is clear that when the wind speed is high, the generation is likewise high, and vice versa. Wind energy is the primary source of RES in a microgrid that is optimized for balancing energy supply, demand, and storage to ensure efficient energy management.

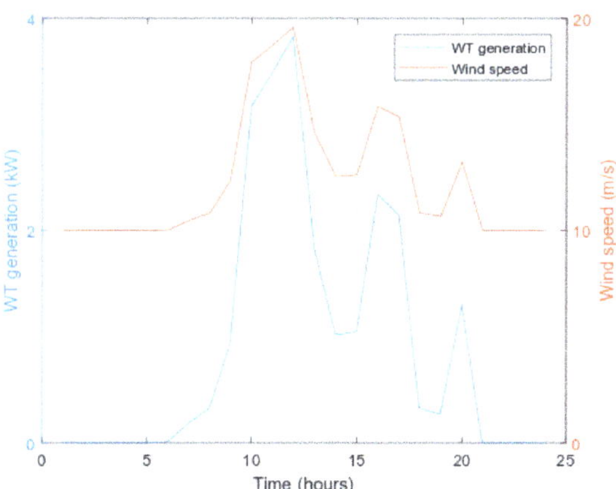

Figure 12. Wind electricity generation profile.

However, a comparison of expected and measured electrolyzer voltage and accompanying power indicates that the slight under-prediction of temperature has virtually no relevance from the perspective of energy system modeling. For instance, we examine the one-day simulation described in our proposed study. By mixing hydrogen and oxygen, a hydrogen tank generates electricity, heat, and water. Frequently, fuel cells and batteries are combined. Both of these devices convert the energy generated by a chemical reaction into electrical energy. The hydrogen fuel cell, on the other hand, will keep producing energy as long as fuel (hydrogen) is available.

Figure 13 illustrates the net electricity demand and imported electricity for serving smart home loads. Figure 13 clearly demonstrates that our proposed scheme shifts the load via scheduling from high price hours to low price hours, and serves the net load from the microgrid. Furthermore, the proposed scheme acquires power from the external grid during the hours when power rates are low. This figure also depicts the RES curve, which ensures that RES are serving the whole day to satisfy the user's electricity demand.

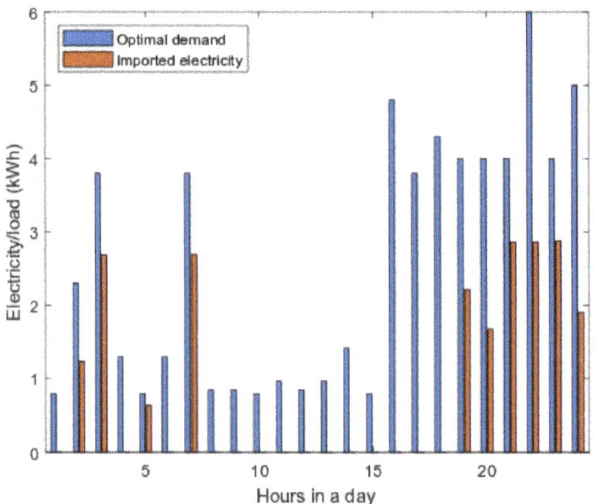

Figure 13. Hourly optimal demand and imported electricity.

Net electricity demand, imported electricity from WT, and purchased electricity from the power grid considering with/without EVs discharging capabilites are illustrated in Figure 14. It is obvious from results that the proposed model minimizes the energy purchased from the power grid during high price hours, importing energy from WT and enabling the discharging mode of EVs to ensure reliable power supply to the load. In contrast, smart homes import more energy in cases when EVs are driving or empty because EVs act as mobile storage.

The mobile storage EVs charging/discharging and electricity generation of the microgrid, including all sources, is depicted in Figure 15. The figure clearly shows that EVs charging mode is enabled when the microgrid generation is maximum and EVs have lower energy, and discharging mode is enabled when microgrid generation is minimum or there is no other source of energy available. In the first hour, the battery is fully charged, in 10–15 h timeslots it is charging, and 2–3 and 16–21 timeslots, it is discharging.

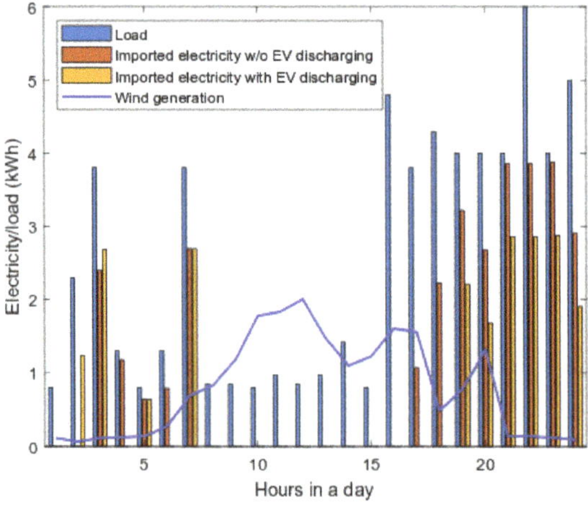

Figure 14. Electricity purchase with/without discharging EVs.

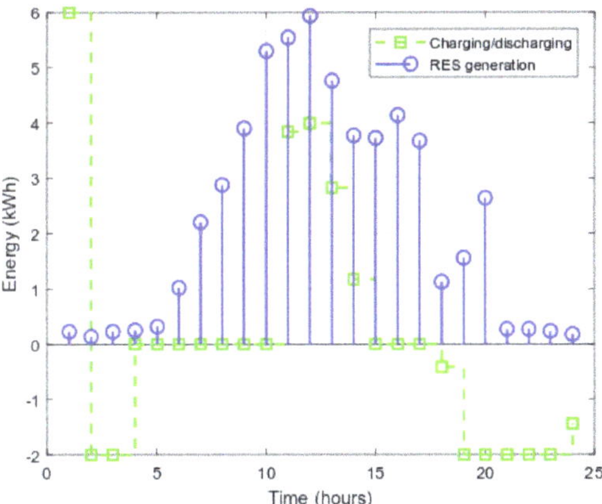

Figure 15. EVs charging and discharging behavior with mircogrid electricity generation.

The EVs and BES participate in the microgrid with RES to significantly contribute in energy cost and PAR minimization as shown in Figure 16. It is obvious from the results that users utilize electricity from BES or EVs rather than importing electricity from the power grid during situations when electricity demand is at peak or high price hours, or when generation from the microgrid is scarce. This behavior highly contributes to energy cost reduction and PAR alleviation.

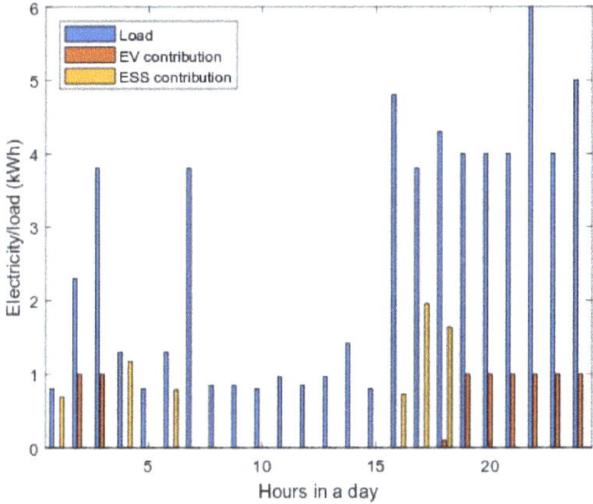

Figure 16. EVs and BES hourly share for serving load.

The MGT is employed instead of a diesel generator to reduce carbon emissions and lower peak load. MGT will only operate when the user demand exceeds a certain threshold value. From Figure 17 it is evident that MGT only functions during peak hours to fulfil customer demand, and so PAR is lowered.

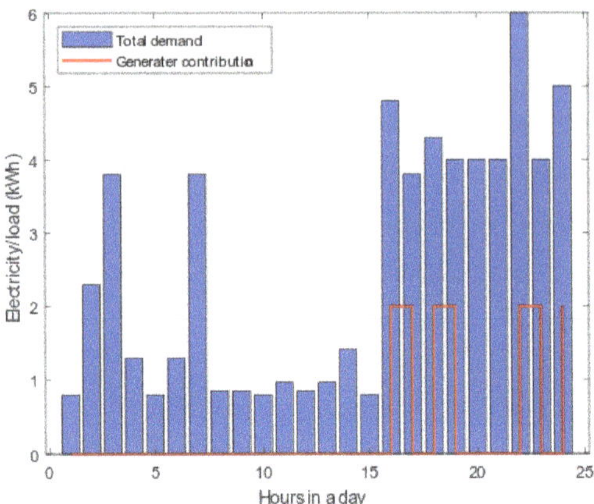

Figure 17. Optimal demand and MGT contribution.

Analysis is conducted for 24 h operation of the microgrid system to evaluate carbon emissions. In the microgrid, the MGT used in place of diesel generator because a diesel generator releases a large carbon footprint with RES. Thus, our proposed scheme with the use of the microgrid significantly minimizes (up to 25%) carbon emission; this is shown in Figure 18.

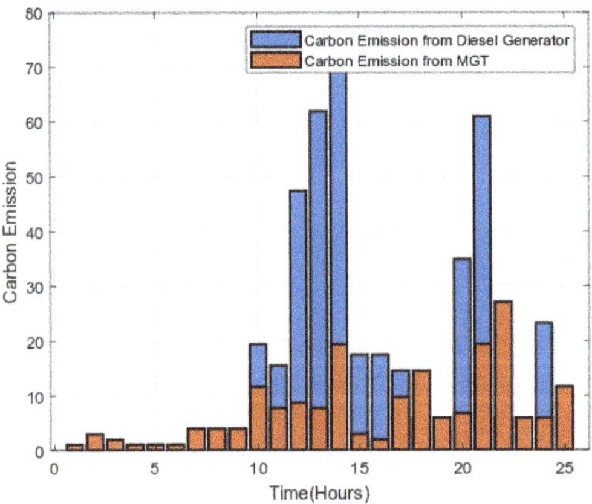

Figure 18. Hourly carbon emissions.

7. Conclusions

This work presents efficient energy management scheme to schedule operation of appliances and charging/discharging of EVs in the presence of RTP signal under utility with and without a microgrid. A prediction model, ANN-mEDE, is developed for accurate electricity generation estimation of the microgrid to contribute to efficient energy management. Then, an ACO algorithm is developed for the proposed scheme to solve energy management problems via scheduling with and without microgrid scenarios. The aims and objectives of solving the energy management problem are maintaining a balance between demand and supply for reducing energy cost, PAR, carbon emission, and user discomfort to facilitate both parties, utility providers and end-users, at the same time. To endorse the developed model, simulations are conducted in comparison with the existing scheme based on MILP and cases without scheduling in with and without microgrid scenarios. The results depict that the proposed scheme reduced energy cost, carbon emission, and PAR by 35%, 25.01%, and 40.12% in scenario I; by 55.05%, 45.5%, and 42.4% in scenario II, respectively.

Author Contributions: Conceptualization, Data curation, Methodology, Resources, Validation, Writing—original draft, Writing—review & editing, F.R.A.; Conceptualization, Data curation, Methodology, Resources, Validation, Writing—original draft, Writing—review & editing, Formal analysis, Funding acquisition, Investigation, Project administration, Supervision, and Visualization, G.H.; Writing—review & editing, Formal analysis, Funding acquisition, Investigation, Project administration, Supervision, and Visualization, I.K.; Writing—review & editing, Formal analysis, Funding acquisition, Investigation, Visualization, S.K., Writing—review & editing, Formal analysis, Funding acquisition, Investigation, Visualization and Investigation, H.I.A.; Writing—review & editing, Formal analysis, Funding acquisition, Investigation, Visualization and Investigation, F.A. and G.R. All authors have read and agreed to the published version of the manuscript.

Funding: The article processing charge is supported by the Taif University Researchers Supporting Project Number (TURSP-2020/264), Taif University, Taif, Saudi Arabia.

Acknowledgments: The authors would like to acknowledge the support from Taif University Researchers Supporting Project Number (TURSP-2020/264), Taif University, Taif, Saudi Arabia.

Conflicts of Interest: The authors declare no conflict of interest.

Nomenclature

Abbreviations	Explanation
ACO	Ant Colony Optimization
AN	Artificial Neuron
ANN	Artificial Neural Network
CE	Carbon Emission
DR	Demand Response
DSM	Demand Side Management
DES	Distributed Electric System
EMC	Energy Management controller
EST	Earliest Starting Time
EDE	Enhanced Differential Evolution
EV	Electric Vehicle
ESS	Energy Storage Systems
ECA	Electrically Controllable Appliances
FA	Firefly Algorithm
IoT	Internet of Things
GHG	Green House Gases
GA	Genetic Algorithm
LOT	Length of Time
LFT	Latest Finishing Time
MGT	Micro Gas Turbine
MILP	Mixed Integer Linear Programming
MINLP	Mixed Integer Non-Linear Programming
MAPE	Mean Absolute Percentage Error
MSE	Mean squared Error
MG	Micro Grid
MPC	Model Predictive Control
MPPT	Maximum Power Point Tracking
NRMSE	Normalized Root Mean Square Error
OCA	Optically Controllable Appliances
PAR	Peak to Average Ratio
PSO	Particle swarm Optimization
PHEV	Plug in Hybrid Electrcic Vehicle

RES	Renewable Energy Sources
RTP	Real Time Pricing
SG	Smart Grids
TCA	Thermostatically Controllable Appliances
WT	Wind Turbine
m	Generation from RES
t	Time interval
Constants	**Explanation**
$T_{\min}^{ac}$	AC lower bound
$T_{\min}^{fridge}$	Fridge lower bound
$T_{\min}^{heater}$	Heater lower bound
$T_{\min}^{freezer}$	Freezer lower bound
$T_{\max}^{ac}$	AC upper bound
$T_{\max}^{fridge}$	Fridge upper bound
$T_{\max}^{heater}$	Heater upper bound
$T_{\max}^{freezer}$	Freezer upper bound
$V^{cut-out}$	Wind cut-out speed
V^{cut-in}	Wind cut-in speed
P_t^{pv}	Electricity consumed per hour from PV
ρEV_a^e	Energy storage level at EV arrival time
ρEV_d^e	Energy storage level at EV departure time
$EV_{\max}^e$	Maximum EV discharging limit
$EV_{\min}^e$	Minimum EV discharging limit
$P^{pv}(t)$	PV contribution in electricity generation
$P^{wt}(t)$	WT contribution in electricity generation
$P^{hyd}(t)$	Hydrogen tank contribution in electricity generation
$P^{elec}(t)$	Electrolyzer contribution in electricity generation
Γ_{ECA}	ECA scheduling
Γ_{OCA}	OCA scheduling
Γ_{TCA}	TCA scheduling
$\Phi(t)$	Hourly imported electricity
EV_a^e	EV arrival time
η^{pv}	Efficiency of Solar panel
η^{ESS}	Efficiency of ESS
Variables	**Explanation**
P^{wt}	Generation from WT
V_t^{wt}	Wind Speed
ρ	Air Density
P_t^{pv}	Hourly produced energy by PV
A^{pv}	Area of Solar panel
$Irr(t)$	Solar Radiation
$Temp(t)$	Outside Temperature
SE	Stored Energy (Ah)
$ES^{ch}(t)$	Charging Status of ESS at time t
$ES^{dis}(t)$	Discharging Status of ESS at time t

References

1. Ackermann, T.; Andersson, G.; Söder, L. Distributed generation: A definition. *Electr. Power Syst. Res.* **2001**, *57*, 195–204. [CrossRef]
2. Gungor, V.C.; Sahin, D.; Kocak, T.; Ergut, S.; Buccella, C.; Cecati, C.; Hancke, G.P. Smart grid and smart homes: Key players and pilot projects. *IEEE Ind. Electron. Mag.* **2012**, *6*, 18–34. [CrossRef]
3. Zakariazadeh, A.; Jadid, S.; Siano, P. Smart microgrid energy and reserve scheduling with demand response using stochastic optimization. *Int. J. Electr. Power Energy Syst.* **2014**, *63*, 523–533. [CrossRef]
4. Ahmad, T.; Zhang, H.; Yan, B. A review on renewable energy and electricity requirement forecasting models for smart grid and buildings. *Sustain. Cities Soc.* **2020**, *55*, 102052. [CrossRef]
5. Flores, J.T.; Celeste, W.C.; Coura, D.J.C.; Rissino, S.D.D.; Rocha, H.R.O.; Moraes, R.E.N. Demand planning in smart homes. *IEEE Lat. Am. Trans.* **2016**, *14*, 3247–3255. [CrossRef]
6. Zhao, X.; Gao, W.; Qian, F.; Ge, J. Electricity cost comparison of dynamic pricing model based on load forecasting in home energy management system. *Energy* **2021**, *229*, 120538. [CrossRef]

7. Rocha, H.R.O.; Honorato, I.H.; Fiorotti, R.; Celeste, W.C.; Silvestre, L.J.; Silva, J.A.L. An Artificial Intelligence based scheduling algorithm for demand-side energy management in Smart Homes. *Appl. Energy* **2021**, *282*, 116145. [CrossRef]
8. Kim, B.-G.; Zhang, Y.; Van Der Schaar, M.; Lee, J.-W. Dynamic pricing and energy consumption scheduling with reinforcement learning. *IEEE Trans. Smart Grid* **2015**, *7*, 2187–2198. [CrossRef]
9. Shirazi, E.; Jadid, S. Cost reduction and peak shaving through domestic load shifting and DERs. *Energy* **2017**, *124*, 146–159. [CrossRef]
10. Silvente, J.; Papageorgiou, L.G. An MILP formulation for the optimal management of microgrids with task interruptions. *Appl. Energy* **2017**, *206*, 1131–1146. [CrossRef]
11. Tushar, M.H.K.; Assi, C.; Maier, M.; Uddin, M.F. Smart microgrids: Optimal joint scheduling for electric vehicles and home appliances. *IEEE Trans. Smart Grid* **2014**, *5*, 239–250. [CrossRef]
12. Umetani, S.; Fukushima, Y.; Morita, H. A linear programming based heuristic algorithm for charge and discharge scheduling of electric vehicles in a building energy management system. *Omega* **2017**, *67*, 115–122. [CrossRef]
13. Wang, X.; Palazoglu, A.; El-Farra, N.H. Operational optimization and demand response of hybrid renewable energy systems. *Appl. Energy* **2015**, *143*, 324–335. [CrossRef]
14. Aslam, S.; Khalid, A.; Javaid, N. Towards efficient energy management in smart grids considering microgrids with day-ahead energy forecasting. *Electr. Power Syst. Res.* **2020**, *182*, 106232. [CrossRef]
15. Pal, S.; Thakur, S.; Kumar, R.; Panigrahi, B. A strategical game theoretic based demand response model for residential consumers in a fair environment. *Int. J. Electr. Power Energy Syst.* **2018**, *97*, 201–210. [CrossRef]
16. Thomas, D.; Deblecker, O.; Ioakimidis, C.S. Optimal operation of an energy management system for a grid-connected smart building considering photovoltaics' uncertainty and stochastic electric vehicles' driving schedules. *Appl. Energy* **2018**, *210*, 1188–1206. [CrossRef]
17. Melhem, F.Y. Optimization Methods and Energy Management in "Smart Grids". Ph.D. Thesis, Université Bourgogne Franche-Comté, Bourgogne Franche-Comté, France, 2018.
18. Giaouris, D.; Papadopoulos, A.I.; Seferlis, P.; Papadopoulou, S.; Voutetakis, S.; Stergiopoulos, F.; Elmasides, C. Optimum energy management in smart grids based on power pinch analysis. *Chem. Eng.* **2014**, *39*, 55–60.
19. Zafar, R.; Mahmood, A.; Razzaq, S.; Ali, W.; Naeem, U.; Shehzad, K. Prosumer based energy management and sharing in smart grid. *Renew. Sustain. Energy Rev.* **2018**, *82*, 1675–1684. [CrossRef]
20. De Angelis, F.; Boaro, M.; Fuselli, D.; Squartini, S.; Piazza, F.; Wei, Q. Optimal home energy management under dynamic electrical and thermal constraints. *IEEE Trans. Ind. Inform.* **2012**, *9*, 1518–1527. [CrossRef]
21. He, M.-F.; Zhang, F.-X.; Huang, Y.; Chen, J.; Wang, J.; Wang, R. A distributed demand side energy management algorithm for smart grid. *Energies* **2019**, *12*, 426. [CrossRef]
22. Subha, S.; Nagalakshmi, S. Design of ANFIS controller for intelligent energy management in smart grid applications. *J. Ambient Intell. Humaniz. Comput.* **2021**, *12*, 6117–6127. [CrossRef]
23. Xing, X.; Xie, L.; Meng, H. Cooperative energy management optimization based on distributed MPC in grid-connected microgrids community. *Int. J. Electr. Power Energy Syst.* **2019**, *107*, 186–199. [CrossRef]
24. Yuan, D.; Lu, Z.; Zhang, J.; Li, X. A hybrid prediction-based microgrid energy management strategy considering demand-side response and data interruption. *Int. J. Electr. Power Energy Syst.* **2019**, *113*, 139–153. [CrossRef]
25. Ouammi, A.; Achour, Y.; Zejli, D.; Dagdougui, H. Supervisory model predictive control for optimal energy management of networked smart greenhouses integrated microgrid. *IEEE Trans. Autom. Sci. Eng.* **2019**, *17*, 117–128. [CrossRef]
26. Mbungu, N.T.; Bansal, R.C.; Naidoo, R.M.; Bettayeb, M.; Siti, M.W.; Bipath, M. A dynamic energy management system using smart metering. *Appl. Energy* **2020**, *280*, 115990. [CrossRef]
27. Bingham, R.D.; Agelin-Chaab, M.; Rosen, M.A. Whole building optimization of a residential home with pv and battery storage in the Bahamas. *Renew. Energy* **2019**, *132*, 1088–1103. [CrossRef]
28. Shakeri, M.; Shayestegan, M.; Reza, S.S.; Yahya, I.; Bais, B.; Akhtaruzzaman, M.; Sopian, K.; Amin, N. Implementation of a novel home energy management system (hems) architecture with solar photovoltaic system as supplementary source. *Renew. Energy* **2018**, *125*, 108–120. [CrossRef]
29. Nemati, M.; Braun, M.; Tenbohlen, S. Optimization of unit commitment and economic dispatch in microgrids based on genetic algorithm and mixed integer linear programming. *Appl. Energy* **2018**, *210*, 944–963. [CrossRef]
30. Silva, B.N.; Han, K. Mutation operator integrated ant colony optimization based domestic appliance scheduling for lucrative demand side management. *Future Gener. Comput. Syst.* **2019**, *100*, 557–568. [CrossRef]
31. Hafeez, G.; Alimgeer, K.S.; Wadud, Z.; Khan, I.; Usman, M.; Qazi, A.B.; Khan, F.A. An innovative optimization strategy for efficient energy management with day-ahead demand response signal and energy consumption forecasting in smart grid using artificial neural network. *IEEE Access* **2020**, *8*, 84415–84433. [CrossRef]
32. Shuja, S.M.; Javaid, N.; Khan, S.; Akmal, H.; Hanif, M.; Fazalullah, Q.; Khan, Z.A. Efficient scheduling of smart home appliances for energy management by cost and PAR optimization algorithm in smart grid. In Proceedings of the Workshops of the International Conference on Advanced Information Networking and Applications, Matsue, Japan, 27–29 March 2019; Springer: Cham, Switzerland, 2019; pp. 398–411.

33. Ayub, S.; Ayob, S.M.; Tan, C.W.; Ayub, L.; Bukar, A.L. Optimal residence energy management with time and device-based preferences using an enhanced binary grey wolf optimization algorithm. *Sustain. Energy Technol. Assess.* **2020**, *41*, 100798. [CrossRef]

34. Shuja, S.M.; Javaid, N.; Rafique, M.Z.; Qasim, U.; Khan, R.F.M.; Butt, A.A.; Hanif, M. Towards efficient scheduling of smart appliances for energy management by candidate solution updation algorithm in smart grid. In Proceedings of the International Conference on Advanced Information Networking and Applications, Matsue, Japan, 27–29 March 2019; Springer: Cham, Switzerland, 2019; pp. 67–81.

35. Liu, Y.; Yang, C.; Jiang, L.; Xie, S.; Zhang, Y. Intelligent edge computing for IoT-based energy management in smart cities. *IEEE Netw.* **2019**, *33*, 111–117. [CrossRef]

36. Choi, J.S. A hierarchical distributed energy management agent framework for smart homes, grids, and cities. *IEEE Commun. Mag.* **2019**, *57*, 113–119. [CrossRef]

37. Ding, Y.; Xie, D.; Hui, H.; Xu, Y.; Siano, P. Game-Theoretic Demand Side Management of Thermostatically Controlled Loads for Smoothing Tie-line Power of Microgrids. *IEEE Trans. Power Syst.* **2021**, *36*, 4089–4101. [CrossRef]

38. Longe, O.M.; Ouahada, K.; Rimer, S.; Harutyunyan, A.N.; Ferreira, H.C. Distributed Demand Side Management with Battery Storage for Smart Home Energy Scheduling. *Sustainability* **2017**, *9*, 120. [CrossRef]

39. Ali, S.; Khan, I.; Jan, S.; Hafeez, G. An Optimization Based Power Usage Scheduling Strategy Using Photovoltaic-Battery System for Demand-Side Management in Smart Grid. *Energies* **2021**, *14*, 2201. [CrossRef]

40. Nadeem, F.; Aftab, M.A.; Hussain, S.M.; Ali, I.; Tiwari, P.K.; Goswami, A.K.; Ustun, T.S. Virtual power plant management in smart grids with XMPP based IEC 61850 communication. *Energies* **2019**, *12*, 2398. [CrossRef]

41. Zhao, C.; Chen, J.; He, J.; Cheng, P. Privacy-preserving consensus-based energy management in smart grids. *IEEE Trans. Signal Process.* **2018**, *66*, 6162–6176. [CrossRef]

42. Je, S.-M.; Huh, J.-H. Estimation of future power consumption level in smart grid: Application of fuzzy logic and genetic algorithm on big data platform. *Int. J. Commun. Syst.* **2021**, *34*, e4056. [CrossRef]

43. Taheri, S.S.; Seyedshenava, S.; Mohadesi, V.; Esmaeilzadeh, R. Improving Operation Indices of a Micro-grid by Battery Energy Storage Using Multi Objective Cuckoo Search Algorithm. *Int. J. Electr. Eng. Inform.* **2021**, *13*, 132–151.

44. Chen, J.; Zhu, Q. A Stackelberg game approach for two-level distributed energy management in smart grids. *IEEE Trans. Smart Grid* **2017**, *9*, 6554–6565. [CrossRef]

45. Förderer, K.; Ahrens, M.; Bao, K.; Mauser, I.; Schmeck, H. Towards the modeling of flexibility using artificial neural networks in energy management and smart grids: Note. In Proceedings of the Ninth International Conference on Future Energy Systems, Karlsruhe, Germany, 12–15 June 2018; pp. 85–90.

46. Misra, S.; Mondal, A.; Banik, S.; Khatua, M.; Bera, S.; Obaidat, M.S. Residential energy management in smart grid: A Markov decision process-based approach. In Proceedings of the 2013 IEEE International Conference on Green Computing and Communications and IEEE Internet of things and IEEE Cyber, Physical and Social Computing, Beijing, China, 20–23 August 2013; pp. 1152–1157.

47. Andrade, I.; Pena, R.; Blasco-Gimenez, R.; Riedemann, J.; Jara, W.; Pesce, C. An Active/Reactive Power Control Strategy for Renewable Generation Systems. *Electronics* **2021**, *10*, 1061. [CrossRef]

48. El-Zonkoly, A.M. Optimal energy management in smart grids including different types of aggregated flexible loads. *J. Energy Eng.* **2019**, *145*, 04019015. [CrossRef]

49. Hafeez, G.; Islam, N.; Ali, A.; Ahmad, S.; Usman, M.; Alimgeer, K.S. A modular framework for optimal load scheduling under price-based demand response scheme in smart grid. *Processes* **2019**, *7*, 499. [CrossRef]

50. Campagna, N.; Caruso, M.; Castiglia, V.; Miceli, R.; Viola, F. Energy Management Concepts for the Evolution of Smart Grids. In Proceedings of the 2020 8th International Conference on Smart Grid (icSmartGrid), Paris, France, 17–19 June 2020; pp. 208–213.

51. Sharifi, A.H.; Maghouli, P. Energy management of smart homes equipped with energy storage systems considering the PAR index based on real-time pricing. *Sustain. Cities Soc.* **2019**, *45*, 579–587. [CrossRef]

52. Rahim, S.; Javaid, N.; Ahmad, A.; Khan, S.A.; Khan, Z.A.; Alrajeh, N.; Qasim, U. Exploiting heuristic algorithms to efficiently utilize energymanagement controllers with renewable energy sources. *Energy Build.* **2016**, *129*, 452–470. [CrossRef]

53. Neural Network Models. Available online: https://otexts.com/fpp2/nnetar.html (accessed on 7 September 2020).

54. Ahmad, A.; Javaid, N.; Guizani, M.; Alrajeh, N.; Khan, Z.A. An accurate and fast converging short-term load forecasting model for industrial applications in a smart grid. *IEEE Trans. Ind. Inform.* **2017**, *13*, 2587–2596. [CrossRef]

55. Niedźwiecki, M.; Ciołek, M. Identification of nonstationary multivariate autoregressive processes-comparison of competitive and collaborative strategies for joint selection of estimation bandwidth and model order. *Digit. Signal Process.* **2018**, *78*, 72–81. [CrossRef]

56. Solar Resource Data. Available online: https://pvwatts.nrel.gov/pvwatts.php (accessed on 12 October 2020).

57. Gensler, A.; Henze, J.; Sick, B.; Raabe, N. Deep Learning for solar power forecasting—An approach using AutoEncoder and LSTM Neural Networks. In Proceedings of the 2016 IEEE International Conference on Systems, Man, and Cybernetics (SMC), Budapest, Hungary, 9–12 October 2016; pp. 002858–002865.

58. Pulipaka, S.; Kumar, R. Comparison of som and conventional neural network data division for pv reliability power prediction. In Proceedings of the 2017 IEEE International Conference on Environment and Electrical Engineering and 2017 IEEE Industrial and Commercial Power Systems Europe (EEEIC/I&CPS Europe), Milan, Italy, 6–9 June 2017; pp. 1–5.

59. Alzahrani, A.; Shamsi, P.; Dagli, C.; Ferdowsi, M. Solar irradiance forecasting using deep neural networks. *Procedia Comput. Sci.* **2017**, *114*, 304–313. [CrossRef]
60. Engelbrecht, A.P. *Computational Intelligence: An Introduction*; John Wiley & Sons: Hoboken, NJ, USA, 2007.
61. Amjady, N.; Keynia, F.; Zareipour, H. Short-term load forecast of microgrids by a new bilevel prediction strategy. *IEEE Trans. Smart Grid* **2010**, *1*, 286–294. [CrossRef]
62. Hu, Y.-L.; Chen, L. A nonlinear hybrid wind speed forecasting model using LSTM network, hysteretic ELM and Differential Evolution algorithm. *Energy Convers. Manag.* **2018**, *173*, 123–142. [CrossRef]
63. Lin, K.-P.; Pai, P.-F.; Ting, Y.-J. Deep belief networks with genetic algorithms in forecasting wind speed. *IEEE Access* **2019**, *7*, 99244–99253. [CrossRef]
64. Wang, Y.; Shen, Y.; Mao, S.; Cao, G.; Nelms, R.M. Adaptive learning hybrid model for solar intensity forecasting. *IEEE Trans. Ind. Inform.* **2018**, *14*, 1635–1645. [CrossRef]
65. Ullah, K.; Hafeez, G.; Khan, I.; Jan, S.; Javaid, N. A multi-objective energy optimization in smart grid with high penetration of renewable energy sources. *Appl. Energy.* **2021**, *299*, 117104. [CrossRef]
66. Chang, G.; Lu, H.; Chang, Y.; Lee, Y. An improved neural network-based approach for short-term wind speed and power forecast. *Renew. Energy* **2017**, *105*, 301–311. [CrossRef]
67. Griesshaber, W.; Sick, F. *Simulation of Hydrogen–Oxygen–Systems with PV for the Self-Sufficient Solar House*; FhG-ISE: Freiburg im Breisgau, Germany, 1991. (In Germany)
68. Havre, K.; Borg, P.; Tommerberg, K. Modeling and control of pressurized electrolyzer for operation in stand alone power systems. In Proceedings of the Second Nordic Symposium on Hydrogen and Fuel Cells for Energy Storage, Helsinki, Finland, 19–20 January 1995; pp. 63–78.
69. Vanhanen, J. On the Performance Improvements of Small-Scale Photovoltaic-Hydrogen Energy Systems. Ph.D. Thesis, Helsinki University of Technology, Espoo, Finland, 1996.
70. Hug, W.; Divisek, J.; Mergel, J.; Seeger, W.; Steeb, H. Highly efficient advanced alkaline electrolyzer for solar operation. *Int. J. Hydrog. Energy* **1992**, *17*, 699–705. [CrossRef]
71. Zhou, L.; Zhou, Y. Determination of compressibility factor and fugacity coefficient of hydrogen in studies of adsorptive storage. *Int. J. Hydrog. Energy* **2001**, *26*, 597–601. [CrossRef]
72. Nascimento, M.A.R.; Rodrigues, L.O.; Santos, E.C.; Gomes, E.E.B.; Dias, F.L.G.; Velásques, E.I.G.; Carrillo, R.A.M. Micro gas turbine engine: A review. *Prog. Gas Turbine Perform.* **2013**, *5*, 107–141.
73. Aslam, S.; Javaid, N.; Asif, M.; Iqbal, U.; Iqbal, Z.; Sarwar, M.A. A mixed integer linear programming based optimal home energy management scheme considering gridconnected microgrids. In Proceedings of the 14th International Wireless Communications & Mobile Computing Conference (IWCMC), Limassol, Cyprus, 25–29 June 2018; pp. 993–998.
74. Shirazi, E.; Jadid, S. Optimal residential appliance scheduling under dynamic pricing scheme via hemdas. *Energy Build.* **2015**, *93*, 40–49. [CrossRef]
75. Aslam, S.; Herodotou, H.; Mohsin, S.M.; Javaid, N.; Ashraf, N.; Aslam, S. A survey on deep learning methods for power load and renewable energy forecasting in smart microgrids. *Appl. Energy* **2021**, *144*, 110992.

 *sustainability*

Article

An Incentive Based Dynamic Pricing in Smart Grid: A Customer's Perspective

Thamer Alquthami [1,*], **Ahmad H. Milyani** [1], **Muhammad Awais** [2] **and Muhammad B. Rasheed** [3,4,*]

1. Electrical and Computer Engineering Department, King Abdulaziz University, Jeddah 21589, Saudi Arabia; ahmilyani@kau.edu.sa
2. Department of Technology, The University of Lahore, Lahore 54000, Pakistan; m.awais.qureshi27@gmail.com
3. Department of Electronics and Electrical Systems, The University of Lahore, Lahore 54000, Pakistan
4. Computer Engineering Department, University of Alcalá, 28805 Alcalá de Henares, Spain
* Correspondence: tquthami@kau.edu.sa (T.A.); Muhammad.rasheed@uah.es (M.B.R.)

Abstract: Price based demand response is an important strategy to facilitate energy retailers and end-users to maintain a balance between demand and supply while providing the opportunity to end users to get monetary incentives. In this work, we consider real-time electricity pricing policy to further calculate the incentives in terms of reduced electricity price and cost. Initially, a mathematical model based on the backtracking technique is developed to calculate the load shifted and consumed in any time slot. Then, based on this, the electricity price is calculated for all types of users to estimate the incentives through load shifting profiles. To keep the load under the upper limit, the load is shifted in other time slots in such a way to facilitate end-users regarding social welfare. The user who is not interested in participating load shifting program will not get any benefit. Then the well behaved functional form optimization problem is solved by using a heuristic-based genetic algorithm (GA), wwhich converged within an insignificant amount of time with the best optimal results. Simulation results reflect that the users can obtain some real incentives by participating in the load scheduling process.

Keywords: demand side management; demand response; load scheduling; real time pricing; genetic algorithm; dynamic incentives

 **check for updates**

Citation: Alquthami, T.; Milyani, A.H.; Awais, M.; Rasheed, M.B. An Incentive Based Dynamic Pricing in Smart Grid: A Customer's Perspective. *Sustainability* **2021**, *13*, 6066. https://doi.org/10.3390/su13116066

Academic Editors: Sheraz Aslam, Herodotos Herodotou and Nouman Ashraf

Received: 20 April 2021
Accepted: 21 May 2021
Published: 27 May 2021

Publisher's Note: MDPI stays neutral with regard to jurisdictional claims in published maps and institutional affiliations.

1. Introduction and Background

The smart grid (SG) is an emerging paradigm shift in power distribution systems that aims to improve itself using various information and communication technologies. It comprises various intelligent controlling and decision-making systems, which manage electricity generation, transmission and distribution through two-way communication mechanisms [1,2]. In addition, SG allows the integration of distributed and renewable generation facilities to cope with various uncertainties (i.e., energy deficits, blackouts, high peaks) that are caused by the energy demand variations and intermittent nature of renewables [3], which helps minimize carbon emissions. This is due to the fact that distributed renewable energy resources can provide power during stand alone or independent/island mode to manage power demand with reduced emission. For this purpose, intelligent autonomous mechanism, blockchain technology and artificial intelligence are key technologies being widely adopted in recent decades. Furthermore, as distributed generation and renewable sources can play a key role in managing energy demand, distribution systems can use the flexibility of variable energy resources to improve the underlying capacity of low voltage distribution networks, which can also be referred to as "active distribution systems". Due to the developments in energy storage technologies, as well as the strong need to reduce transportation-related costs and emissions [3], the focus on electric vehicles has been increasing. Electric vehicles need to be recharged at charging stations. Thus, based on the aforementioned justifications, EVs integration has a threefold set of objectives:

(i) it can be used for transportation purposes replacing the traditional vehicles due to a large amount of carbon emissions and abrupt fluctuations in fuel prices; (ii) EVs can also be used for the transmission of electricity; (iii) the integration of EVs can help with alleviating the high and rebound peaks majorly caused by high energy extracted during low pricing hours.

Recent advancement in communication and control technologies has envisioned the DR programs as an important tool for load management and scheduling. In [4], the authors used an opportunistic load scheduling technique, which is based on optimal stopping theory. Each load is assigned a time factor to decide its priority and then threshold criteria is calculated to find optimal time slot for each individual load. As it is a pure threshold policy. Therefore, this policy would be unfavorable when energy consumption and market spot pricing trends are dynamic. A similar work is proposed [5] to schedule residential loads in conjunction with on-site renewable energy and storage system. A mixed integer linear programming algorithm is used to solve cost minimization and comfort maximization problems. A slightly different work is presented [6] for real-time pricing demand response with fault-tolerant and flexible user enrolment to predict dynamic pricing. This scheme is relatively efficient in terms of less computational overhead and transmission delay. In [7], a two layered model for a hybrid energy system is proposed with a demand biding strategy. Before scheduling the load, a demand bidding based on Nash equilibrium theory is used to find the optimal pricing involving different stakeholder. Then, a coordinated multiagent framework is used to ensure the stability of this system with an event triggered mechanism. The work presented in [8] is devoted to modeling a locational marginal pricing based demand response using a monotonously decreasing linear function. In [9], a price based demand response program to schedule the residential load is presented and solved by using a decision support system. The load is first predicted and then scheduled based on market price to minimize overall cost. A novel pricing scheme considering residential demand response, renewable energy and power losses is presented for peak load alleviation [10]. All the generation and demand facilities actively collaborate in a distributed manner to find the best optimal price without affecting their objectives., where demand response DR allows the potential users to shift their consumption level or curtail some portion of the load in response to time-varying dynamic pricing such as time of use (TOU), real-time pricing (RTP), day-ahead pricing (DAP), and critical peak pricing (CPP) [11]. In [12], a multiagent based strategy to integrate the flexibility potential of industrial and residential demand is presented. A particular focus is on considering the cement and metal smelting industry, where residential and industrial demand is fulfilled through renewable and grid energy sources, respectively. In [13], a multiagent based load scheduling scheme is proposed, which utilizes optimal stopping theory to obtain the optimal scheduling instants. To optimally utilize the grid and renewable energy resources, a heuristic algorithm is used to solve the load scheduling problem [14]. Furthermore, to maximize user comfort in terms of scheduling delay, the load is modeled to prioritize comfort over cost and vice versa. To model the electricity prices, the authors first used a data mining approach to find the load consumption patterns from historical data using density based clustering with noise [15]. Then, a mixed integer nonlinear programming technique is applied to find electricity prices over the given time slots. To analyze the performance of this model, the online network enabled optimization system (NEOS) is used. The work presented in [16] is devoted to analyzing the advantages and disadvantages of active demand response programs in relation to residential load management. This work also analyzed the features of the energy system and highlights the key issues of decentralized energy resources. To reduce the peak load demand, the work presented in [17] is devoted to modeling the load and heating ventilation air-conditioning, particularly using demand response. In response to electricity prices and energy demand requirements, the dynamic demand response controller (DDRC) adjusts the control set points after every 15 min and shifts the load in case demand exceeds the predefined threshold. The proposed model is designed in the MATLAB/SIMULINK environment which is connected with the EnergyPlus model via a building control virtual test-bed. In [18], the authors proposed a renewable energy buying back scheme with a dynamic pricing strategy for energy efficiency in the smart grid. Here, the dynamic pricing is formulated as a convex optimization duel problem and a day-ahead tune dependent pricing in a distributed manner is proposed to ensure user privacy. This work is designed in such a way to provide the benefits to users and utility, dynamically. This is to note that the benefits offered by the DR program should not be underestimated due to their great impact on power system reliability and stability through its load management and optimization. However, the discussed benefits could be undermined due to the negligible impact of small power consumption loads on power system stability. Additionally, another drawback associated with the improper implementation of DR programs is the restriction of some high power rating loads being non-shiftable or non-interruptible [13]. On the other hand, EVs are becoming one of the potential candidates both in household and charging lots that draw a large amount of electric power, causing serious concerns from energy management perspectives [16]. There might be several reasons that exacerbate the utility's DR and create rebound peaks due to massive power that is drawn. Firstly, the aggregated charging loads of EVs can serve as a great resource of demand and capacity to obtain a noticeable impact on DR programs if there exists coordination among EV charging loads. Secondly, there should be greater flexibility on shifting and interruption offered by EV charging operators through the coordination with DR. This could manage EV charging load demand while fulfilling the key concerns of DR. These parameters are addressed and modeled in the work proposed in [19].

2. Related Work

Various price-based DR optimization techniques for optimally control residential loads have been reported in the literature. In [20], a new architecture based on technical building systems is presented, which is suitable for nearly zero energy building. To achieve this objective, the active load profiles of all the participants are obtained and then a building demand response program is used to model and manage the load for peak alleviation and flatten the load profile. These include model-based optimization methods [21], heuristic optimization techniques [22,23], where residential load is modeled with load classification [22] and customer preferences [23]. A stochastic optimization algorithms taking into consideration the inherent uncertainties in appliance scheduling time and electricity pricing [24]. To handle these uncertainties, a stochastic technique involving energy consumption adaptation variable and load consumption patterns is used. In [25], a probabilistic demand response program is proposed to model the load demand of residential sector. The main objective is to analyze the operational objectives used to balance the total cost. The stochastic optimization techniques provide the statistical results of the energy consumption of residential appliances. However, these algorithms are unable to guarantee an efficient DR policy from a day ahead perspective. On the other hand, model-based DR programs can guarantee the DR policy if the accurate load and patterns are available. However, a big challenge is how to get a correct estimate of the energy demand and usage patterns of residential loads. Because the energy consumption patterns cannot be fixed due to variable habits of energy consumers. In the third category, where home appliances are scheduled in response to dynamic prices using heuristic optimization methods. It is mentioned that optimal results in terms of cost and peak to average ratio reductions can be achieved with compromise on discomfort. Because, unlike electricity cost minimization, the comfort/social welfare maximization are two different objectives and cannot be achieved at the same time.

Moreover, the model-based DR schemes are developed based on simplified energy consumption models. In [26], the energy consumption of a residential house is modelled using simplified conduction heat transfer equations. In [27], the authors use a quasi-steady state approach for the estimation of a day-ahead electricity demand in the DR market. In [2], the authors adopt a model predictive control strategy that employs the model of building dynamics based on a thermal resistance-capacitance network. A low voltage residential load model based on price based DR has been proposed in [28]. In this work, high-resolution load models were developed by combing Monte Carlo Markov chain based bottom-up demand models, time-variant load model, shot water demand models and discrete state-space representation of thermal loads. Then, a price based DR program is modelled to control the working of all these loads in response to consumers. The model is useful for predicting the distributed impact of introducing dynamic pricing in the system. In [29], the authors consider the problems of considering dynamic pricing in the network. They have used a Stackelberg game approach to model the interactions between end-users and electricity producers. Then, a comprehensive characterization of the tradeoffs between consumer surplus and net profit is obtained. They have also analyzed the effects of renewable energy integration and distributed storage in the system. It is concluded that all benefits go to energy retailers when the capacity of renewable energy is small. These are the main conditions where renewable energy systems are used. In another similar work, the authors consider the problem of variable pricing and load consumption, where a utility or energy retailer acts as an agent between retailer and consumers [30], whereas variable pricing is somehow useful for the electricity retailer in stabilizing the electric network. However, its implementation seems difficult due to the lack of information about the consumers and associated uncertainties. Similarly, consumers are also more likely to face the difficulties in establishing their schedules for loads scheduling due to variable pricing. However, this problem can be tackled by using a reinforcement learning technique without background information of consumers and retailers. In [31], the authors propose a demand-side management (DSM) technique that takes into consideration user preferences. The authors identified a trade-off between cost reduction and comfort maximization and developed a Game theoretic-based algorithm to overcome this trade-off. In [32], the authors proposed a new pricing mechanism for low and high energy consumption users based on a time of use pricing policy. The price signal is further divided into different blocks and forecasted load is scheduled based on these prices to curtail the cost of end users. Here, an artificial neural network, due to its efficiency, is adopted to forecast the short term load in a day-ahead fashion. However, a few other algorithms to forecast the short and long term load are also being widely used by different authors [33]. In another similar approach, [34], the consumer's behaviours on each other's DSM decisions have been accounted for. Then a non-cooperative game strategy is used, where each user will decide whether to participate in the DR program or not. Here, to minimize the electricity cost is the main objective. While in traditional game-theoretic approaches, it is being considered that consumers are free in taking their decisions. In [35], the authors show that incoordination between consumer and electricity distributor may create the chances of high and rebound peaks leading to catastrophic behavior of the electric grid. In this regard, the authors propose a system-wide framework to coordinate the DR of end-users in a smart grid. The key objective of this framework is to provide monetary benefits, privacy and comfort to end-users.

3. Motivation

With population growth and advancements in information and communication technologies (ICTs), the power demand has also been increased. Consequently, scientists and researchers have been working to find new energy sources and energy management mechanisms to meet the growing power demand. In this regard, the SG vision has come to incorporate distributed and renewable energy resources, advance metering infrastructure (AMI) [36] and communication protocols, DR programs [20,21], and optimization techniques to efficiently manage the power demand. In accordance with this, the residential customers have been provided with the facility to curtail their load or reschedule the working slots in response to the DR programs and time-varying prices found in the literature [6–10]. Consequently, the utility obtained the benefits regarding of grid stability, while users can gain get the reduced bill as these pricing mechanisms are designed in such a way that the hourly price factor is calculated based on aggregated load consumed/demanded in a particular region which is operated under the same DSO. In almost all the techniques, DSM algorithms are used to schedule the load for energy management by taking into consideration the load demand, market clearing price, dynamic consumption trends and optimization algorithms. To achieve the effective outcomes of the DR based load scheduling programs, the end users are being offered various incentives in terms of bill reduction, uninterruptible supply of power, and so forth. However, the users are given the opportunity to decide whether they are willing to participate or not. Unlike the participating customers, the other customers would not be able to receive any benefit until they agreed to participate in DR programs by accepting terms and conditions. Furthermore, it is also understood that there could be multiple energy retailers being operated in some specific geographic region with dynamic and time varying price signals. The participating users have to manage their load according to the available price signal without knowing which other users are impacting on their benefits. For example, the high, medium and low energy consumers are always charged electricity bills according to the consumed load based on DR pricing policy. Here, the price signal would remain the same even if the customers are consuming low or high power over a given time period. Consequently, the those customers who have maintained a stable load profile and are consuming relatively less power are not getting the actual benefits or incentives in participating in DR and load scheduling programs. This is due to the discriminatory price signal being provided by the market retailer because it is difficult to provide a separate price signal to every user, which is a limitation of price based DR programs. Thus, by consideration of this problem and the underlying limitation of the DR program, this work proposes a new mechanism to provide some incentives to the customers maintaining a balanced load profile.

4. Contribution

This mechanism must possess the characteristics of a fair distribution of prices and incentives without any discrimination. By considering the aforementioned limitations and drawbacks, this paper further investigates the price based DR programs and introduces a novel mechanism for load management and price calculation with dynamic incentives. By taking into account the existing mechanisms (i.e., TOU, RTP) for calculating electricity prices, we have designed a new and slightly different method to calculate electricity prices and incentives based on load consumption and shifting patterns over given time period. For this purpose, the profiles of each user are obtained via smart meters using advanced metering infrastructure (AMI) [18] and are maintained just like the charging management system (CMS) in the electric vehicle scenario [28]. Then, by using the proposed mechanism, the price is calculated whether the user will get the incentives or overpriced signal, which is independent of each user and load. One of the unique and novel aspects of the proposed mechanism is that these incentives are non-discriminatory and independent from other users participating in a load scheduling based incentivise mechanism. These prices rather depend upon individual consumption trends and the RTP signal over the given time period. Hence, the proposed mechanism is designed to facilitate the users in reducing high and rebound peaks by incorporating the upper limit of energy consumption during the scheduling process. However, this may be inconvenient for those users who intend to use energy as per their defined schedules to improve the social welfare level. Consequently, the utility may receive extra benefits in terms of system stability through reliable energy generation and supply systems. Finally, the developed system model is tested and validated using the proposed algorithm which presents the idea of calculating dynamic incentives instead of static and predefined incentives.

5. Characterizing DR

As understood, electricity cannot be stored at large scale power systems due to physical limitations and constraints. Therefore, it is strongly recommended that the difference between electricity generation and consumption must be minimal. Moreover, the marginal cost of electricity in a day-ahead market is extremely dynamic due to variable and unexpected energy consumption patterns, whereas the electricity cost varies over the given period (i.e., 60 min, 30 min, 15 min) of fixed intervals, and consumers receive retail electricity prices reflecting average energy generation trends including transmission and distribution losses. Therefore, the frequent disconnections between short term marginal and long term marginal costs can cause serious concerns regarding real-time energy generation and consumption. This is because consumers do not have the information about the marginal cost of supplied electricity. This results in an inefficient participation in DR programs. Consequently, the consumers receive little or no incentive for participating in DR programs. From the above-mentioned discussion, it can be concluded that a flat rate tariff encourages the consumers to overuse electricity during low pricing hours. As a result, electricity tariffs may have higher values due to excessive energy consumption during particular hours. Therefore, utilities have to fulfil energy demand by turning on extra generators leading to costly tariffs. In conclusion, inefficient utilization of DR programs and calculation of electricity tariffs in a day-ahead market may affect a consumer's objectives. As a response, consumers may be discouraged to adopt DR programs, which can create serious concerns regarding grid stability in the long term. So, it is desired to adopt some real-time mechanisms that could be feasible for both utilities and consumers, without disturbing/violating their objectives.

6. System Model

In this section, we describe the proposed system model used to calculate incentive based electricity prices. Let $i \in N$ denote the loads l_i with time deadline denoted by $t \in T$ such that $t = 1, 2, 3, ...T$ and $j \in M$ denotes index of users who are participating in load scheduling program and getting incentives. The energy demand of each load i is denoted by $d_{l_i}(t)$. Let $\phi(t)$ denoted the global pricing policy (i.e., RTP) obtained from the electricity market via AMI. Let β be the binary decision variable and represents the ON/OFF status of load. Each load has a predefined start τ_i^s and end time τ_i^e, which lies within scheduling time interval T. Without losing generality, all customers have the equal opportunity to schedule their loads within the given time frame. Consequently, the cost-sensitive customers' $\Bbbk_1$ can get some incentives by shifting some load from on-peak to off-peak hours. In contrast, the customers $\Bbbk_3$ who are not willing to participate in DR programs may receive higher prices, which are calculated based on energy price and demand during critical hours. The details of $\Bbbk_1$ and $\Bbbk_3$ are given in the next section.

6.1. Previous Model

In traditional schemes being used to calculate energy consumption prices, where all users receive global prices irrespective of individualized demands, this may result in re-bound peaks, that is, due to demand shifting from critical hours to off-peak hours [1,37,38]. In consequence, the non-homogeneous cost may be charged to customers, which participated in DSM programs. In addition, exposing end-users to the wholesale electricity market (i.e., real-time and day-ahead prices) has to lead the users to shift their load to a low pricing area, which eventually increases demand during this time span [38]. This would eventually lead to higher peaks. It is also demonstrated in [37] that this may create rebound peaks due to demand shifting, and hence, DSMs need to be studied in accordance with homogeneous and non-homogeneous energy users with the objective of peak shaving and fair price distribution among all types of consumers. Similarly, the work reported in [39] provided the concept of user aware price policies, which is based on individualized demand profiles obtained from smart grid communication network. Although the concept is novel, which helps with alleviating non-homogeneous price penalties on potential users. However, in reality, energy consumption has a dynamic trend which is based on customers' daily life activities. Eventually, this mechanism [39] may create homogeneous cost profiles for a given time. A general energy price is calculated using Equation (1).

$$p_i = \sum_{i \in N} \sum_{t \in T} \left\{ \left[(\beta \times \ell_i(t)) \times d_{\ell_i}(t) \right] \times \phi(t) \right\}, \forall i \in T, t \in T, \tag{1}$$

where $\phi(t)$ is the real time electricity price obtained from a day-ahead market, β is a binary decision variable which denotes, that is, ON/OFF states of connected loads $\ell_i(t)$ during time t, and is expressed as;

$$\beta = \begin{cases} 1; & \textit{If load is ON} \\ 0; & \textit{If load is OFF.} \end{cases} \tag{2}$$

As the utility pricing model is based on some specified time interval, which is considered 24 h equally divided on all time slots. Therefore, t_s and t_e are used to denote start and end time intervals of ℓ_i over the given price $\phi(t)$, respectively, whereas the ϕ is obtained from the electricity retail market, which is changed due to load trends and consumer demand over the given period $[t \in T]$, while τ_i denotes utility cycles (i.e., number of time slots) of all loads subject to $(\beta\ell_i(t))$. The following sections discuss different cases related to electricity price calculation methods.

6.2. Demand Aware Prices (Case-1)

To cater for the aforementioned uncertainty, the proposed work provides a mathematical mechanism to calculate electricity prices on the basis of energy consumption profiles of all users/units which are obtained from smart grid communication network. For this purpose, there need to be community networks to handle individual price profiles of all users in accordance with the individualized energy consumption trends. The electricity price (p'_i) is calculated by using the proposed mechanism, which depends on the load consumption of each customer, and is expressed through Equation (3):

$$p'_i(t) = \sum_{i \in N} \sum_{t \in T} \left\{ \left[(\beta \times \ell_i(t)) \times d_{\ell_i}(t) \right] \times \phi'(t) \right\}, \forall i \in T, t \in T, \tag{3}$$

where $\phi'(t)$ denotes homogeneous price policies which further depends on $\Bbbk_1$ variable denoting the actual electricity price each customer will be charged, giving the variation in energy consumption of ℓ_i over give time t. The $\Bbbk_1$ can be calculated on the bases of $\ell_i(t)$ and $\phi(t)$ and expressed in Equation (4):

$$\Bbbk_1(t) = \sum_{i \in N} \sum_{t \in T} \left\{ \left[((\beta \times \ell_i(t)) \times d_{\ell_i}(t)) \times \phi(t) \right] \times (d_i)^{-1/2} \right\}, \forall i \in T, t \in T, \tag{4}$$

$$p'_i(t) = \sum_{i \in N} \sum_{t \in T} \left\{ \left[(\beta \times \ell_i(t)) \times d_{\ell_i}(t) \right] \times \Bbbk_1(t) \right\}, \forall i \in T, t \in T. \tag{5}$$

Equation (5) denotes the electricity price (p'_i) in the proposed case and the cost minimization objective function can be described through Equation (6):

$$min \sum_{i \in N} \sum_{t \in T} \Phi_i(t) \tag{6}$$

$$s.t: \ \forall i \in N : \Phi_i = \left(\sum_{i \in N} \sum_{t \in T} p'_i(t). \right) \tag{7}$$

6.3. An Incentive Based Price Overview (Case-2)

In this case, we first identify the total users consuming high power, contributing towards a high electricity tariff. Eventually, the other users who have a balanced power consumption profile may be affected in terms of bearing high prices. This section aims to devise a mechanism to fairly design customized price profiles for all users. For this purpose, the pricing tariff for the customers consuming high power needs to be calculated, which would be used in calculating incentives for the customers showing balanced power consumption trends. After calculating incentives and penalties for respective customers, we will then compare the cost values of both traditional and proposed mechanisms for validation purpose. Using Equation (8), we can identify the consumers who have drawn more power leading to adding more to the electricity price being provided.

$$p'_i(t) - p_i(t) = 0, \forall i \in T, t \in T. \tag{8}$$

Now, we can design the customized electricity tariff for high consumption users. Meanwhile, we also calculate the incentives to those customers who willingly maintain stable load profile to seek incentives from utility.

6.4. An Incentive Based Price Calculation

This section discusses the incentive-based price policies designed for those customers who participated in the energy management process $j \in M$ to reduce peak load demand. In response, the participating customers have been given incentives in the form of a bill reduction. In contrast, the customers who do not want to compromise on their comfort level and have not yet participated in energy management programs are charged prices in accordance with load demand and non-discriminatory penalties. These penalties have been calculated to preserve utility revenue. Initially, we have to calculate the electricity cost of two types of customers: (i) the customers with surplus electricity cost being charged after participation in the energy management (EM) process; and (ii) the customers with reduced electricity costs after participation in the EM process. For this purpose, we first calculate the reduced cost as expected by the M customers, which can be calculated through Equation (9):

$$p'_i(t) = \sum_{j \in M} \sum_{i \in N} \sum_{t \in T} \left\{ \left[(\beta \times \ell_{i,j}(t)) \times d_{\ell_{i,j}}(t) \right] \times \Bbbk_2(t) \right\}, \forall i \in, j \in J, T, t \in T \tag{9}$$

$$\Bbbk_2(t) = \sum_{i \in N} \sum_{j \in M} \sum_{t \in T} \left\{ \left[((\beta \times \ell_{i,j}(t)) \times d_{\ell_{i,j}}(t)) \phi'(t) \right] \times (d_{i,j})^{-1/2} \right\}, \forall i \in T, t \in T, \tag{10}$$

where Equation (10) denotes the fraction by which the customers j consumes more energy, and $\phi'(t)$ is the price, which is expected by customers j in order to reduce their electricity cost. The actual electricity price $p''_i(t)$ which is obtained after load scheduling is expressed through Equation (11):

$$p''_i(t) = \sum_{i \in N} \sum_{j \in M} \sum_{t \in T} \left\{ \left[(\beta \times \ell_{i,j}(t)) \times d_{\ell_{i,j}}(t) \right] \times \phi''(t) \right\}, \forall i \in T, j \in M, t \in T, \tag{11}$$

where $\phi''(t)$ depicts the electricity price being charged to j customers. Now, we can calculate the surplus electricity price being charged to j customers, based on which the comfort cost is calculated for i consumers. Once this cost is calculated, the extra cost charged to j customers would be adjusted in order to provide incentives in the form of electricity bill reduction. The total amount of electricity cost for i can be calculated through Equation (12):

$$\overline{p_j}(t) = \sum_{i \in N} \sum_{j \in M} \sum_{t \in T} \left\{ \left[\left((\beta \times \ell_{i,j}(t)) \times d_{\ell_{i,j}}(t) \times (\Bbbk_1(t) + \alpha) \right) \right] \times (d_{i,j})^{-1/2} \right\}, \forall i \in N, j \in M, t \in T, \tag{12}$$

where $\overline{p_i}$ denotes the actual price charged to M customers and $\alpha = \Bbbk_2 - \Bbbk_3$ is the approximate cost difference between scheduled and unscheduled cases. $\Bbbk_3$ can be through Equation (13):

$$\Bbbk_3(t) = \sum_{i \in N} \sum_{j \in M} \sum_{t \in T} \left\{ \left[(\beta \times \ell_{i,j}(t)) \times d_{\ell_{i,j}}(t) \times \phi'(t) \right] \times (d_{i,j})^{-1/2} \right\}, \forall i \in N, j \in M, t \in T. \tag{13}$$

Now, the total energy consumption cost $\hat{\Phi}_i(t)$ over time t is calculated by using Equation (14);

$$\hat{\Phi}_{i,j}(t) = \left\{ \overline{p_i} \sum_{i\in N} \sum_{t\in T}(t) + \left[\sum_{i\in N} \sum_{j\in M} \sum_{t\in T} \left\{ \left[(\beta \times \ell_{i,j}(t)) \times d_{\ell_{i,j}}(t) \right] \times \Bbbk_3(t) \right\} \times (d_{i,j})^{-1/2} \right] \times \sum_{i\in N} \sum_{j\in M} \sum_{t\in T} d_{i,j}(t) \right\}, \quad (14)$$
$$\forall i \in N, j \in M, t \in T.$$

Now, we can calculate incentives for M customers who took part in the energy management program. Let $\mathbb{E}_{\overline{p_i}}$ denote the monetary incentives given to those customers who have participated in the energy management process to flatten the peak demand during critical hours. From the above Equations (1)–(12), we can calculate $\mathbb{E}_{\overline{p_i}}$ as given below;

$$\mathbb{E}_{\overline{p_i}} = \sum_{i\in N} \sum_{j\in M} \sum_{t\in T} \left[\hat{\Phi}_{i,j}(t) - \left\{ \Bbbk_3(t) \times \left((\beta \times \ell_{i,j}(t)) \times d_{\ell_{i,j}}(t) \right) \times \phi'(t) \right\} \right], \forall i \in N, j \in M, t \in T. \quad (15)$$

Now, the final objective function Equation (16) which is the incentive maximization and subject to the constraints of Equations (17)–(20) is written as:

$$min \ \sum_{i\in N} \sum_{j\in M} \sum_{t\in T} \left(\mathbb{E}_{\overline{p}_{i,j}}(t) \right) \quad (16)$$

$$\mathbb{E}_{\overline{p_i}} \leq (d_{\ell_i} \times \phi) : \forall i \in N. \quad (17)$$

Equation (16) refers to the cost minimization objective function and Equation (17) denotes that the total cost obtained from the proposed mechanism should always be less than the traditional mechanism. Otherwise, the proposed mechanism is not an optimal one.

$$\hat{\Phi}_i = u_r : \forall i \in N \quad (18)$$

$$p_{\ell_{i,j}^{un}} = p_{d_{i,j}^{sch}} : \forall i \in N. \quad (19)$$

Equation (18) shows the total cost of load consumed, which should be equal to the cost/utility revenue. Otherwise, the mismatch could ultimately affect the objectives. Equation (19) denotes that the unscheduled and scheduled load before and after the scheduling process must be equal. Otherwise, the proposed mechanism could be based on energy conservation/reduction instead of load scheduling or management.

$$\overline{p_{d_{i,j}}} < p_{d_{i,j}} < \underline{p_{d_{i,j}}} : \forall i \in N \quad (20)$$

$$\ell_i \leq \beta \times \ell_{uti}(t) : \forall i \in N, t \in T. \quad (21)$$

Equation (20) shows the upper and lower limits on the load demand and Equation (21) shows that the load demand must not exceed the utility capacity.

7. Proposed Algorithm

In the proposed work, we used GA to solve the incentive-based load scheduling problem. Although, the consideration of heuristic algorithms is gaining popularity due to their ability to obtain the best optimal results even when other mathematical algorithms fail due to the diverse nature of control parameters and inherent complexity while designing or formulating the underlying problem, mathematically. That is why we have considered and used GA due to its ability to obtain global optimal results in all situations with a high convergence rate. Moreover, the crossover and mutation operators further allow obtaining the best results even if some random parameters affect the performance of the results. Generally, convergence is one of the important parameters when dealing with successive series of experiments in iterative methods [36,40]. Generally, the GA works by considering some initial parameters with random initial population, fitness function, and values of other control parameters. Initially, the objective/cost function is evaluated as per given control values such as cost, incentives and delay in the proposed work. Once the fitness is calculated and the best results are obtained, the algorithm stops by following some predefined criteria or based on an initial number of iterations. Then, the best optimal results are saved and crossover and mutation operators are applied to evaluate the fitness function based on a new population. The probabilities of crossover P_c and P_m are set in such a way to obtained global optimal results. However, to avoid premature convergence, we have compared the best convergence results with the newly obtained results in each iteration and used a Sigma scaling factor so that premature results should be discarded. We have also tested the convergence against different control values to assess the performance under the proposed model. It is also worth mentioning here that the performance of any optimization algorithm also depends on the well behaved functional form of a developed mathematical model. Otherwise, the optimal results may take more time or face complexities in achieving them. Table 1 provides the loads with demand and working hours, while the Table 2 gives the values of control variables used to find the optimal scheduling patterns and incentives. Max. generation is selected as 800 so as to have the sufficient space for optimal results. The population size is chosen as 400 in order to avoid the solution being premature. The P_c and P_m values are selected and set after monitoring the convergence results. We observed through experiments that these values are the most suitable for fast convergence. The Algorithm 1 explains the working steps involved in getting incentives based price profiles. In step-4, the initial random population is generated which is evaluated in step-5 using Equation (7). Step-6 is involved in calculating the power using RTP and proposed mechanisms. Equations (9), (11), (12), (14) and (15) are involved for the calculation. Steps 1–8 show that the algorithm moves to the next step if an optimal solution is obtained. Otherwise, the control will move to step-1 for new calculation and evaluation. Steps 13–14 compare the optimal solutions obtained from Equation (16) and Equation (6). After saving the results, the crossover and mutation operators are applied and the control moved to the step-1. The same process repeats over a 24 h time period.

Table 1. Load consumption and duty cycle requirements.

l_i	Working Hours	d_{l_i} (kW)
l_1	20	2.5
l_2	24	3
l_3	5	2
l_4	7	2.5
l_5	8	3.5
l_6	8	3

Table 2. Control parameters of GA [36,40].

Parameters	Values
Number of loads	6
Number of users	3
Max. generation	800
Population size	400
Probability of crossover	0.9
Probability of mutation	0.003

Algorithm 1 Steps involved in calculating incentives using GA.

Require: $l_i, d_{l_i}, \phi, popsize, P_c, P_m,$

1: **for** $t = 1$ **to** T **do**
2:    **for** $i = 1$ **to** N **do**
3:       **for** $j = 1$ **to** M **do**
4:          generate initial population
5:          evaluate Equation (7)
6:          calculate $p'(t), p''(t), \overline{p_j}, \hat{\Phi}_{i,j}, \mathbb{E}_{\overline{p}_{i,j}}$ using Equations (9), (11), (12), (14) and (15)
7:          evaluate Equation (16)
8:          **if** *best* $==$ *min.* **then**
9:            save result
10:         **else**
11:            move to step-1
12:         **end if**
13:         **if** *min.Equation* (16) $\leq$ *min.Equation* (6) **then**
14:            save the results
15:         **else**
16:            go to step-4
17:         **end if**
18:         do selection process
19:         do P_c
20:         do P_m
21:         move to step-1
22:       **end for**
23:    **end for**
24: **end for**

---

8. Results and Discussion

Figure 1 shows the electricity price signal used in the proposed work. Unlike other pricing signals being widely used in the literature and real-time works such as RTP, TOU, DAP, CPP, the proposed price signal is dynamic with changing values at every instant of time. This price signal is considered to deeply analyze the realistic cost and incentive profiles of all the consumers instead of developing the load scheduling techniques of algorithms based on DAP, which is known in advance to the users and energy management controllers. Furthermore, these price signals elucidate the realistic changing behaviours of price with respect to load. Figure 2 gives the convergence profile of GA based on the control and load scheduling parameters. It is clear from the figure that GA converges within 400 iterations and results with the best optimal cost in each iteration are obtained. Figure 3 give and comparison of the price profiles between the RTP signal and the proposed pricing signals being calculated using proposed incentive based model. It can be seen from this figure that the proposed price profiles differ from RTP, showing that the proposed mechanism calculates the price signal based on load demand, scheduling capacity and incentives, respectively. Figure 1 shows a comparison of different patterns of price profiles using RTP and proposed price (PP). From first sight, it is very clearly visible that the price profiles using PP has variations when compared with RTP obtained from a day-ahead market. These variations are due to various incentives provided to different consumers based on load shifting and taking part in the load scheduling mechanism. For example, if any user is willingly participating in load scheduling and maintain a consumption level under the upper threshold limit as compared to the other consumers, he gets incentives in electricity tariffs of that particular hour, instead of for the whole day. Similarly, if any user is consuming relatively more power and exceeding the upper threshold limit, then s/he has to be charged the extra price due to creating trouble for the electricity retailer of the producer.

This is because the electricity producer or distributor has to maintain a balance between generation and demand. Figure 4 provides the energy consumption profiles of all the units using RTP. t can be seen that scheduled load profiles show remarkable trends reflecting the scheduling capability of the proposed mechanism based on GA algorithm, which is in terms of load scheduling and peak management as well. Similarly, Figure 5 shows the cost profiles of all the units using RTP. These variations provide an overview of the scheduling mechanism designed to lower the overall cost and providing incentives to the users. It is also seen from the profiles that scheduled cost is lower than unscheduled cost. Here, the maximum peak is around 350$ and the minimum peak is around 20$. In contrast, Figure 6 provides the cost profiles of all the units using PP. Here, we can see that the maximum peak is around 370$ and the minimum peak is around 10$. It reflects that those customers who are willingly participating in the proposed scheduling mechanism are getting the incentives in terms of reduced cost. On the other hand, the other customers who are not interested in participating in load management programs are getting a comparatively higher price and cost.

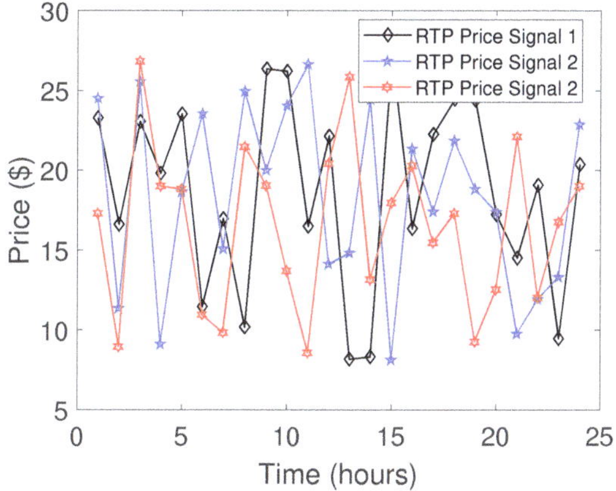

Figure 1. Real time electricity price signals over time *t*.

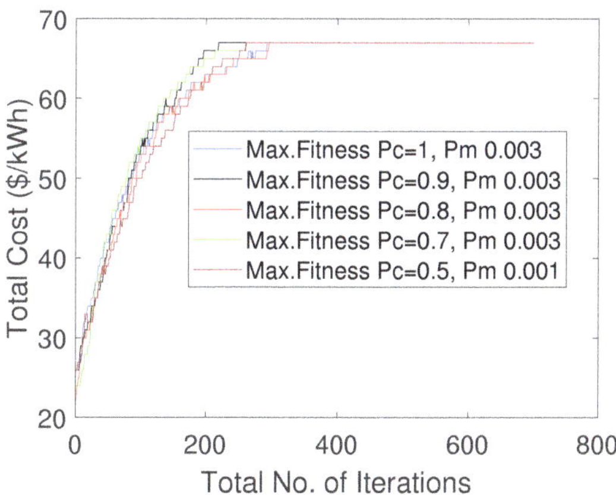

Figure 2. A comparison between total cost and fitness value over different values of control parameters.

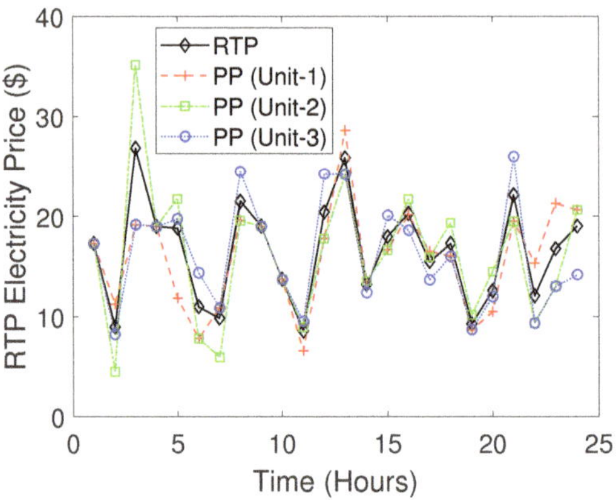

Figure 3. A comparison between real time electricity price and the proposed method for different units over time *t*.

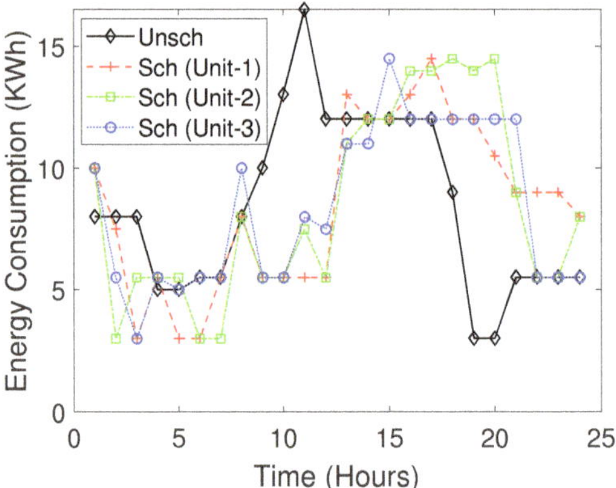

Figure 4. A comparison of energy consumption using real time electricity and proposed price signals for different units over time *t*.

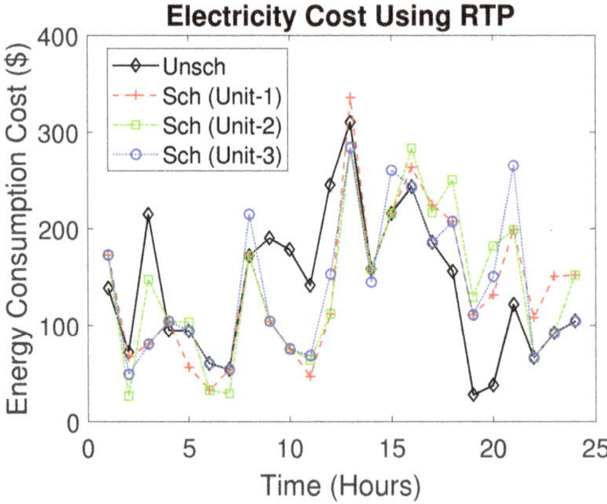

Figure 5. Electricity cost incurred for different units over time *t* using real time price signal.

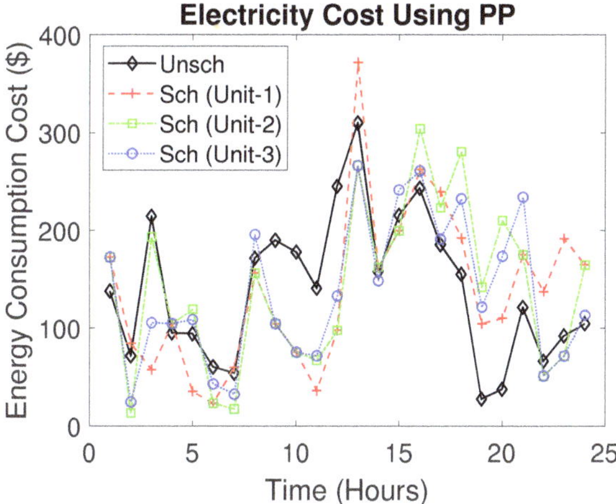

Figure 6. Electricity cost incurred for different units over time t using proposed method.

9. Conclusions

Price based DR programs through AMI are designed to encourage users to participate in load scheduling processes to balance generation and demand capacity without considering costlier generation. However, these programs faced difficulties in attracting a large number of customers to participate. This is due to the inherent limitations caused by its discriminatory nature if applied in different geographical regions being operated under the same distributed system operator. To handle this limitation, this work has proposed an incentive-based load scheduling mechanism using a real-time pricing policy to facilitate energy retailers' and end-users' particularity. The incentives, in terms of dynamic pricing profiles using load consumption trends, are calculated. However, as these incentives are calculated based on the dynamic price profiles of different customers, we first obtained these profiles based on load consumption trends. In response, each customer received a different price signal, which is dynamically changed with load consumption reflecting incentives or an overpriced tariff. To analyze the impact of the proposed mechanism, we provided a system model to calculate these prices and incentives and formulate an objective function. Then, a well behaved functional form of optimization function is solved by using a heuristic based GA and obtained the energy and load consumption profiles. The results are then compared with the unscheduled profiles without incentives. It is clear from the results that participating customers are able to get dynamic pricing signal based incentives without affecting the electricity bills of other customers. In the future, we intend to extend and implement this model in designing electricity prices in peer-to-peer energy transactions using blockchain technology.

Author Contributions: T.A., M.B.R. and M. Awais did the mathematical modelling and implementation. T.A., A.H.M. check and verify the results. T.A., A.H.M. and M.B.R. supervise the project. All authors have read and agreed to the published version of the manuscript.

Funding: This project was funded by Deanship of Scientific Research (DSR) at King Abdulaziz University, Jeddah, under grant number (RG-13-135-41). The authors therefore acknowledge with thanks DSR for their technical and financial support. This project has also received funding from the European Union Horizon 2020 research and innovation programme under the Marie Sklodowska-Curie grant agreement No 754382, GOT ENERGY TALENT. The content of this [report/study/article/publication] does not reflect the official opinion of the European Union. Responsibility for the information and views expressed herein lies entirely with the author(s).

Institutional Review Board Statement: Not applicable.

Informed Consent Statement: Not applicable.

Data Availability Statement: No new data were created or analysed in this study. Data sharing is not applicable to this article.

Acknowledgments: The authors thank the referees for careful reading and useful comments that helped to improve the paper.

Conflicts of Interest: The authors declare no conflict of interest.

Nomenclature

t	index of time
i	index of loads
j	index of customers
a_r	number of arrival requests
d_{l_i}	load demand of ith load
l_i	load index for ith load
p_i	energy consumption price of ith load
β	binary decision variable
$\overline{p_i}$	actual price charged to M customers
$\hat{\Phi}_{i,j}$	total energy consumption cost
u_r	utility revenue
$p_{d_{i,j}^{sch}}$	scheduled load of jth customer
$p_{d_{i,j}}$	min. limit on load demand of jth customer
$\overline{\phi(t)}$	electricity price over time t
τ_i	utility cycle of ith load
p'	electricity price using proposed method
κ_1	actual electricity price phase-1
Φ_i	total electricity cost of ith load
κ_2	actual electricity price phase-2
p''	actual electricity price after incentives
$\phi 1$	electricity price after scheduling
α	approximate cost difference
$\mathbb{E}_{\overline{p}_i}$	incentives for customer
$p_{\ell_{i,j}^{un}}$	unscheduled load jth customer
$\overline{p_{d_{i,j}}}$	max. limit on load demand of jth customer
l_{uti}	utility load

References

1. Baimel, D.; Tapuchi, S.; Baimel, N. Smart grid communication technologies-overview, research challenges and opportunities. In Proceedings of the 2016 International Symposium on Power Electronics, Electrical Drives, Automation and Motion (SPEEDAM), Capri, Italy, 22–24 June 2016; pp. 116–120. [CrossRef]
2. Siroky, J.; Oldewurtel, F.; Cigler, J.; Privara, S. Experimental analysis of model predictive control for an energy efficient building heating system. *Appl. Energy* **2011**, *88*, 3079–3087. [CrossRef]
3. Konda, S.R.; Mukkapati, B.; Panwar, L.K.; Panigrahi, B.K.; Kumar, R. Dynamic Energy Balancing Cost Model for Day Ahead Markets with Uncertain Wind Energy and Generation Contingency under Demand Response. *IEEE Trans. Ind. Appl.* **2018**, *54*, 4908–4916. [CrossRef]
4. Peizhong, Y.; Xihua, D.; Abiodun, I.; Chi, Z.; Shufang, L. Real-time oppertunistic scheduling for residential demand response. *IEEE Trans. Smart Grid* **2013**, *4*, 174–227.
5. Aslam, S.; Khalid, A.; Javaid, N. Towards efficient energy management in smart grids considering microgrids with day-ahead energy forecasting. *Electr. Power Syst. Res.* **2020**, *182*, 106232. [CrossRef]
6. Xue, K.; Yang, Q.; Li, S.; Wei, D.S.L.; Peng, M.; Menmon, I.; Hong, P. PPSO: A Privacy-Preserving Service Outsourcing Scheme for Real-Time Pricing Demand Response in Smart Grid. *IEEE Internet Things J.* **2019**. [CrossRef]
7. Zhang, H.; Yue, D.; Dou, C.; Li, K.; Xie, X. Event-Triggered Multiagent Optimization for Two-Layered Model of Hybrid Energy System With Price Bidding-Based Demand Response. *IEEE Trans. Cybern.* **2021**. [CrossRef] [PubMed]
8. Ding, T.; Qu, M.; Amjady, N.; Wang, F.; Bo, R.; Shahidehpour, M. Tracking Equilibrium Point under Real-time Price-Based Residential Demand Response. *IEEE Trans. Smart Grid* **2020**. [CrossRef]
9. Ozturk, Y.; Senthilkumar, D.; Kumar, S.; Lee, G. An intelligent home energy management system to improve demand response. *IEEE Trans. Smart Grid* **2013**, *4*, 694–701. [CrossRef]

10. Nguyen, D.H.; Narikiyo, T.; Kawanishi, M. Optimal Demand Response and Real-Time Pricing by a Sequential Distributed Consensus-Based ADMM Approach. *IEEE Trans. Smart Grid* **2018**, *9*, 4964–4974. [CrossRef]

11. Gholian, A.; Mohsenian-Rad, H.; Hua, Y. Optimal Industrial Load Control in Smart Grid. *IEEE Trans. Smart Grid* **2016**, *7*, 2305–2316. [CrossRef]

12. Golmohamadi, H.; Keypour, R.; Bak-Jensen, B.; Pillai, J.S.R. A multi-agent based optimization of residential and industrial demand response aggregators. *Int. J. Electr. Power Energy Syst.* **2019**, *107*, 472–485. [CrossRef]

13. Rasheed, M.B.; Javaid, N.; Malik, M.S.A.; Asif, M.; Hanif, M.K.; Chaudary, M.H. Intelligent Multi-Agent Based Multilayered Control System for Opportunistic Load Scheduling in Smart Buildings. *IEEE Access* **2019**, *7*, 23990–24006. [CrossRef]

14. Asgher, U.; Rasheed, M.B.; Al-Sumaiti, A.; Rahman, A.; Ali, I.; Alzaidi, A.; Alamri, A. Smart Energy Optimization Using Heuristic Algorithm in Smart Grid with Integration of Solar Energy Sources. *Energies* **2018**, *11*, 3494. [CrossRef]

15. Yang, J.; Zhao, J.; Wen, F.; Dong, Z. A Model of Customizing Electricity Retail Prices Based on Load Profile Clustering Analysis. *IEEE Trans. Smart Grid* **2018**. [CrossRef]

16. Darby, S.J. Load management at home: Advantages and drawbacks of some active demand side options. *J. Power Energy* **2012**, *227*, 9–17. [CrossRef]

17. Yoon, J.H.; Baldick, R.; Novoselac, A. Dynamic demand response controller based on rea-time retail price for residential buildings. *IEEE Trans. Smart Grid* **2014**, *5*, 121–129. [CrossRef]

18. Chiu, T.C.; Shih, Y.Y.; Pang, A.C.; Pai, C.W. Optimized Day-Ahead Pricing with Renewable Energy Demand-Side Management for Smart Grids. *IEEE Internet Things J.* **2017**, *4*, 374–383. [CrossRef]

19. Ridoy, D.; Wang, Y.; Putrus, G.; Kotter, R.; Marzband, M.; Herteleer, B.; Warmerdam, J. Multi-objective techno-economic-environmental optimisation of electric vehicle for energy services. *Appl. Energy* **2020**, *257*, 113965.

20. Martirano, L.; Habib, E.; Parise, G. Demand Side Management in Microgrids for Load Control in Nearly Zero Energy Buildings. *IEEE Trans. Ind. Appl.* **2017**, *53*, 1769–1779. [CrossRef]

21. Adika, C.O.; Wang, L. Demand-side bidding strategy for residential energy management in a smart grid environment. *IEEE Trans. Smart Grid* **2014**, *5*, 172–1733. [CrossRef]

22. Rasheed, M.B.; Javaid, N.; Ahmad, A.; Khan, Z.A.; Qasim, U.; Alrajeh, N. An efficient power scheduling scheme for residential load management in smart homes. *Appl. Sci.* **2015**, *5*, 1134–1163. [CrossRef]

23. Rasheed, M.B.; Javaid, N.; Ahmad, A.; Jamil, M.; Khan, Z.A.; Qasim, U.; Alrajeh, N. Energy optimization in smart homes using customer preference and dynamic pricing. *Energies* **2016**, *9*, 593. [CrossRef]

24. Chen, X.; Wei, T.; Hu, S. Uncertainty-aware household appliance scheduling considering dynamic electricity pricing in smart home. *IEEE Trans. Smart Grid* **2013**, *4*, 932–941. [CrossRef]

25. Harsha, P.; Sharma, M.; Natarajan, R.; Ghosh, S. A framework for the analysis of probabilistic demand response schemes. *IEEE Trans. Smart Grid* **2013**, *4*, 2274–2284. [CrossRef]

26. Black, J.W. Integrating Demand into the U.S. Electric Power System: Technical, Economic, and Regulatory Frameworks for Responsive Load. Ph.D. Dissertation, Engineering Systems Division, Massachusetts Institute of Technology: Cambridge, MA, USA, 2005.

27. Zhou, Z.; Zhao, F.; Wang, J. Agent-based electricity market simulation with demand response from commercial buildings. *IEEE Trans. Smart Grid* **2011**, *2*, 580–588. [CrossRef]

28. McKenna, K.; Keane, A. Residential load modeling of price-based demand response for network impact studies. In Proceedings of the 2016 IEEE Power and Energy Society General Meeting (PESGM), Boston, MA, USA, 17–21 July 2016; p. 1. [CrossRef]

29. Jia, L.; Tong, L. Dynamic Pricing and Distributed Energy Management for Demand Response. *IEEE Trans. Smart Grid* **2016**, *7*, 1128–1136. [CrossRef]

30. Kim, B.G.; Zhang, Y.; Schaar, M.v.; Lee, J.W. Dynamic Pricing and Energy Consumption Scheduling With Reinforcement Learning. *IEEE Trans. Smart Grid* **2016**, *7*, 2187–2198. [CrossRef]

31. Yaagoubi, N.; Mouftah, H.T. User-Aware Game Theoretic Approach for Demand Management. *IEEE Trans. Smart Grid* **2015**, *6*, 716–725. [CrossRef]

32. Aurangzeb, K.; Aslam, S.; Mohsin, S.M.; Alhussein, M. A Fair Pricing Mechanism in Smart Grids for Low Energy Consumption Users. *IEEE Access* **2021**, *9*, 22035–22044. [CrossRef]

33. Aslam, S.; Herodotou, H.; Mohsin, S.M.; Javaid, N.; Ashraf, N.; Aslam, S. A survey on deep learning methods for power load and renewable energy forecasting in smart microgrids. *Renew. Sustain. Energy Rev.* **2021**, *144*, 110992. [CrossRef]

34. Wang, Y.; Saad, W.; Mandayam, N.B.; Poor, H.V. Load Shifting in the Smart Grid: To Participate or Not? *IEEE Trans. Smart Grid* **2016**, *7*, 2604–2614. [CrossRef]

35. Safdarian, A.; Fotuhi-Firuzabad, M.; Lehtonen, M. Optimal Residential Load Management in Smart Grids: A Decentralized Framework. *IEEE Trans. Smart Grid* **2016**, *7*, 1836–1845. [CrossRef]

36. Katoch, S.; Chauhan, S.S.; Kumar, V. A review on genetic algorithm: Past, present, and future. *Multimed Tools Appl.* **2021**, *80*, 8091–8126. [CrossRef] [PubMed]

37. Li, Y.; Ng, B.L.; Trayer, M.; Liu, L. Automated residential demand response: Algorithmic implications of pricing models. *IEEE Trans. Smart Grid* **2012**, *3*, 1712–1721. [CrossRef]

38. Zhao, Z.; Lee, W.C.; Shin, Y.; Song, K.-B. An optimal power scheduling method for demand response in home energy management system. *IEEE Trans. Smart Grid* **2013**, *4*, 1391–1400. [CrossRef]

39. Al-Rubaye, S.; Al-Dulaimi, A.; Mumtaz, S.; Rodriguez, J. Dynamic Pricing Mechanism in Smart Grid Communications Is Shaping Up. *IEEE Commun. Lett.* **2018**, *22*, 1350–1353. [CrossRef]
40. Houck, C.R.; Joines, J.; Kay, M.G. A genetic algorithm for function optimization: A Matlab implementation. *Ncsu-ie tr* **1995**, *95*, 1–10.

Article

Study of an Optimized Micro-Grid's Operation with Electrical Vehicle-Based Hybridized Sustainable Algorithm

Muhammad Shahzad Nazir [1,*], Zhang Chu [1], Ahmad N. Abdalla [2], Hong Ki An [3], Sayed M. Eldin [4], Ahmed Sayed M. Metwally [5], Patrizia Bocchetta [6] and Muhammad Sufyan Javed [7,*]

1. Faculty of Automation, Huaiyin Institute of Technology, Huai'an 223003, China
2. Faculty of Information and Electronic Engineering, Huaiyin Institute of Technology, Huai'an 223003, China
3. Department of Transportation Engineering, Huaiyin Institute of Technology, Huai'an 223003, China
4. Center of Research, Faculty of Engineering, Future University in Egypt, New Cairo 11835, Egypt
5. Department of Mathematics, College of Science, King Saud University, Riyadh 11451, Saudi Arabia
6. Dipartimento di Ingegneria dell'Innovazione, Università del Salento, Via Monteroni, 73100 Lecce, Italy
7. School of Physical Science and Technology, Lanzhou University, Lanzhou 730000, China
* Correspondence: msn_bhutta88@yahoo.com (M.S.N.); safisabri@gmail.com (M.S.J.)

Abstract: Recently, the expansion of energy communities has been aided by the lowering cost of storage technologies and the appearance of mechanisms for exchanging energy that is driven by economics. An amalgamation of different renewable energy sources, including solar, wind, geothermal, tidal, etc., is necessary to offer sustainable energy for smart cities. Furthermore, considering the induction of large-scale electric vehicles connected to the regional micro-grid, and causes of increase in the randomness and uncertainty of the load in a certain area, a solution that meets the community demands for electricity, heating, cooling, and transportation while using renewable energy is needed. This paper aims to define the impact of large-scale electric vehicles on the operation and management of the microgrid using a hybridized algorithm. First, with the use of the natural attributes of electric vehicles such as flexible loads, a large-scale electric vehicle response dispatch model is constructed. Second, three factors of micro-grid operation, management, and environmental pollution control costs with load fluctuation variance are discussed. Third, a hybrid gravitational search algorithm and random forest regression (GSA-RFR) approach is proposed to confirm the method's authenticity and reliability. The constructed large-scale electric vehicle response dispatch model significantly improves the load smoothness of the micro-grid after the large-scale electric vehicles are connected and reduces the impact of the entire grid. The proposed hybridized optimization method was solved within 296.7 s, the time taken for electric vehicle users to charge from and discharge to the regional micro-grid, which improves the economy of the micro-grid, and realizes the effective management of the regional load. The weight coefficients λ_1 and λ_2 were found at 0.589 and 0.421, respectively. This study provides key findings and suggestions that can be useful to scholars and decisionmakers.

Keywords: microgrid; sustainable society; electric vehicles; flexible load; optimization

Citation: Nazir, M.S.; Chu, Z.; Abdalla, A.N.; An, H.K.; Eldin, S.M.; M. Metwally, A.S.; Bocchetta, P.; Javed, M.S. Study of an Optimized Micro-Grid's Operation with Electrical Vehicle-Based Hybridized Sustainable Algorithm. *Sustainability* **2022**, *14*, 16172. https://doi.org/10.3390/su142316172

Academic Editors: Herodotos Herodotou, Sheraz Aslam and Nouman Ashraf

Received: 5 November 2022
Accepted: 1 December 2022
Published: 3 December 2022

Publisher's Note: MDPI stays neutral with regard to jurisdictional claims in published maps and institutional affiliations.

1. Introduction

The concept of smart, environmentally friendly, and sustainable cities is crucial to assessing how well nations have advanced their civilizations and development [1–3]. The goal of developed nations' research and development efforts is to create greener cities and communities that enhance the state of the environment worldwide and reduce pollution from human activity [4]. To accomplish a comprehensive energy solution, it is crucial to control the demand for and distribution of produced energy [3,4]. Furthermore, it is also necessary to implement various forms of renewable energy technology in cities and societies [5]. To enable sustainable energy for cities, a mix of several renewable energy sources, such as solar, wind, geothermal, tidal, etc., is necessary [6]. Intelligent energy management strategies can be implemented at all levels, starting at home and extending to every nook and cranny of the city, including transportation, schools, hospitals, factories, streets, etc. [7]. The increasing penetration of renewables has driven power systems to operate closer to their stability boundaries, increasing the risk of instability [8]. With the upcoming dynamic power generation in many countries, the installed capacity of power generation can gradually and effectively use energy to promote energy conservation, which plays an important role in achieving sustainable energy development [8,9]. The authors discussed the rapid development of power grid technology in the mix with the electric vehicles (EVs) industry (V2G) [10,11]. Using large-scale EV charging piles in the area to realize vehicle network interaction allows large-scale EVs and EVs to take part in economic optimization management, while the use of an energy storage system allows users to create energy arbitrage by discharging during price peaks and charging during off-peak periods if a variable energy price is considered [9].

The control methods of the microgrid (MG) are more diversified, and the development of safety emergency response capabilities has become a current research hotspot. In terms of reducing the valley gap, the literature uses the temporal and spatial characteristics of EVs to construct an orderly charging and discharging load for EVs and a real-time electricity price response model [12]. EVs and other power generation equipment can take part in the economic dispatch of the MG. To study the different strategies between EV power stations and MG, an economic dispatch optimization model was constructed [13]. To solve the increase in the popularity of the complex EV access point network, it has been proposed large-scale EVs be connected to the network, and there is a good deal of optimization scheduling methods. EVs are effectively used to optimize charging and reduce system load peaks and valleys [14]. However, the economic dispatch model of the literature mentioned above considered three factors of an MG, while the user benefits and safety of MG operation do not cogitate the performance of MG management and the participation of EV users [8,15].

These days, smart parking lots are becoming more and more popular since they offer a workable solution to power outages [16]. Systems for managing energy can benefit from heuristic algorithms since they speed up decision-making and develop a novel heuristic algorithm for MG energy management [17]. The principle behind this algorithm is to avoid wasting the available renewable potential at each period. Model predictive control is used to reduce the output power loss caused by converter failure, panel shading, and dirt buildup on wind and PV panels [18].

Authors discussed the dimensional optimization algorithm for optimizing scheduling problems, such as the endless combination of particle swarm optimization (PSO) algorithm and differential evolution algorithm (DE), and the random particle swarm algorithm (RDPSO) [19,20]. Authors proposed that WOANN predicts the required control gain parameters of the hybrid renewable energy systems to maintain the power flow, based on the active and reactive power variation on the load side [19]. The imperialist competition algorithm (ICA) combined mutation, destruction, and selection of a variety of different operators with PSO and other methods studied [21,22]. However, these methods have some shortcomings in finding the optimal solution and the best ability to overcome them [23]. Based on the above analysis, starting from the management side of the MG operator, we comprehensively considered the three factors of MG operation safety, environmental governance, and user participation. A preliminary EV participation in the MG operation management optimization model was established to realize the operating and management costs, environmental pollution control costs, and the lowest mid-term cost [24]. The multi-agent chaotic particle swarm optimization (MACPSO) algorithm combined with the chaotic particle swarm optimization algorithm and chaotic particle swarm optimization (CPSO) algorithm was used to solve the problem [25]. The demand and response characteristics of each power generation unit, large-scale EVs, and electric load in the region were different, and the constraint conditions of each power generation unit presented nonlinear characteristics [26]. A regional MG under the constraints of nonlinear equations is one difficulty for researchers [27]. Furthermore, achieving the best states of management cost, environmental pollution, and load fluctuation variance are another topic of discussion [26]. Authors used the penalty function method to deal with the relevant constraints; this method adds a penalty term to generate a new objective function [20]. Energy conservation has become a long-term strategic policy for global economic and social development [27]. The enhancement of energy management can improve energy efficiency and promote energy conservation and emissions reduction [28]. However, integrating renewable energy and a flexible load makes the integrated energy system a complex dynamic with high uncertainty, bringing great challenges to modern energy management [29]. With the increasing number of vehicle charging piles installed in recent years, load peak periods are brought to the station area, resulting in an insufficient capacity of the station area, in turn resulting in an overload of the distribution transformer in the station area, increased loss of lines and transformers, and other problems [30]. A good auxiliary power supply is the key to the coordinated development between vehicle charging and the power grid in a smart MG [28]. The wind and solar energy generation system can transform the natural resources of the station area into a stable power supply. Authors proposed an energy management method for a grid-connected wind-solar storage MG system with multiple types of energy storage [31]. Authors have researched optimal energy scheduling of MG considering EV charging load [32]. According to the two operation modes of the MG, namely grid-connected and isolated islands and the different access modes of EVs, the MG operation control strategy including EVs was customized [21]. To investigate if solar energy and wind energy are naturally complementary, an energy storage system and an optimized battery energy storage control strategy were combined to put forward a hybrid landscape storage system control strategy considering the charging effect of batteries [33]. The author discussed the operation energy management strategy of the isolated grid of an MG containing hybrid energy storage [33]. However, none of these explored strategies were studied with regard to their application in smart stations/MGs and EVs.

This paper proposes large-scale EVs involved with MG operation and its management optimization method. This method first makes full use of the EV natural flexible load property, and the response of the large-scale EV scheduling model is constructed. Then, considering the system's operation, user participation, and environmental governance, an optimization model is established. The system operation management cost of the MG, environmental pollution control cost, and load fluctuation variance are integrated to achieve an optimal system. Finally, by comparing the optimization results of multiple scenarios, it is verified that the model can realize effective load management in the region and reduce the management cost of MGs and environmental pollution treatment costs to support a healthy society.

The rest of the study is organized as follows, Section 2 presents the related work, Section 3 describes the proposed model, Section 4 explains the results and discussion and the conclusion is presented in Section 5.

2. Related Work

2.1. Model of Large-Scale EVs

Assume that some users of EVs in the area respond to the dispatching information of the regional grid to charge from and discharge to the MG. In contrast, some users do not respond to dispatching information and randomly access charging [21]. The dispatching structure diagram is shown in Figure 1.

According to the operating characteristics of the EVs in the area, the regional response-dispatching EV cluster is divided into a charging cluster and a discharging cluster [34]. The charging and discharging responsiveness of EV users at time t expressed as:

$$\varphi_d(t) = \frac{N_d(t)}{N \times 100\%} \tag{1}$$

$$\varphi_c(t) = \frac{N_c(t)}{N \times 100\%} \tag{2}$$

$$\varphi(t) = \varphi_d(t) + \varphi_c(t) \tag{3}$$

where N is the total number of EVs in the area; $N_d(t)$ is the number of EVs that respond to discharge information; $N_c(t)$ is the number of EVs that respond to charging; $\varphi_d(t)$ is the discharge responsiveness of EV users; $\varphi_c(t)$ is the user's charging responsiveness; $\varphi(t)$ is the user's responsiveness. When a user responds to charging, $\varphi_d(t) = 0$, and when the user responds to discharge, $\varphi_c(t) = 0$.

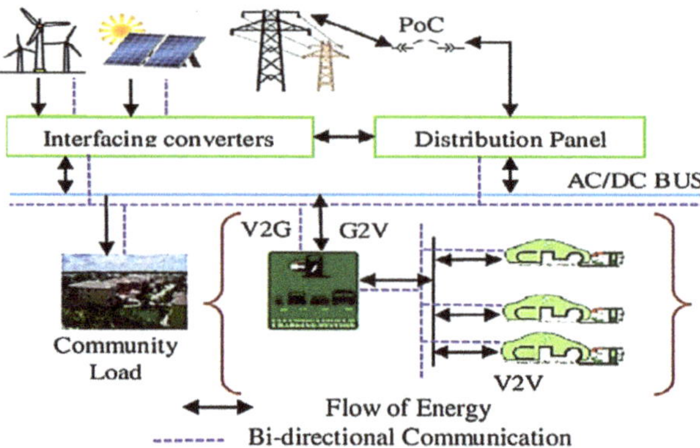

Figure 1. Layout of large-scale EV response to MG dispatch.

2.2. EV Disorderly Charging Model

It is assumed that the charging power of an EV is equal to the power of the connected charging piles and limited by the power of the charging piles installed in the area [35]. The charging power of different types of charging piles is inconsistent. According to the maximum state of charge of the i-th EV $SOC_{i,max}$, the state of charge SOC_i at the beginning of charging, the power of the connected charging pile $P_{i,ch}$, the power battery capacity C, and the charging efficiency η_{CEV}, the charging duration of EVs obtained in the formula is as follows [36]:

$$t_{c,i} = \frac{(SOC_{i,\max} - SOC_i)C}{\eta_{CEV}P_{ch,i}} \tag{4}$$

The calculation expression is:

$$P_{uno}(t) = (1 - \varphi(t))\sum_{i=1}^{N} P_{ch,i}(t).\alpha_{tcp,t}.\alpha_{state,t} \tag{5}$$

where $P_{uno}(t)$ represents the total disordered charging load in the area at time t, t $1, 2, 3, \ldots,$ 24; N is the number of EVs; $P_{ch,i}(t)$ is the connection at the charging time t, $\alpha_{state,i}$ represents the state of the charging pile, $\alpha_{state,i} = 1$ represents the charging state, and $\alpha_{tcp,i}$ represents the parking time; $t_{p,i}$ is greater than the required charging duration. $t_{c,i}$ is used to calculate the charging power that is $\alpha_{tcp,i}$ 1 and the total disordered charging load in the t period area.

2.3. Model of Charging, Discharging, and User Response

EV users respond to dispatched charging and discharging, while users responding to the dispatching information of the regional MG and connect to the regional MG for controllable charging and discharging in an orderly manner. Suppose that when the user of the i-th EV responds to the scheduling information, they also provide feedback on important parameters such as the state of charge of the power battery SOC_i, the parking time $t_{p,i}$, the next trip, and the unit power consumption of the EV to the area of the MG to calculate the energy consumption for the rest of the user's journey $SOC_{rest,i}$ according to the feedback parameter information, the state of charge of comprehensive user anxiety $SOC_{anx,i}$, and to protect the battery reserve with capacity margin of not less than 20% to calculate the charging and discharging time of the EV.

The charging duration is consistent with Equation (6), and the discharge duration is expressed as [37]:

$$t_{d,i} = \frac{(SOC_i - SOC_{i,rest} - SOC_{anx,i} - 20\%)C}{\eta_{dEV}P_{dis,i}} \tag{6}$$

where η_{dEV} is the discharge efficiency of EV; $P_{dis,i}$ represents the discharge power of the connected charging pile.

The total charge and discharge power $P_{resEV}(t)$ of EVs in the area at time t can be obtained by:

$$P_{disLoad}(t) = \varphi_{dis}(t)\sum_{i=1}^{N} P_{dis,i}(t).\alpha_{tdp,t}.\alpha_{state,t} \tag{7}$$

$$P_{cLoad}(t) = \varphi_c(t)\sum_{i=1}^{N} P_{ch,i}(t).\alpha_{tcp,t}.\alpha_{state,t} \tag{8}$$

$$P_{resEV}(t) = P_{disLoad}(t) + P_{cLoad}(t) \tag{9}$$

where $\alpha_{tdp,i}$ means that the parking time $t_{p,i}$ is greater than the continuous discharge time $t_{d,i}$ to calculate the discharge power, that is, $\alpha_{tdp} = 1$, according to the discharge of the connected charging pile power.

3. Proposed Model

3.1. Objective Function

The operating and management costs of the regional MG and pollutant control costs can be collectively referred to as the total operating and management costs of the regional MG, defined as:

$$minF \quad \lambda_1(F1 + F2) + \lambda_2 F3 \tag{10}$$

where λ_1 and λ_2 are weighting factors, where $\lambda_1 + \lambda_2 = 1$.

The integrated operation and management cost of an MG with large-scale EVs mainly includes the economic operation cost of the MG and the incentive cost for EVs that respond to dispatching to take part in the dispatch as follows:

$$F_1 = \sum_{i=1}^{T} \{[C_n(P_{DG}(n,t)) + C_{w,n}(P_{DG}(n,t))]\} + C_{grid}(t)P_{grid}(t) + C_{excit}(t) + C_{dc}(t) \tag{11}$$

where T is a dispatch cycle; N_{DG} is the type of power generation unit installed in the area; $C(P_{DG}(n,t))$ is the power generation cost of the nth type of power generation unit; $C_{W,n}(P_{DG}(n,t))$ is the maintenance cost of the nth type of power generation unit; $P_{DG}(n,t)$ is the power generation of the n type of power generation unit; $c_{grid}(t)$ and $P_{grid}(t)$ are the agreement points of the MG and the power grid company at time t, respectively; $C_{excit}(t)$ is the cost of incentivizing EVs to participate in dispatch; $C_{dc}(t)$ is the difference between the operation purchase of electricity from users (discharge) and the sale of electricity to EVs (charging), an additional fee is required.

The power generation cost of a distributed generation unit is defined as:

$$C_n(P_{DG}(n,t)) = \alpha_n[P_{DG}(n,t)]^2 + \beta_n P_{DG}(n,t) + \gamma_n \tag{12}$$

where α_n, β_n, and γ_n are cost constants, which are related to the type of distributed generation (DG) unit.

The maintenance cost of the power-generating unit is approximately proportional to the power generated by the power-generating unit.

$$C_{w,n}(P_{DG}(n,t)) = \lambda_n^m(P_{DG}(n,t)) \tag{13}$$

where λ_n^m represents the maintenance coefficient of n types of distributed power generation units. Different types of distributed power generation units have different maintenance coefficients. For different types of maintenance coefficients, please refer to the literature.

To use the capacity of the EV's power battery, the regional MG adopts certain incentives for car owners to attract EV owners to actively respond to the dispatching information of the regional MG, charge/discharge the regional MG, and take incentive measures.

The cost calculation expression is:

$$C_{excit}(t) = \varphi(t) \sum_{i=1} (SOC_{i,\max} - SOC_{i,\min}) C \rho_{excit}(t) \wedge_i (t) \tag{14}$$

where $\rho_{excit}(t)$ represents the unit incentive cost at time t, and its value is calculated by referring to the distributed generation kilowatt-hour subsidy standard [38]; $\wedge_i(t)$ is the connection state of the charging pile at time t, $\wedge_i(t) = 1$ means the charging pile is connected, $\wedge_i(t) = 0$ means the charging pile is not connected.

The additional cost added by the difference in the electricity price during the charging and discharging period of the user can be calculated by:

$$C_{dc}(t) = \sum_{i=1}^{T} \sum_{i=1}^{N} \{\varphi_{dis}(t) P_{dis,i}(t) p(t) \triangle_{d,t} - (1 - \varphi_{dis}(t) P_{ch,i}(t) c(t) \Delta_{c,i})\} \tag{15}$$

where T is a dispatch period; N is the total number of EVs in the area; $p(t)$ is the on-grid price of the user for discharging into the regional MG at time t; $c(t)$ is the charging price of the user at time t; $\Delta t_{c,i}$ and $\Delta t_{d,i}$, respectively, represent the continuous charging and discharging time of EV users to the regional MG. The pollutant penalty costs for MG operation are as follows:

$$F_2 = \sum_{t=1}^{T} \left| \sum_{n=1}^{N_{DG}} \sum_{m=1}^{M} C_m \alpha_{m,n} P_{DG}(n,t) + \sum_{m=1}^{M} C_m \alpha_{grid,m} P_{grid}(t) \right| \tag{16}$$

where M is the type of pollutant, and the power generation process mainly considers NOx, SO$_2$, and carbon emissions; C_m represents the cost per kilogram of treating these m types of pollutants; $\alpha_{n,m}$ represents the first type of power generation unit produced. The emission coefficient of m-type gas pollutants; $P_{DG}(n,t)$ is the power generation of the nth type of power generation unit; $\alpha_{grid,m}$ represents the emission coefficient of m-type gas pollutants generated when the public large-scale power grid transmits electric energy; $P_{grid}(t)$ represents the power flowing in both directions between the regional MG and the large public grid.

The power supplied by the grid to the MG is positive, and the power supplied by the regional MG to the large power grid is negative due to the load fluctuation and due to improving the security and stability of the MG's economic operation.

$$F_3 = \frac{1}{T} \sum_{t=1}^{T} \left(P_{load}(t) + P_{uno}(t) + P_{resEV}(t) - \sum_{n=1}^{N_{DG}} P_{DG}(n,t) - P_{av}(t) \right)^2 \tag{17}$$

where $P_{load}(t)$ represents the basic electricity load in the MG in the period t; $P_{uno}(t)$ represents the disorderly charging load of EVs in the period t; $P_{resEV}(t)$ represents the EVs response charge and discharge in the period t load.

3.2. Constraints

To achieve a balanced state of power on the supply and demand side in regional MGs, the power balance constraint can be expressed as:

$$P_{load}(t) + (P_{uno}(t) + P_{resEV}(t)) = \sum_{n=1}^{N_{DG}} P_n(t) - P_{bss}(t) + P_{grid}(t) \tag{18}$$

where $P_n(t)$ is the power supply of the nth type of power generation unit in the area; $P_{bss}(t)$ is the energy storage system's total charge and discharge power; greater than 0 means discharging, and less than 0 means charging.

EV charging and discharging state constraints:

$$SOC_i(t+1) = SOC(t)_i + \left(\frac{\eta_{c,i}P_{c,i}(t)}{C}\alpha_{state,i}\cdot \triangle_{c,t} + \frac{P_{dis,i}(t)}{\eta_{d,i}C}\alpha_{state,i}\cdot\triangle_{d,t} \right) \tag{19}$$

where $SOC_i(t+1)$ and $SOC_i(t)$ are, respectively, the state of charge of the ith EV power battery during $(t+1)$ and t periods; $\eta_{c,i}$, $\eta_{d,i}$ represents the charge and discharge efficiency of EV; $\Delta_{c,t}$ and $\Delta_{d,t}$ represent the duration of charge and discharge.

To meet the user's next trip needs, the user can set the desired power battery power.

$$SOC_i(t_{leave}) \geq SOC_{desired,i} \tag{20}$$

where $SOC_{desired,i}$ represents the state of charge of the power battery expected by the user when leaving the charging pile; $SOC_i(t_{leave})$ represents the actual state of charge of the power battery when the EV leaves the charging pile.

3.3. Hybridized Algorithm

3.3.1. The GSA Algorithm

The GSA acts on agents as objects whose actions are recorded by the masses [39]. Objects are to display a solution or a portion of a solution. The gravitational pull attracts items to themselves, causing a worldwide movement toward objects with larger masses [40]. Because the heavier masses have higher fitness criteria, achieving a worthy ideal answer is more difficult.

The position is defined with N as:

$$X_i = (x_i^1 \ldots x_i^d \ldots x_i^n) \; for\, i = 1, 2, \ldots, N \tag{21}$$

First, the agents of the solution have given a solution based on Newton's gravitational theory [41]. The gravitational force is calculated as follows:

$$F_{ij}^d(t) = G(t)\frac{M_i(t)xM_j(t)}{R_{ij}(t)+\varepsilon}\left(x_j^d(t) - x_i^d(t)\right) \tag{22}$$

The Euclidian distance can be written as:

$$R_{ij}(t) = \|X_i(t), X_j(t)\|_2 \tag{23}$$

The total force acting can be presented as:

$$F_i^d(t) = \sum_{j\in kbest, j\neq i}^{N} rand_j F_{ij}^d(t) \tag{24}$$

Moreover, $a_i^d(t)$ can be presented as:

$$a_i^d(t) = \frac{F_i^d(t)}{M_{ii}(t)} \tag{25}$$

Furthermore, a technique based on this concept can be described to obtain an agent's subsequent speed and location. An agent's subsequent speed can be represented as a function of its current velocity plus its current acceleration. As a result, the improved location and speed provides:

$$v_i^d(t+1) = rand_i \times v_i^d(t) + a_i^d(t) \tag{26}$$

$$x_i^d(t+1) = x_i^d(t) + v_i^d(t+1) \tag{27}$$

To appropriately regulate the search procedure, the gravitational constant (G) is set arbitrarily at the start and gradually decreases over time as follows:

$$G(t) = G(G_0, t) \tag{28}$$

$$G(t) = G_0 e^{-\alpha \frac{t}{T}} \tag{29}$$

The masses of the agents can be determined via fitness evaluation. The greater an agent's act mass, the more significant that agent is to obtaining the answer. According to Newton's laws of gravity and motion, a hefty mass has greater pull-on power and moves slower. The masses can be described as follows:

$$M_{ai} = M_{pi} = M_{ii} = M_i, \ \ i = 1, 2, \ldots, N$$

$$m_i(t) = \frac{fit_i(t) - worst(t)}{best(t) - worst(t)} \tag{30}$$

$$M_i(t) = \frac{m_i(t)}{\sum\limits_{j=1}^{N} m_j(t)} \tag{31}$$

3.3.2. The RFR Algorithm

The fitness function for the GSA algorithm must be established to assess the benefits and drawbacks of the RFR model for each node [42]. The following new particle will result from the crossing:

$$X_{inew}^k = r X_i^k + (1 - r) X_j^k \tag{32}$$

$$V_{inew}^k = \frac{V_i^k + V_j^k}{|V_i^k| + |V_j^k|} |V_j^k| \tag{33}$$

where the random number r value is between 0 to 1; the velocities of particles are V_i and V_j, X_i and X_j; X_i and V_i are the new positions and velocities of the different particles, while X_i replaces them.

The method of dynamically adjusting the inertia factor is used for great particles.

In the initial phase, w is nominated to develop global searchability. The slighter w stood in the later phase to attain a more sophisticated search. The efficient formulation of the inertia factor is presented in Equation (33).

$$w(t) = (w_1 - w_2) \times (T - t)/T + w_2 \tag{34}$$

The following was chosen for the mean square of the residual:

$$R_{RF}^2 = 1 - \frac{MSE_{ooB}}{\sigma_y^2} \tag{35}$$

where the predicted cost of variance is $\sigma^2 y$; residual error $R^2{}_{RF}$ is the mean square. Each input feature's significance can be determined by random forest as:

$$f_i = \frac{\sum_{j \in feature i} n_j}{\sum_k n_k} \tag{36}$$

$$n_k = w_k M_k - w_{k1} M_{k1} - w_{k2} M_{k2} \tag{37}$$

where node k importance is n_k; the feature division is the node with the feature I as n_j; the number of samples in node k is w_k, w_1, and w_2, the ratio and its sub-nodes to all the samples, respectively; the node k mean square errors are M_k, M_1, and M_2 and its sub-nodes.

Figure 2 shows the multi-objective model is weighted into a comprehensive single-objective model. There may be errors in weighting to overcome subjective experience. This paper uses the entropy method for weighting, forming a weighted single-objective optimization model, and implementing specific steps.

(1) Take the objective function $F_i(i = 1, 2, \ldots, n)$ as the optimization target for the single-objective solution.
(2) According to step (1), a single objective function value and a comprehensive, objective function value can be obtained.
(3) According to the single objective function value and the comprehensive objective function, the objective function value is unified and dimensionless, and the preprocessed objective function value set is obtained.
(4) Apply the entropy weight method to obtain the weight coefficient of the objective function.

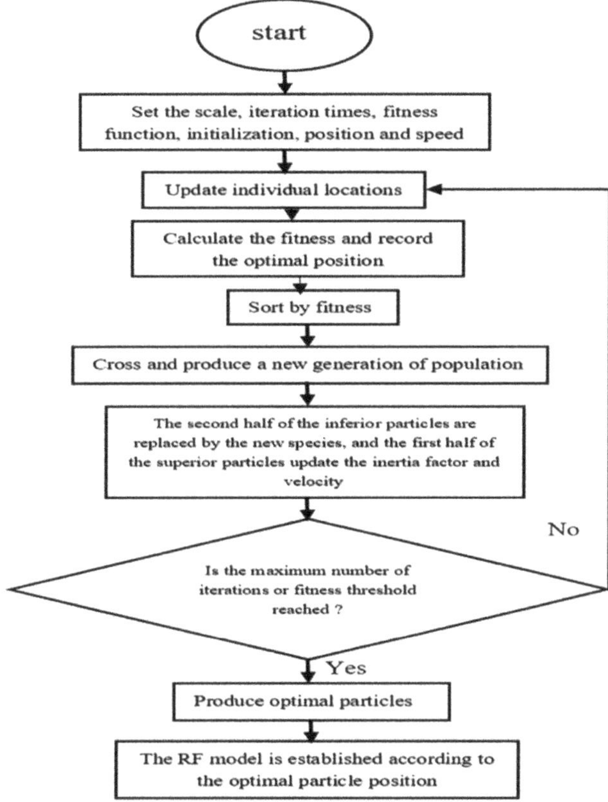

Figure 2. MG operation-based hybridized GSA-RFR algorithm.

4. Result and Discussion

4.1. Model Parameter

Figure 3 shows a selection of the typical wind and solar power forecast curves in the MG of a park. The charging piles are installed in the park and can be divided into three levels. Table 1 lists the allowable charging power of each level of charging piles. The three-level charging pile proportions are 50%, 30%, and 20%. Due to the limitations of the actual park's MG installed capacity and transmission power lines, it is assumed that there are 100 EVs in the park. Tables 2–4 show the parameters of the power generation units installed in the park, unit power generation costs and maintenance costs, environmental governance cost coefficients, and time-of-use electricity prices of the park [43]. Assuming that all EVs in the park are of the same type, the charging and discharging power and the charging and discharging efficiency are the same. The EV's parameters are shown in Table 5. The GSA-RFR algorithm parameters; population size is 300, the maximum iteration time $MaxTime = 350$, the upper limit of inertia weight $w_{max} = 0.9$, the lower limit of inertia weight $w_{min} = 0.4$, the initial self-learning factor $c_i = 2.5$, termination self-learning factor $c_{1f} = 0.5$, initial social learning factor $c_i = 20.5$, termination social learning factor $c_{2f} = 2.5$, environment size $l_{size} = 20$, optimal chaotic environment size $h_{size} = 3$, the search radius $r = 0.5$, the number of iterations of optimal chaos $h_{cir} = 10$.

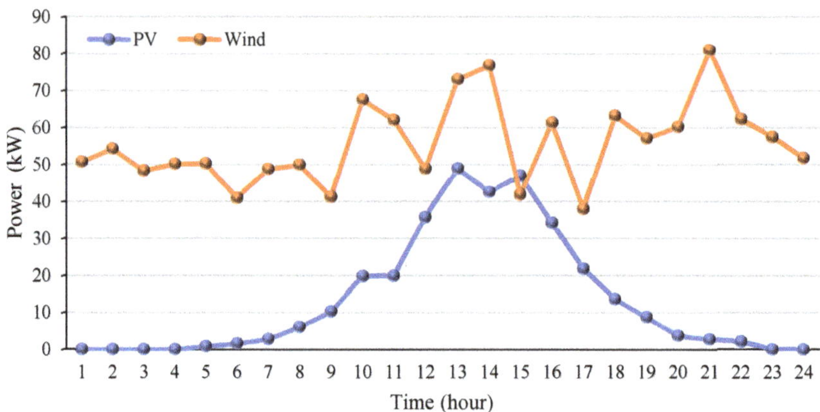

Figure 3. Typical wind and PV power generation forecast curve in the area.

Table 1. Charging power rating of piles.

Cases	Case1	Case2	Direct Current
Power (kW)	1.4–3	5–10	25–180

Table 2. Parameters of each power generation unit, generation cost, and maintenance cost.

Resource	Power Range (kW)	Generation Cost (USD/kWh)	Maintenance Cost (USD/kWh)
BSS	−150~150	0.68	0.08439
WT	0.100~0.39	0.009	6
PV	0.75~0.56	0.001	2
MT	20~150	0.41	0.0401

Table 3. Environmental costs and pollutant emission coefficient.

Contaminant Type	Control Costs (USD/kg)	Pollutant Emission Factor (kWh)			
		Grid	MT	WT	PV
CO_2	0.21	889	724	0	0
SO_2	14.842	1.8	0.0036	0	0
NOx	62.94	41.6	0.2	0	0

Table 4. Local time-of-use electricity price list.

	Peak	Normal	Valley
Period (Time)	**10:00–15:00** **19:00–22:00**	**7:00–9:00** **16:00–18:00**	**1:00–6:00** **23:00–24:00**
Electricity price (USD/kWh)	1.56	0.7	0.43
Purchase price (USD/kWh)	0.75	0.43	0.14

Table 5. Electric vehicle parameters.

Category	Battery Capacity (kWh)	SOC Limit (%)	Charge and Discharge Power Limit (kW)	Charge and Discharge Efficiency (%)
Value	52.5	20/90	50	0.95

4.2. Effectiveness of the Large-Scale EV Response Scheduling Model

To verify the effectiveness of a large-scale EV response scheduling model in this paper in reducing load volatility, three different EV charging and discharging models are selected: (1) user autonomous charging model; (2) orderly charging and discharging model based on peak-valley time-of-use electricity prices; (3) response scheduling model, using three different charging and discharging models, the total electricity load, including EVs, and the charging and discharging loads were calculated, and the user response degree was set to 100%. The operating results are shown in Figure 4.

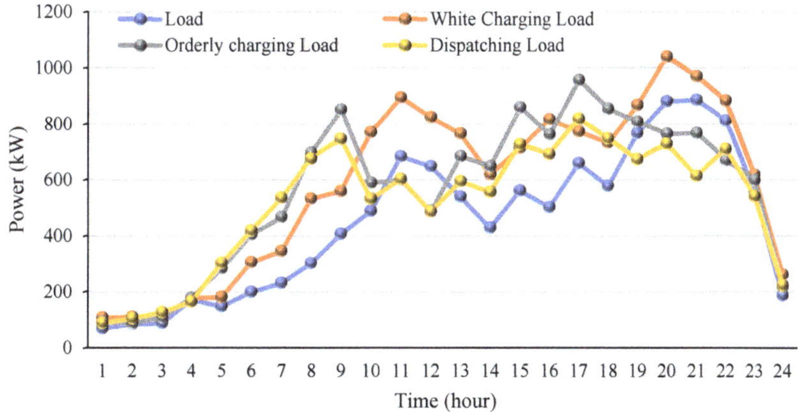

Figure 4. The operation results of MG with different loads.

In Figure 4, the orderly charge and discharge model based on the peak-valley time-of-use electricity price is shown. The performance of orderly charge and discharge according to the peak-valley time-of-use electricity price information is reduced by 11.6%. Meanwhile, this causes EV users to charge and discharge the MG to generate new electricity (peaks), which makes a new impact on the MG. The response scheduling model guides EVs to charge from and discharge to the MG based on peak-valley time-of-use electricity prices and regional load information. Compared with the user-autonomous charging model and the orderly charge–discharge model of peak-valley time-of-use electricity prices, the variance of load fluctuations is reduced by 25.3% and 15.5%. The response scheduling model improves the smoothness of the regional load.

4.3. Weight Coefficients and Impact of Different Optimization Results

To analyze the influence of different weight coefficients on the optimization results, the table compares the optimization results under five groups of different weight coefficients. Among them, the weight coefficient λ_1 in each group has a value of 0, 0.3, 0.5, 0.8, and 1; the weight coefficient λ_2 is 1, 0.7, 0.5, 0.2, 0, respectively. According to the optimization results, the statistics of the integrated operation and management costs and load fluctuation variance of the MG in a dispatch period are shown in Table 6.

Table 6. Comparison of optimization results of different weight coefficients.

Weight Coefficient	MG Operation Costs (USD)	Environmental Pollution Costs (USD)	Load Fluctuation Variance
1	94,099.872	10,735.098	12.56
2	42,739.869	8585.706	86.68
3	36,717.197	8081.684	118.62
4	31,347.512	7632.539	126.76
5	31,190.257	6976.039	285.38

The MG operation and management, environmental pollution control, load fluctuation variance, and other indicators of cost reach the best operating state to select the appropriate weighting coefficients. Table 6 shows that setting different weight coefficients has an impact on the optimization results. When the weight coefficient λ_1 increases, the MG operating costs and environmental pollution cost gradually decrease, and the load fluctuation variance increases with the decrease in the weight coefficient λ_2. Table 6 shows the maximum value of MG operation management cost, environmental pollution control cost, and load fluctuation variance as USD 94,099.872 and USD 10,735.098. The minimum values are USD 31,190.257, USD 6976.039, and USD 12.56, respectively. The entropy weight method is used to calculate the weight coefficients of each objective function, and the weight coefficients λ_1 and λ_2 are 0.589 and 0.421, respectively. Applying the obtained weight coefficients to weigh different objective functions and solving the proposed model, the MG operation management, environmental pollution costs and load fluctuation are USD 31,984.413, USD 76,695.169, and USD 120.236, respectively.

4.4. Controllable Power Generation Unit Output with Different Responsiveness

The output of controllable generating units with different responsiveness were studied, and the user responsiveness in the selected area is 0%, 50%, 80%, and 100%, respectively. Figure 5 shows the operation results, and the responses of different users in a dispatch period were counted. The operation management, load fluctuation and the costs of the MG with a high degree are shown in Table 6.

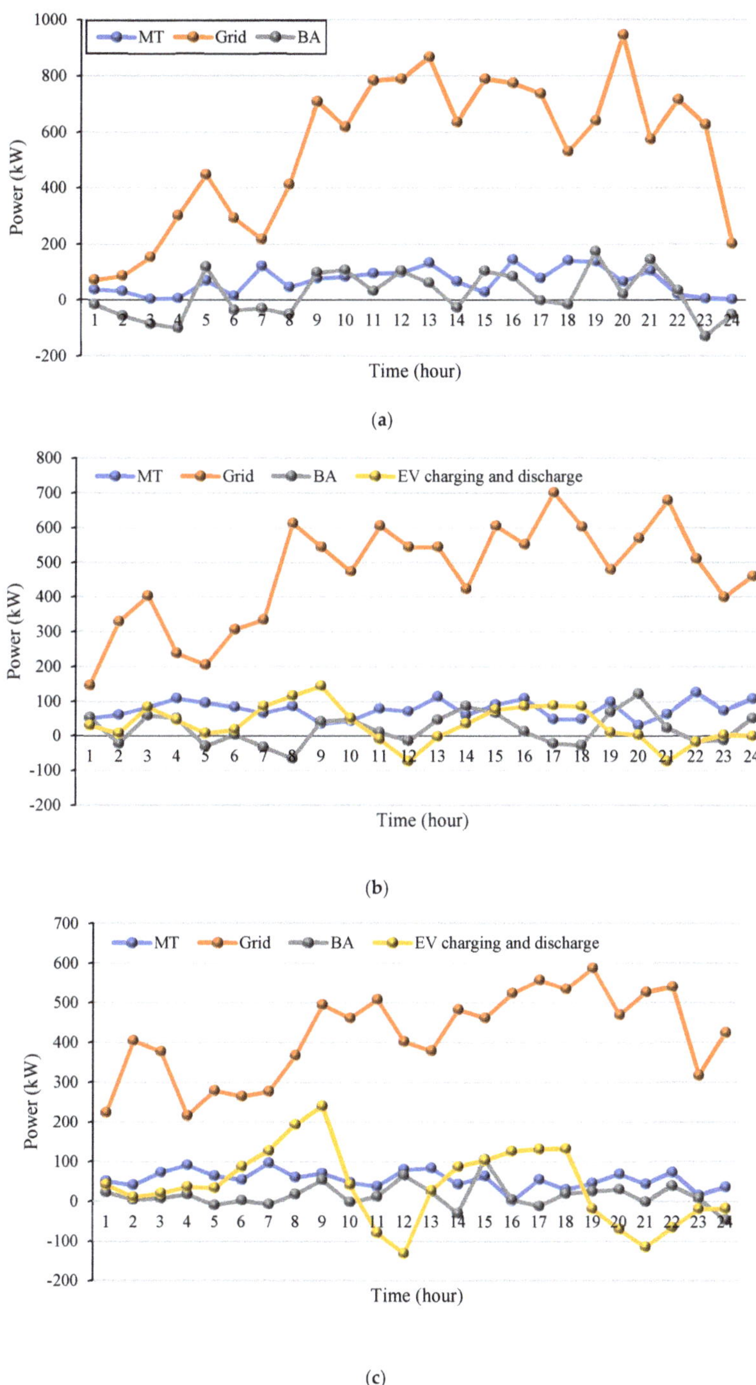

(**a**)

(**b**)

(**c**)

Figure 5. *Cont.*

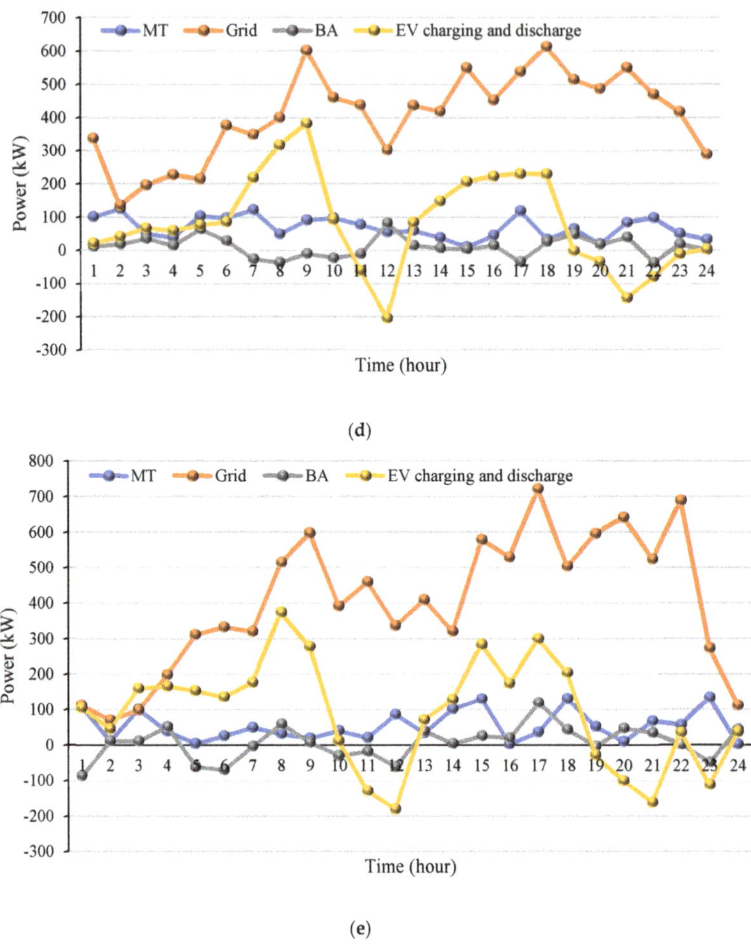

(d)

(e)

Figure 5. The planned output of controllable power generation units with different user responsiveness; (**a**) 0% of the planned output of controllable generating units, (**b**) 30% of the planned output of controllable power generation units, (**c**) 50% of the planned output of controllable power generation units, (**d**) 80% of the planned output of controllable power generation units, (**e**) 100% controllable power generation unit planned output.

Figure 6 shows the power supply change curve of the large-scale public power grid. During the period 8:00–20:00 large-scale EVs are connected to the MG in disorder, and the power demand of the MG during the peak period of electricity load is sharp. Increasing the amount of power supplied by the large public grid to the MG is likely to cause the large public grid to be overloaded. Figure 6 and Table 7 show the statistical data that when the user responsiveness is between 30% and 80%, as the user responsiveness increases, the operation and management costs of the MG and the environmental pollution control costs are reduced correspondingly, and the variance of load fluctuations is first. Considering the tendency to increase after decreasing when the user response rate is 100%, large-scale EVs are connected to the regional MG for charging and discharging. Due to the limitation of the installed capacity of the regional MG-distributed energy, the demand for electricity in the region increases, leading to the microgrid's related costs. Correspondingly, at the same time, the variance of load fluctuations increases compared with the variances of load fluctuations with a responsiveness of 50% and 80%.

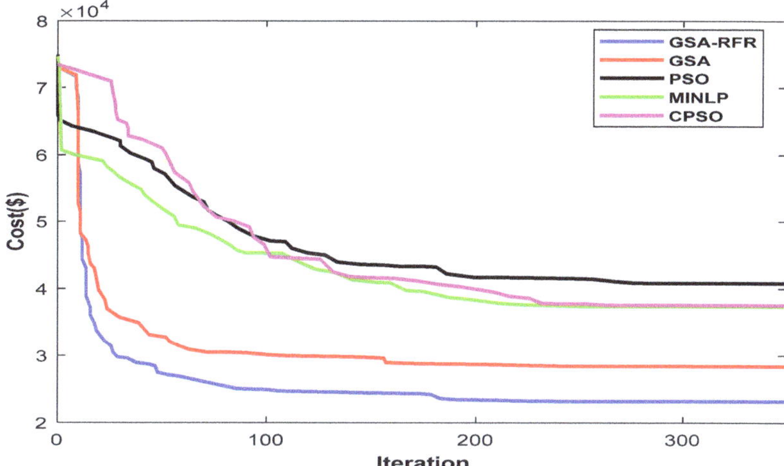

Figure 6. Convergence curves of different optimization algorithms.

Table 7. Comparison of MG indicators for different user responsiveness.

Responsiveness	MG Operating Costs (USD)	Environmental Pollution Treatment Costs (USD)	Load Fluctuation
0%	45,454.215	9467.724	253.71
30%	37,967.285	8760.162	144.13
50%	31,690.206	7881.186	114.24
80%	28,655.759	7836.853	130.19
100%	35,750.162	7972.188	196.15

4.5. Comparison of Different Optimization Algorithms

A comparison is made to verify the effectiveness of the GSA-RFR algorithm in solving high-dimensional, non-continuous problems with the multi-constrained optimization problems (MINLP) method, GSA algorithm, CPSO algorithm, and multi-agent PSO algorithm to solve the economic optimization model put forth in this study and contrast the GSA-RFR algorithm's solution outcomes with the optimization outcomes of various algorithms. The maximum number of iterations is 350, and the population size is 300. Figure 6 depicts the convergence curves of various algorithms.

The PSO method has the fastest convergence speed, as shown in Figure 6, but it is prone to premature phenomena and cannot converge upon the global extreme point. Although the solution time of the MINLP method is not much different from that of the PSO algorithm, its optimization accuracy is better than that of the PSO algorithm. In the CPSO algorithm, the chaotic search strategy is added to improve the global search ability of the algorithm to a certain extent, thereby avoiding falling into local extreme points. Still, its convergence accuracy needs to be improved. The CPSO algorithm requires more iterations and time to achieve convergence in the solution process, but it improves the optimization results. Compared with several optimization methods, the GSA-RFR algorithm integrates a multi-agent system and a chaotic search mechanism to increase its time consumption, but its optimization effect is the best.

Table 8 shows that when the economic optimization model is established, the optimization result's operating cost, environmental pollution control cost, and load fluctuation variance are minimized. The GSA-RFR algorithm demonstrates good performance in solving high-dimensional, non-continuous, and multi-constrained optimization problems.

Table 8. Comparison of optimization results of different optimization algorithms.

Algorithms	MG Operation Costs (USD)	Environmental Pollution Control Expenses (USD)	Variance of Load Fluctuation	Solution Running Time (s)
GSA-RFR	48,974.386	8986.562	236.45	296.7
GSA	45,499.685	8654.639	214.32	307.5
MINLP	43,875.754	8579.546	196.53	314.3
CPSO	32,497.179	7608.743	184.65	375.6
PSO	28,785.478	6937.591	132.36	509.8

5. Conclusions

The goal of developed nations' research and development efforts is to create greener cities and communities that enhance the state of the environment worldwide and reduce environmental pollution. EVs will play a critical role in energy systems over the coming years, due to their environmental friendliness and capacity to reduce/absorb superfluous power from renewable energy sources. Meanwhile, a large-scale EV charging pile of regional power grids increases the randomness and uncertainty of the load in the concerned area. The proposed study constructed a large-scale EV response dispatch model that significantly improves the load smoothness of the MG after large-scale EVs are connected, and reduces the impact of the entire MG. The weight coefficients λ_1 and λ_2 were determined as 0.589 and 0.421, respectively, the controllable power generation output scheme unit was best observed, and the operational management, environmental pollution control, and variance of load fluctuations costs were interestingly observed as lowest at USD 31,983.813, USD 76,695.169, and USD 120.236, respectively. The proposed hybridized optimization method directs EV users to charge from and discharge to the regional MG with the presence of renewable energy resources (wind and PV), which improves the economics of the MG and realizes the operation management and environmental pollution regularized to establish a friendly society.

In the future, studies on different scenarios, including the maximum renewable model, the uncoordinated charging model, the load levelling model, and the charging-discharging model, can be used to further enhance EV demand. Additionally, the effects of various electric vehicle (EV) charging/discharging strategies on the costs associated with operation and the removal of pollutants in remote micro-grid (MG) modes are also relevant areas for future study.

Author Contributions: Methodology, writing—original draft preparation, M.S.N.; formal analysis, writing, review and editing, Z.C.; A.N.A., investigation, writing, review and editing, H.K.A., S.M.E., P.B., M.S.J., and A.S.M.M. All authors have read and agreed to the published version of the manuscript.

Funding: This work was funded by the Researchers Supporting Project No.(RSP-2021/363), King Saud University, Riyadh, Saudi Arabia.

Institutional Review Board Statement: Not applicable.

Informed Consent Statement: Not applicable.

Data Availability Statement: Data will be made available on request.

Acknowledgments: The authors would like to acknowledge the technical and related facilities from affiliated institutes/universities. This work was funded by the Researchers Supporting Project No.(RSP-2021/363), King Saud University, Riyadh, Saudi Arabia.

Conflicts of Interest: The authors declare no conflict of interest.

Nomenclature

MG	micro-grid
EVs	electric vehicles
MINLP	multi-constrained optimization problems
PSO	particle swarm optimization
DE	differential evolution algorithm
ICA	imperialist competition algorithm
MACPSO	multi-agent chaotic particle swarm optimization
CPSO	chaotic particle swarm optimization
DG	distributed generation
$N_d(t)$	number of EVs that respond to discharge information
$N_c(t)$	number of EVs that respond to charging information
$SOC_{i,max}$	the state of charge
$P_{i,ch}$	the power battery capacity
η_{CEV}	charging efficiency
$P_{uno}(t)$	represents the total disordered charging
$P_{resEV}(t)$	EVs response charge and discharge
$P_{load}(t)$	basic electricity load

References

1. Bibri, S.E. Data-Driven Smart Eco-Cities of the Future: An Empirically Informed Integrated Model for Strategic Sustainable Urban Development. *World Futures* **2021**, 1–44. [CrossRef]
2. Li, J.; Sun, W.; Song, H.; Li, R.; Hao, J. Toward the construction of a circular economy eco-city: An emergy-based sustainability evaluation of Rizhao city in China. *Sustain. Cities Soc.* **2021**, *71*, 102956. [CrossRef]
3. Mignoni, N.; Scarabaggio, P.; Carli, R.; Dotoli, M. Control frameworks for transactive energy storage services in energy communities. *Control Eng. Pract.* **2023**, *130*, 105364. [CrossRef]
4. Scarabaggio, P.; Carli, R.; Dotoli, M. Noncooperative Equilibrium Seeking in Distributed Energy Systems Under AC Power Flow Nonlinear Constraints. *IEEE Trans. Control Netw. Syst.* **2022**, 1–12. [CrossRef]
5. Nazir, M.S.; Mahdi, A.J.; Bilal, M.; Sohail, H.M.; Ali, N.; Iqbal, H.M.N. Environmental impact and pollution-related challenges of renewable wind energy paradigm—A review. *Sci. Total Environ.* **2019**, *683*, 436–444. [CrossRef]
6. Ma, H.; Zhang, C.; Peng, T.; Nazir, M.S.; Li, Y. An integrated framework of gated recurrent unit based on improved sine cosine algorithm for photovoltaic power forecasting. *Energy* **2022**, *256*, 124650. [CrossRef]
7. Aziz, S.; Peng, J.; Wang, H.; Jiang, H. ADMM-Based Distributed Optimization of Hybrid MTDC-AC Grid for Determining Smooth Operation Point. *IEEE Access* **2019**, *7*, 74238–74247. [CrossRef]
8. Egbue, O.; Uko, C. Multi-agent approach to modeling and simulation of microgrid operation with vehicle-to-grid system. *Electr. J.* **2020**, *33*, 106714. [CrossRef]
9. Yao, M.; Molzahn, D.K.; Mathieu, J.L. An optimal power-flow approach to improve power system voltage stability using demand response. *IEEE Trans. Control Netw. Syst.* **2019**, *6*, 1015–1025. [CrossRef]
10. Rodrigues, Y.R.; de Souza, A.Z.; Ribeiro, P.F. An inclusive methodology for Plug-in electrical vehicle operation with G2V and V2G in smart microgrid environments. *Int. J. Electr. Power Energy Syst.* **2018**, *102*, 312–323. [CrossRef]
11. Suresh, V.; Bazmohammadi, N.; Janik, P.; Guerrero, J.M.; Kaczorowska, D.; Rezmer, J.; Jasinski, M.; Leonowicz, Z. Optimal location of an electrical vehicle charging station in a local microgrid using an embedded hybrid optimizer. *Int. J. Electr. Power Energy Syst.* **2021**, *131*, 106979. [CrossRef]
12. Rajamand, S. Vehicle-to-Grid and vehicle-to-load strategies and demand response program with bender decomposition approach in electrical vehicle-based microgrid for profit profile improvement. *J. Energy Storage* **2020**, *32*, 101935. [CrossRef]
13. Anastasiadis, A.G.; Konstantinopoulos, S.; Kondylis, G.P.; Vokas, G.A. Electric vehicle charging in stochastic smart microgrid operation with fuel cell and RES units. *Int. J. Hydrogen Energy* **2017**, *42*, 8242–8254. [CrossRef]
14. Sattarpour, T.; Nazarpour, D.; Golshannavaz, S. A multi-objective HEM strategy for smart home energy scheduling: A collaborative approach to support microgrid operation. *Sustain. Cities Soc.* **2018**, *37*, 26–33. [CrossRef]
15. Tidjani, F.S.; Hamadi, A.; Chandra, A.; Saghir, B.; Mounir, B.; Garoum, M. Energy management of micro grid based electrical vehicle to the building (V2B). In Proceedings of the 2019 7th International Renewable and Sustainable Energy Conference (IRSEC), Agadir, Morocco, 27–30 November 2019.
16. Sadeghian, O.; Oshnoei, A.; Mohammadi-Ivatloo, B.; Vahidinasab, V.; Anvari-Moghaddam, A. A comprehensive review on electric vehicles smart charging: Solutions, strategies, technologies, and challenges. *J. Energy Storage* **2022**, *54*, 105241. [CrossRef]

17. Adil, M.; Mahmud, M.P.; Kouzani, A.Z.; Khoo, S. Energy trading among electric vehicles based on Stackelberg approaches: A review. *Sustain. Cities Soc.* **2021**, *75*, 103199. [CrossRef]
18. Kumar, A.; Jha, B.K.; Das, S.; Mallipeddi, R. Power flow analysis of islanded microgrids: A differential evolution approach. *IEEE Access* **2021**, *9*, 61721–61738. [CrossRef]
19. Venkatesan, K.; Govindarajan, U. Optimal power flow control of hybrid renewable energy system with energy storage: A WOANN strategy. *J. Renew. Sustain. Energy* **2019**, *11*, 015501. [CrossRef]
20. Ajithapriyadarsini, S.; Mary, P.M.; Iruthayarajan, M.W. Automatic generation control of a multi-area power system with renewable energy source under deregulated environment: Adaptive fuzzy logic-based differential evolution (DE) algorithm. *Soft Comput.* **2019**, *23*, 12087–12101. [CrossRef]
21. Garcia-Guarin, J.; Rodriguez, D.; Alvarez, D.; Rivera, S.; Cortes, C.; Guzman, A.; Bretas, A.; Aguero, J.R.; Bretas, N. Smart microgrids operation considering a variable neighborhood search: The differential evolutionary particle swarm optimization algorithm. *Energies* **2019**, *12*, 3149. [CrossRef]
22. Deepa, S.; Selladurai, R.; Chelladurrai, C. Cost minimization in a MicroGrid connected with Wind and PV generations using a hybrid Cat Swarm optimization and micro Differential Evolution. In Proceedings of the 2019 9th International Conference on Power and Energy Systems (ICPES), Perth, WA, Australia, 10–12 December 2019.
23. Moradi, M.H.; Abedini, M.; Hosseinian, S.M. Improving operation constraints of microgrid using PHEVs and renewable energy sources. *Renew. Energy* **2015**, *83*, 543–552. [CrossRef]
24. Ikeda, S.; Ooka, R. Application of differential evolution-based constrained optimization methods to district energy optimization and comparison with dynamic programming. *Appl. Energy* **2019**, *254*, 113670. [CrossRef]
25. Gholami, K.; Jazebi, S. Multi-objective long-term reconfiguration of autonomous microgrids through controlled mutation differential evolution algorithm. *IET Smart Grid* **2020**, *3*, 738–748. [CrossRef]
26. Essiet, I.O.; Sun, Y.; Wang, Z. Optimized energy consumption model for smart home using improved differential evolution algorithm. *Energy* **2019**, *172*, 354–365. [CrossRef]
27. Aziz, S.; Wang, H.; Liu, Y.; Peng, J.; Jiang, H. Variable Universe Fuzzy Logic-Based Hybrid LFC Control With Real-Time Implementation. *IEEE Access* **2019**, *7*, 25535–25546. [CrossRef]
28. Fattahi, A.; Nahavandi, A.; Jokarzadeh, M. A comprehensive reserve allocation method in a micro-grid considering renewable generation intermittency and demand side participation. *Energy* **2018**, *155*, 678–689. [CrossRef]
29. Aslam, S.; Khalid, A.; Javaid, N. Towards efficient energy management in smart grids considering microgrids with day-ahead energy forecasting. *Electr. Power Syst. Res.* **2020**, *182*, 106232. [CrossRef]
30. Gholami, K.; Dehnavi, E. A modified particle swarm optimization algorithm for scheduling renewable generation in a micro-grid under load uncertainty. *Appl. Soft Comput.* **2019**, *78*, 496–514. [CrossRef]
31. Arcos-Aviles, D.; Pacheco, D.; Pereira, D.; Garcia-Gutierrez, G.; Carrera, E.V.; Ibarra, A.; Ayala, P.; Martínez, W.; Guinjoan, F. A Comparison of Fuzzy-Based Energy Management Systems Adjusted by Nature-Inspired Algorithms. *Appl. Sci.* **2021**, *11*, 1663. [CrossRef]
32. Nazari, A.; Keypour, R. Participation of responsive electrical consumers in load smoothing and reserve providing to optimize the schedule of a typical microgrid. *Energy Syst.* **2020**, *11*, 885–908. [CrossRef]
33. Mena, R.; Hennebel, M.; Li, Y.-F.; Zio, E. Self-adaptable hierarchical clustering analysis and differential evolution for optimal integration of renewable distributed generation. *Appl. Energy* **2014**, *133*, 388–402. [CrossRef]
34. Marzband, M.; Fouladfar, M.H.; Akorede, M.F.; Lightbody, G.; Pouresmaeil, E. Framework for smart transactive energy in home-microgrids considering coalition formation and demand side management. *Sustain. Cities Soc.* **2018**, *40*, 136–154. [CrossRef]
35. Garcia-Guarin, J.; Infante, W.; Ma, J.; Alvarez, D.; Rivera, S. Optimal scheduling of smart microgrids considering electric vehicle battery swapping stations. *Int. J. Electr. Comput. Eng.* **2020**, *10*, 5093.
36. Cao, Y.; Tang, S.; Li, C.; Zhang, P.; Tan, Y.; Zhang, Z.; Li, J. An optimized EV charging model considering TOU price and SOC curve. *IEEE Trans. Smart Grid* **2011**, *3*, 388–393. [CrossRef]
37. Tushar, M.H.K.; Zeineddine, A.W.; Assi, C. Demand-side management by regulating charging and discharging of the EV, ESS, and utilizing renewable energy. *IEEE Trans. Ind. Inform.* **2017**, *14*, 117–126. [CrossRef]
38. Bewley, S.K. *The Potential Market Applications of Distributed Generation of Electricity*; Massachusetts Institute of Technology: Cambridge, MA, USA, 2002.
39. Rashedi, E.; Nezamabadi-Pour, H.; Saryazdi, S. GSA: A gravitational search algorithm. *Inf. Sci.* **2009**, *179*, 2232–2248. [CrossRef]
40. Duman, S.; Sönmez, Y.; Güvenç, U.; Yörükeren, N. Optimal reactive power dispatch using a gravitational search algorithm. *IET Gener. Transm. Distrib.* **2012**, *6*, 563–576. [CrossRef]
41. Shin, D.-K.; Lee, J.J. Analysis of asymmetric warpage of thin wafers on flat plate considering bifurcation and gravitational force. *IEEE Trans. on Compon. Packag. Manuf. Technol.* **2014**, *4*, 248–258. [CrossRef]
42. Johannesen, N.J.; Kolhe, M.L.; Goodwin, M. Smart load prediction analysis for distributed power network of Holiday Cabins in Norwegian rural area. *J. Clean. Prod.* **2020**, *266*, 121423. [CrossRef]
43. Midenet, S.; Boillot, F.; Pierrelée, J.-C. Signalized intersection with real-time adaptive control: On-field assessment of CO_2 and pollutant emission reduction. *Transp. Res. Part D Transp. Environ.* **2004**, *9*, 29–47. [CrossRef]

sustainability

Article

Optimal DG Location and Sizing to Minimize Losses and Improve Voltage Profile Using Garra Rufa Optimization

Riyadh Kamil Chillab [1], Aqeel S. Jaber [2,*], Mouna Ben Smida [1] and Anis Sakly [1]

[1] National Engineering School of Monastir (ENIM), University of Monastir, Ibn El Jazzar, Skaness, Monastir 5019, Tunisia
[2] Departments of the Electrical Power Engineering, Al-Ma'moon University College, Baghdad 10013, Iraq
* Correspondence: aqe77el@yahoo.com

Abstract: Distributed generation (DG) refers to small generating plants that usually develop green energy and are located close to the load buses. Thus, reducing active as well as reactive power losses, enhancing stability and reliability, and many other benefits arise in the case of a suitable selection in terms of the location and the size of the DGs, especially in smart cities. In this work, a new nature-inspired algorithm called Garra Rufa optimization is selected to determine the optimal DG allocation. The new metaheuristic algorithm stimulates the massage fish activity during finding food using MATLAB software. In addition, three indexes which are apparently powered loss compounds and voltage profile, are considered to estimate the effectiveness of the proposed method. To validate the proposed algorithm, the IEEE 30 and 14 bus standard test systems were employed. Moreover, five cases of DGs number are tested for both standards to provide a set of complex cases. The results significantly show the high performance of the proposed method especially in highly complex cases compared to particle swarm optimization (PSO) algorithm and genetic algorithm (GA). The DG allocation, using the proposed method, reduces the active power losses of the IEEE-14 bus system up to 236.7873%, by assuming 5DGs compared to the active power losses without DG. Furthermore, the GRO increases the maximum voltage stability index of the IEEE-30 bus system by 857% in case of the 4DGs, whereas GA rises the reactive power of 5DGs to benefit the IEEE-14 bus system by 195.1%.

Keywords: distribution generation; Garra Rufa ptimization; PSO; GA; power system

Citation: Chillab, R.K.; Jaber, A.S.; Smida, M.B.; Sakly, A. Optimal DG Location and Sizing to Minimize Losses and Improve Voltage Profile Using Garra Rufa Optimization. *Sustainability* **2023**, *15*, 1156. https://doi.org/10.3390/su15021156

Academic Editors: Herodotos Herodotou, Sheraz Aslam and Nouman Ashraf

Received: 4 December 2022
Revised: 1 January 2023
Accepted: 4 January 2023
Published: 7 January 2023

1. Summary

The interconnection of generation, transmission, and distribution with a centralized control increases the power system complexity with the increase in the number of nodes and branches. To overcome centralized control and long-distance power transmission, distributed generators (DGs) are among the most common clean energy solutions introduced in the last 20 years [1–4]. Their advantages are not limited only to reduce the complexity and enhance the environment as in the smart city. They are also extended to other indexes, such as the economy, environment, quality, stability, losses, voltage profile, and sensitivity. Those benefits have positively increased with the corresponding proper selection of each benefit. On the other hand, within the context of intelligent optimization and control, the huge thinking of the maximum economic operation and high efficiency, the researchers spotlighted and paid attention to the intelligent behavior of nature [5–9]. The new artificial intelligence (AI) advances in software engineering are related to all scientific topics which provide new opportunities and challenges for scientists for tackling highly complex issues that are difficult to solve with conventional techniques [10,11]. The nature-inspired algorithms ranked the highest in predicting the exact solutions, efficiency, and speed even in multi-objective functions. Since intelligent optimization led the way in engineering science, several studies have been conducted on the topic of determination of DG location and size which are listed in the references [12–23]. For instance, Suresh and Belwin [12] used the Dragonfly algorithm to optimal DG size for multi-objective function. IEEE 15, 33, and 69 examined the algorithm performance. Ogunsina et al. [13] determined the DG effect using the electrical transient and analysis program (ETAP) model for the IEEE 30 bus standard via enhancing the active power loss and voltage profile. The used method was artificially intelligent colony optimization. The optimal size and sitting of DGs were estimated by Marimuthu et al. [14] by a hybrid of particle swarm optimization (PSO) and time-varying acceleration coefficients. The voltage Stability with other four objectives were the goal for enhancing a 69-node power system. Montoya et al. [15] suggested a solution for the DGs allocation by employing a master-slave technique using a modified genetic algorithm (GA) that named the Chu-Beasley genetic algorithm. The master-slave solved the mixed-integer nonlinear identification problem in the complex system and the slave determined the optimal power flow using MATLAB. Another attempt of GA for optimizing the allocation was introduced by Chandel et al. [16]. The differential evolution was the comparison base that was selected to enhance five objective functions of the IEEE 18 standard system. Elhosseny et al. [17] developed a PSO technique for selecting the location and the size of DG. The IEEE14 standard system was implemented to validate the build-up of the PSO method for reducing power losses and improving voltage stability. The power loss was reduced by using the Bat algorithm during the optimal selection of DG size in [18]. The IEEE 33-bus standard was the only system that has been tested to validate the system. Suresh and Edward [19] considered the hybridization of fuzzy and one-rank cuckoo search algorithms as the best method to allocate DG. The power losses and voltage profile were the objectives to improve IEEE 15-bus, 33-bus, and 69-bus test systems. Abedini and Saremi [20] proposed a hybrid of two intelligent methods, PSO and GA, to locate the DG with a fuzzy optimization idea to transfer the multi-objective into a single objective problem. The method was tested using the 52-bus of Hamadan power networks. A fuzzy logic control method with GA has been tested to optimize the D-STATCOM by optimizing the allocation of the DG [21]. A radial distribution standard of 33-bus was the system that examined the Fuzzy-GA method to improve three of the indexes of the power system.

However, each of the optimal algorithms that is used in the literature has its own drawback. For example, GA suffers from premature dependence convergence, slow convergence, and difficulty in parameter determination [24,25]. The last iterations of PSO converge slowly and drop easily into a locally optimal solution [24,26]. The convergence rate of the bee colony is slow convergence, also has the same problem of PSO of the local optimal point. The ant colony convergence is normally slow, and hilly dependent on parameter selection [24]. The modified and the hybrid methods are more complex and increase the complexity of the system [26]. These limitations are due mainly to the use of Garra Rufa optimization (GRO) for estimating the size and location of DGs. The high flexibility feature of GRO may lead to solving the DGs issues. First of all, a simplified view of the DG, the proposed method, and the used systems are introduced. Secondly, applying three factors in one objective function that are active and reactive power losses minimization, and voltage stability index enhancement. The three objectives were weighted depending on the corresponding priority of each index on the power system's economic and quality. Moreover, the proposed method is applied to the 14 and 33-bus bar standard systems with PSO and GA algorithms with five cases of distribution generator numbers.

Finally, the proposed optimization technique is compared with PSO and GA in order to show its tracking performance via the active power losses, reactive power losses, and voltage profile. By examining the optimal allocation of each method, the performance validation is conducted by analyzing the IEEE of 14-bus and 30-bus.

2. Related Work

A population of strings representing several potential solutions makes up the population used by the population-based search technique known as GA [27]. As a result, when it's used to solve challenging optimization issues, GA has latent parallelism that improves its search capabilities and speeds up the discovery of the optima. In order to discover a globally optimal solution, GA is an effective point-based optimization technique that has been widely used in a variety of engineering issue.

Another evolutionary computation method is used to validate the proposed method is (PSO). PSO has been inspired by the behaviors of wildlife like swarming fishes and flocking birds. In several cases, PSO is typically described as a clear, simple-to-use, and computationally efficient technique.

The GA and PSO can be widely used to investigate optimal DG placement. The nonlinear model of the power system, as settled, has problems of irregularity and discontinuity. The objective function organized by the genetic algorithm has a high feature of adaptability compared to PSO [28]. PSO is more effective than GA and has a balanced method to improve local and global exploration capabilities [29,30]. However, some shortcomings have been found in the performance of most of the basic algorithms, such as GA, ABC, ACO, CKO, and PSO [31]. Figures 1 and 2 represent the intelligent algorithms for GA and PSO respectively.

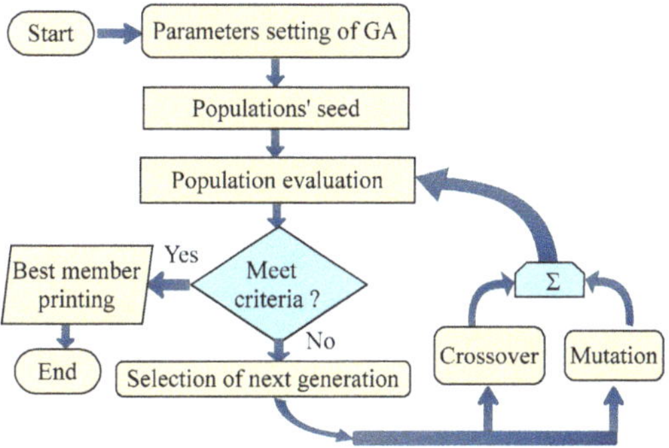

Figure 1. GA algorithm.

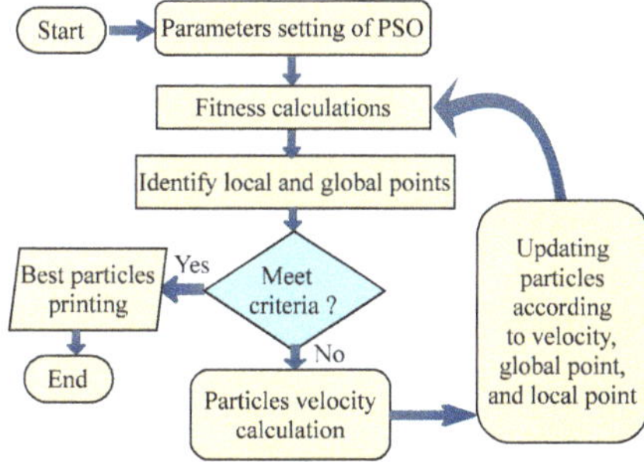

Figure 2. PSO algorithm.

To overcome the drawback of many of the intelligent optimization methods, GRO is one of the most flexible and efficient methods introduced to solve the highly complex issues [32,33]. To understand the GRO mechanism, the algorithm could be simplified in three steps. Step one is initialization, step two is the leaders' crossover, and group followers' crossover is the last step.

2.1. GRO Initialization

The main principle of GRO is to divide the total particles into more than one group, each group has its' own guide to the local and global optimal group points. Moreover, initial assumptions must be assumed in the GRO algorithm such as each fish could be either a guide or a follower according to the corresponding global optimal point of each group. Before the next iteration, a percentage of the followers will move from the weak leader to the stronger one that got the best optimal value. This percentage must be initially assumed. Other initial parameters are the inertia weight (ω) and acceleration coefficients (c_1, c_2). The initialization equation is listed as shown [31].

$$followers\ number = \frac{total\ umber\ of\ partcles - number\ of\ groups}{number\ of\ groups} \tag{1}$$

2.2. GRO Leaders' Crossover

Two types of leaders' crossovers are assumed in the GRO algorithm, firstly, a new leader (guide) is elected for each group. Secondly, select the best leader to lead the number one group which has the maximum number of followers. These steps are considered the basis paving the way for the most important step, which gives flexibility to the method of GRO.

2.3. GRO Followers' Crossover

The flexible movement for the sleeve fish between the groups is more probability to search in the problem space. The highly complex problems can cause disorientation for all the intelligent optimization algorithms that have inflexible nature of moving from one search space to another. This issue occurs due to a large number of ripples and the multiplicity of parameters of complex problems. By the follower crossover between the groups, GRO found a way to keep searching in the wider area space of the problem by applying three steps. First of all, a random number of fish will move to the strong leader from all other groups. Secondly, one step is moving toward each leader which must be done by determining the velocity (v_i) and the position (X_i) using the classical Equations (2) and (3).

$$v_i(t+1) = \omega v_i(t) + c_1 r_1 (p_i(t) - X_i(t)) + c_2 r_2 (G_i(t) - X_i(t)) \tag{2}$$

$$X_i(t+1) = X_i(t) + v_i(t+1) \tag{3}$$

After that, the fitness of the new groups' figures will be recalculated, including all followers and leaders. Equations (4) and (5) represent the novelty steps of GRO.

$$moving\ folwoers_i = integer\ (£ * random) \tag{4}$$

$$followers_{ij} = Max\ ((followers_{ij-1} - moving\ folwoers_i), o) \tag{5}$$

where £ is the highest possible number of moving fish. Figure 3 shows the algorithms flowchart of GRO method.

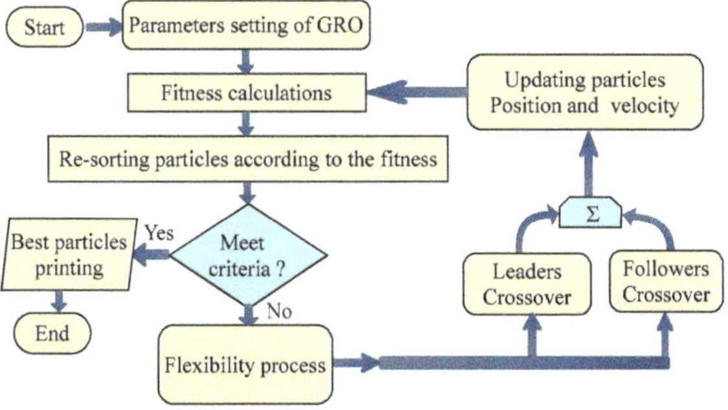

Figure 3. GRO algorithm.

In this paper, particle swarm (PSO) and GA optimization algorithms will be used for comparison with GRO to optimize the size and location of the DGs. The PSO, GA, and GRO basic fundamental equations are implemented without any further modifications according to Kennedy and Eberhart (PSO) [34], Abedini (GA) [20], and Jaber [31].

3. Proposed System Model

As previously mentioned, power system models and their networks comprise a nonlinear high-order system of equations that arise from a large number of parameters.

Thus, prior to solving the DGs problem of estimating their size and location, an obvious objective function is required. The objective function of any DG allocation could be one or more.

According to the power flow using the formulation that contained the different types of power system variables (active reactive power, voltage, and power angle), important indexes have been introduced to improve the power system quality, e.g., power losses, voltage profile, reliability, stability, and economic issues. In this paper, three of the most important challenges in DG topic are selected to determine the size and location of each DG. The following indexes represent the base indexes formulas of the objective function.

3.1. Minimization of Total Active Power Losses

The total losses have a substantial effect on the total power generation, thus could grow the economic and environmental merits. Two power systems are used to estimate the current flow in the lines between the buses of those systems. These currents result in power losses (P_L) that represent the most important objective function which has a mathematical model as (6).

$$P_L = \sum_{line=1}^{N_u} G_{line}\left(V_i^2 + V_s^2 - 2V_i V_s \cos(\alpha_i - \alpha_s)\right) \tag{6}$$

where N_u is the total number of transmission lines in the system, G_{line} is the conductance of the line, V_i and V_s are the magnitudes of the sending end voltages and receiving end voltages of the line, α_i and α_s are angles of the end voltages.

3.2. Minimization of Reactive Power Losses

Q_L are referred to as the complex part of the apparent power losses. The amount of reactive power losses (Q_L) has a significant impact on conductor capacity and voltage profile. In addition, the QL reduction could increase the stability and reliability indexes. The second objective function in this study can be mathematically written as (7).

$$Q_L = \sum_{line=1}^{N_u} B_{line}\left(V_i^2 + V_s^2 - 2V_i V_s \sin(\alpha_i - \alpha_s)\right) \tag{7}$$

where N_u is the total number of transmission lines in the system, B_{line} is the susceptance of the line, V_i and V_s are the magnitudes of the sending end voltages and receiving end voltages of the line, α_i and α_s are angles of the end voltages.

3.3. Voltage Stability Index Improvement

In order to satisfy the modern power system quality, the improvement of voltage stability is essential that it is concerned with the power capability for maintaining acceptable voltages at all buses in both normal and up normal conditions. In this paper, the voltage deviation index is used to estimate the voltage stability index which is based on the power flow calculations [35,36]. The described voltage stability index (VSI) is formulated as follows:

$$\text{VSI} = \sum_{Bus=1}^{N_b} \left(V_{ref} - V_{bus}\right) \tag{8}$$

where N_b is the number of buses; V_{ref} is the reference voltage; and V_{bus} is the bus voltage.

From the other hands, the weights are selected to give the corresponding priority to each impact indices of DGs allocation objective functions [37,38]. The appropriate weight selection also relies on the experience of power system researchers and the heeds of distribution side utilities. Nowadays, total power loss reduction in both of its components is one of the major concerns in the power system operation and control due to its impact on the economy, stability, and environment, while the voltage stability index is less important than the power loss reduction. Hence, the weights for P_L, Q_L, and VSI (w_1, w_2, and w_3) have been taken as 0.50, 0.15, and 0.35, respectively.

Moreover, in order to make the maximum and minimum values and the possible changes of each index as a result of adding the DG homogeneous in terms of units and influence, a corresponding base index was chosen that represents the same of each index without adding the DG. Figure 4 represents the objective function calculation according to the AI algorithm. Furthermore, the optimization problem is given by Equation (9)

$$\text{objective} = \text{minimize}\left(w_1 * \frac{P_L \text{ with } DG}{P_L \text{ without } DG} + w_2 * \frac{Q_L \text{ with } DG}{Q_L \text{ without } DG} + w_3 * \frac{VSI \text{ with } DG}{VSI \text{ without } DG} \right) \qquad (9)$$

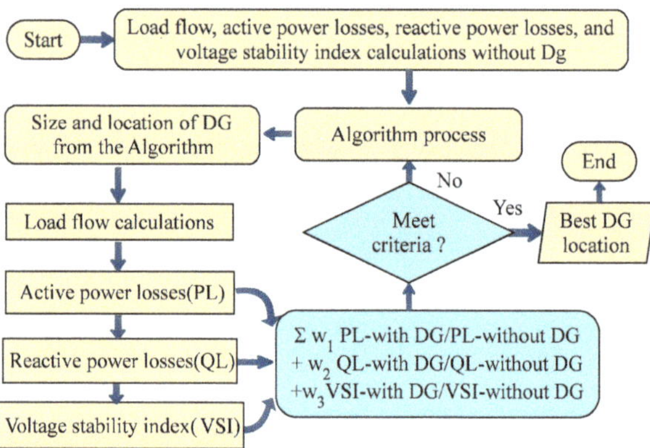

Figure 4. The objective function.

4. Result and Discussion

The discovery of the high efficiency and flexibility of GRO in highly complex issues inspires the authors of this paper to utilize the algorithm to estimate the location and size of DGs in order to reduce active and reactive losses of power and improve voltage stability.

To achieve the results and performances of the prepared scenario, all tested cases and algorithms have been simulated by MATLAB. Additionally, the optimization algorithms are tested with the 30 and 14 bus IEEE standards which are shown in Figure 5.

Besides, for fair comparisons, the objective functions are implemented using GRO, PSO, and GA with the same different numbers of population and iterations depending on the problem case and the system, as shown in Table 1.

Table 1. Algorithm parameters.

System	30 Bus	30 Bus	30 Bus	30 Bus	30 Bus	14 Bus	14 Bus	14 Bus	14 Bus	14 Bus
DGs–number	1	2	3	4	5	1	2	3	4	5
Particles	25	30	40	40	40	20	20	20	20	20
Iterations	30	30	35	40	40	20	20	30	30	30

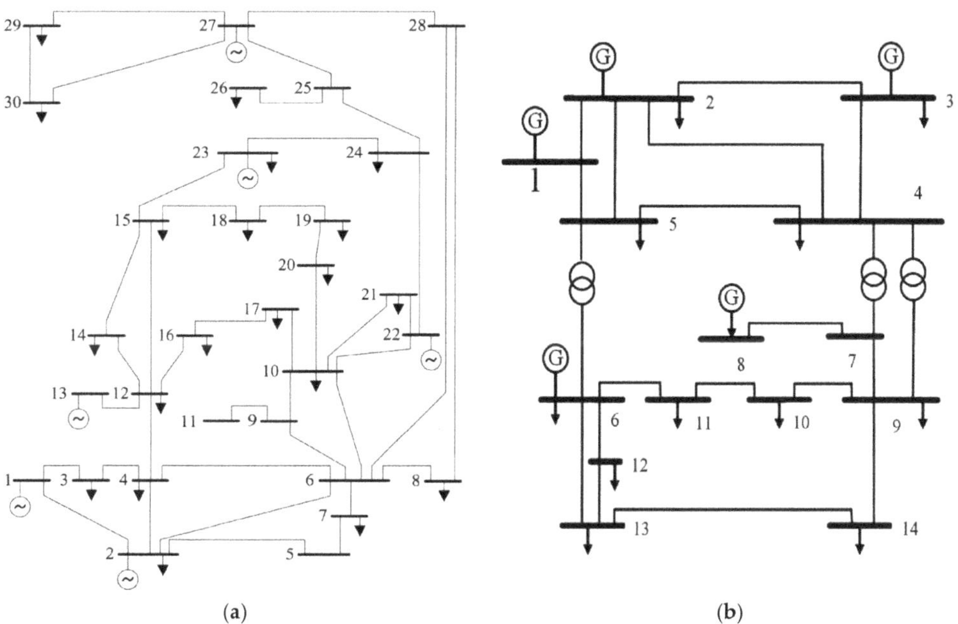

Figure 5. Tested systems (**a**) IEEE 30 bus system (**b**) IEEE 14 bus system.

4.1. Test Case 1: IEEE-14 Bus Standard

Between 1 and 5 DGs are assumed to be added to the selected IEEE systems to find the effectiveness of DG of the objective functions. The addition of different numbers of DG is create multi complexity cases. Furthermore, the load flow calculations for all the cases have been done using the Newton–Raphson method. Table 2 shows the size and location for the three methods.

Table 3 describes some of the important directories for the changes in Loss saving and voltage profile improvements by the three used methods. It can be noted from Table 3 that all three methods have an impact on the apparent power losses and improve the stability voltage index but in different values. In cases of single, two, and three generators, there is no significant noted advance of GRO on the objective value, while the effectiveness of the proposed method is clearly shown in the four and five DGs allocation problem solutions.

4.2. Test Case 2: IEEE-30 Bus Standard

In order to validate the proposed method, the same procedure sequence is followed with a higher complex system of the IEEE system instead of the IEEE-14 bus. The IEEE-30 bus consists of six generators and 41 lines between the 30 buses. Tables 4 and 5 illustrate the DGs locations for the five cases of the number of generators and the three cases of the optimization methods.

The specific underlined values seen in Tables 3 and 5 show the best optimal values of each optimization method for all the classes. To clarify those distinction of these optimization methods, Figures 6–8 illustrate samples of power system improving in power losses and voltage profile.

Table 2. DGs allocation of IEEE-14 bus.

Dg Number	1DG			2DG			3DG			4DG			5DG		
Method	GRO	PSO	GA	GRO	PSO	GA	GRO	PSO	GA	GRO	PSO	GA	GRO	PSO	GA
Location	7	7	8	12,8	9,5	3,7	6,5,4	2,4,5	2,6,13	2,3,12,10	9,2,2,3	2,2,8,11	2,3,10,11,2	2,7,2,5,4	11,10,2,2,3
size	17.771	20.85	20.848	17.381 21.432	20.850 20.010	20.837 20.707	46.174 41.925 45.954	49.685 51.350 50.850	50.789 43.203 49.329	15.652 17.134 13.558 38.021	20.850 8.048 17.471 20.85	1.582 20.849 20.777 20.825	11.352 39.087 25.309 12.051 10.112	6.725 20.211 20.309 20.850 20.850	20.754 20.812 16.147 20.835 20.828

Table 3. IEEE-14 bus objective functions.

Dg Number	1DG			2DG			3DG			4DG			5DG		
Method	GRO	PSO	GA	GRO	PSO	GA	GRO	PSO	GA	GRO	PSO	GA	GRO	PSO	GA
P_{LOSS}	15.055	14.324	13.961	11.695	11.716	11.453	6.7	8.427	7.051	6.795	8.087	7.85	4.034	4.786	4.190
Q_{LOSS}	61.259	58.986	57.859	35.931	35.558	36.541	22.04	26.057	24.758	26.126	27.724	27.383	21.565	19.347	19.281
|VSI|	0.2707	0.2210	0.1916	0.1062	0.0866	0.1330	0.0294	0.0334	0.13685	0.0626	0.0470	0.0994	0.0107	0.2150	0.1990
Objective	0.547	0.545	0.547	0.459	0.471	0.435	0.323	0.358	0.410	0.280	0.400	0.423	0.212	0.361	0.329

Table 4. DGs allocation of IEEE-30 bus.

Dg Number	1DG			2DG			3DG			4DG			5DG		
Method	GRO	PSO	GA	GRO	PSO	GA	GRO	PSO	GA	GRO	PSO	GA	GRO	PSO	GA
Location	5	16	2	19, 14	14, 18	18,14	3,4,9	17,28,13	29,17,8	6,23,9,7	17,27,15,9	17,29,18,10	5,10,21,25,24	28,24,2,16,29	7,11,28,18,12
size	44.641	61.054	70.076	48.608 52.589	47.358 63.882	52.719 40.175	22.123 72.297 101.7	46.051 70.850 58.858	9.562 69.359 67.387	56.195 26.217 25.888 66.511	30.850 30.850 30.850 26.487	30.849 22.015 30.839 30.756	98.468 47.781 35.144 7.556 7.448	98.468 47.781 35.144 7.556 7.448	50.819 42.210 46.125 10.804 46.461

Table 5. IEEE-30 bus objective functions.

Dg Number	1DG			2DG			3DG			4DG			5DG		
Method	GRO	PSO	GA	GRO	PSO	GA	GRO	PSO	GA	GRO	PSO	GA	GRO	PSO	GA
P_{LOSS}	11.671	11.371	11.371	9.268	9.022	8.961	6.708	6.518	6.595	8.697	9.621	8.841	5.531	7.72	6.889
Q_{LOSS}	49.266	48.157	48.157	40.092	39.34	39.081	23.964	23.164	23.446	33.58	39.033	36.448	23.624	32.691	27.842
I VSI I	0.1737	0.2990	0.6672	0.1527	0.1430	0.6922	0.4708	0.5261	0.5632	0.0955	0.3012	0.4576	0.2341	0.4215	0.2812
objective	0.507	0.544	0.685	0.4104	0.398	0.606	0.424	0.438	0.455	0.358	0.475	0.507	0.299	0.453	0.365

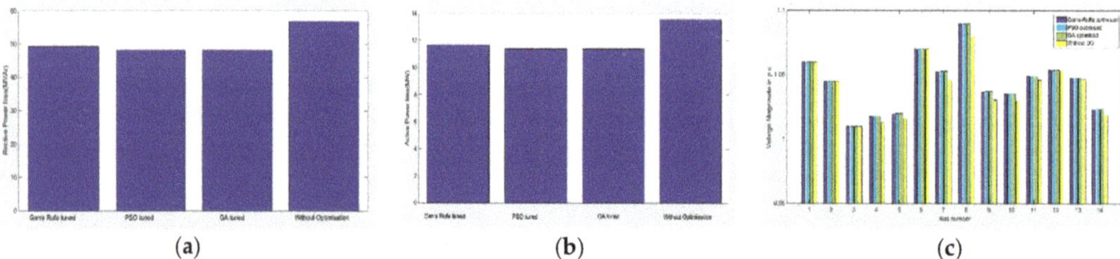

Figure 6. IEEE-14, 1-DG Power system improvement (**a**) reactive power, (**b**) active power, and (**c**) voltage profile.

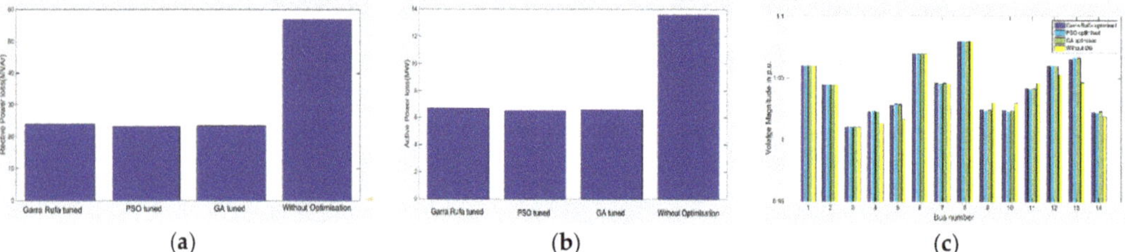

Figure 7. IEEE-14, 3-DG Power system improvement (**a**) reactive power, (**b**) active power, and (**c**) voltage profile.

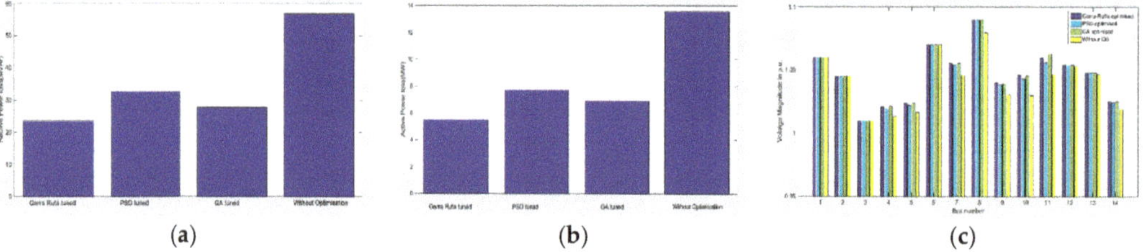

Figure 8. IEEE-14, 5-DG Power system improvement (**a**) reactive power, (**b**) active power, and (**c**) voltage profile.

As observed in Tables 3 and 5, there are loss savings in all three methods. The maximum saving in the active power of the IEEE-14 bus system is 236.7873% in the case of the 5DGs GRO method, and the minimum benefit is –9.7576% in the 1DG GRO method, which means GRO failed in finding an acceptable solution. Moreover, the maximum saving in the reactive power of the IEEE-14 bus system is 195.1% in the case of the 5DGs GA method, and the minimum benefit is −7.1173% in the 1DG GRO method. Furthermore, the maximum saving in the voltage stability index of the IEEE-14 bus system is 515.2% in the case of the 5DGs GRO method, and the minimum benefit is 107.6 in the 1DG GRO method. While the best objective function is 0.212 in the case of the five DG GRO method, the worst value is 0.547 in both GA and GRO single DG.

The maximum saving in the active power of the IEEE-30 bus system is 216.6% in the case of the 5DGs GRO method, and the minimum benefit is 50.07% in the 1DG GRO method. Moreover, the maximum saving in the reactive power of the IEEE-30 bus system is 193.8% in the case of the 3DGs GA method, and the minimum benefit is 39.79% in the 1DG GRO method. Furthermore, the maximum saving in the voltage stability index of the IEEE-30 bus system is 857% in the case of the 4DGs GRO method, and the minimum benefit is 32.04% in the 2DG GA method. While the best objective function is 0.299 in the case of the five DG GRO method, and the worse value is 0.606 in both GA 2DGs.

Of the ten cases that were selected to study the new method which represents a multi-level of complexity, seven of the cases were advanced via the objective function using GRO, while two and one cases are optimized better by using PSO and GA, respectively. Moreover, all the high levels of complexity had a better solution using the proposed method.

Converge mechanism of the proposed optimization methods by searching in several areas in the problem space is the reason behind the overcome of GRO method in most cases. The convergence for two cases of each optimization method is shown in Figures 9 and 10. Additionally, all the other load flow results and figures can be shown in Supplementary Materials.

From Figures 9 and 10, it can be noted that for the assumed iterations and the same number of search particles, GRO converges more effectively than the others (GA, PSO) in terms of the minimum objective function. Moreover, in Figure 10, GRO could find a better solution even after five constant iterations (from 10 to 15). This gives the impression that GRO may be successful in skipping in falling into a single optimal point.

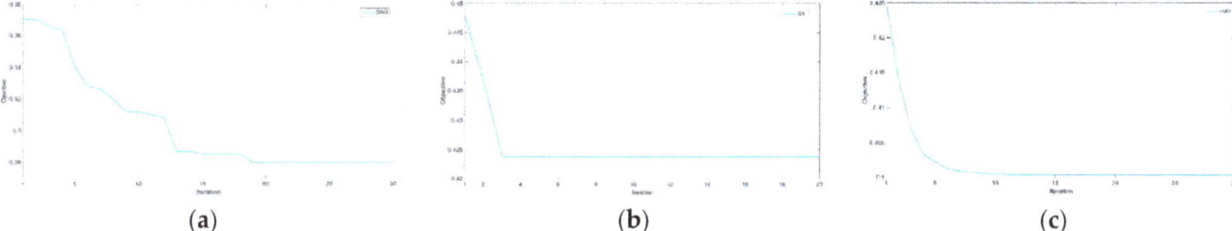

(a)   (b)   (c)

Figure 9. IEEE-14, 4-DG algorithm convergence (**a**) GRO, (**b**) GA, and (**c**) PSO.

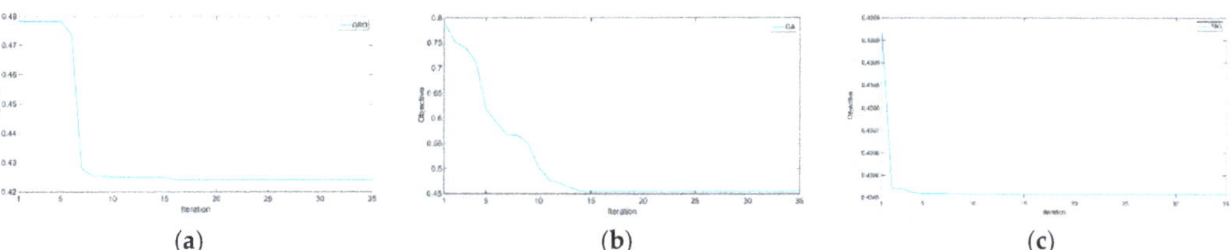

(a)   (b)   (c)

Figure 10. IEEE-30, 3-DG algorithm convergence (**a**) GRO, (**b**) GA, and (**c**) PSO.

5. Conclusions

In this work, the GRO, a recent nature-inspired algorithm, has been utilized to specify DGs' location and size in power network distribution. The proposed method was investigated on two well-known IEEE14 and IEEE30 bus standards by DG, 2DGs, 3DGs, 4DGs, and 5DGs installation and succeeded in terms of loss reduction improvement in voltage stability. Three algorithms, including GA, PSO, and GRO, were applied to solve the assigned issue for comparative aims. As a result of proper DG allocation using the proposed methods, the active and reactive power losses were reduced and the voltage stability index was enhanced up to 236.7873%, 857%, and 195.1% respectively. Moreover, there was a single case where the GRO was unable to find a proper solution, which was reducing the reactive power in a single DG and IEEE 14-bus standard. The GRO mechanism and its superior exploration and exploitation features over other swarm intelligence methods for solving highly complex engineering optimization issues were conveyed. In the end, it is expected that the GRO may provide efficient solutions to existing highly complex power engineering issues, such as load forecasting or load frequency control. On the other hand, the newly proposed method can be modified to be able to solve low-complex problems with more accuracy.

Supplementary Materials: The following supporting information can be downloaded at: https://www.mdpi.com/article/10.3390/su15021156/s1.

Author Contributions: Conceptualization, R.K.C.; Methodology, A.S.; Writing—original draft, A.S.J.; Writing—review & editing, M.B.S. All authors have read and agreed to the published version of the manuscript.

Funding: This research received no external funding.

Institutional Review Board Statement: Not applicable.

Informed Consent Statement: Informed consent was obtained from all subjects involved in the study.

Data Availability Statement: Not applicable.

Conflicts of Interest: The authors declare no conflict of interest.

References

1. Somefun, T.; Popoola, O.; Abdulkareem, A.; Awelewa, A. Review of Different Methods for Siting and Sizing Distributed Generator. *Int. J. Energy Econ. Policy* **2022**, *12*, 16–31. [CrossRef]
2. Suyarov, A.; Hasanov, M.; Boliev, A.; Nazarov, F. Whale Optimization Algorithm for Intogreting Distributed Generators in Radial Distribution Network. 2021. Available online: https://ssrn.com/abstract=3938852 (accessed on 8 September 2021).
3. Abdalla, A.N.; Jing, W.; Nazir, M.S.; Jiang, M.; Tao, H. Socio-economic impacts of solar energy technologies for sustainable green energy: A review. *Environ. Dev. Sustain.* **2022**, 1–38. [CrossRef]
4. Zhu, Y.; Zhou, Y.; Wang, Z.; Zhou, C.; Gao, B. A terminal distribution network black-start optimization method based on pruning algorithm considering distributed generators. *Energy Rep.* **2021**, *8*, 237–244. [CrossRef]
5. Abidi, M.H.; Alkhalefah, H.; Moiduddin, K.; Al-Ahmari, A. Novel improved chaotic elephant herding optimization algorithm-based optimal defense resource allocation in cyber-physical systems. *Soft Comput.* **2022**, 1–16. [CrossRef]
6. Hussein, B.M.; Jaber, A.S. Unit commitment based on modified firefly algorithm. *Meas. Control.* **2020**, *53*, 320–327. [CrossRef]
7. Zhang, X.; Li, Y.; Fan, Y. Regularization Cuckoo Search Algorithm for Multi-Parameter Optimization of the Multi-Laminated Controlled Release System. *Axioms* **2022**, *11*, 500. [CrossRef]
8. Aderibigbe, M.; Adoghe, A.; Agbetuyi, F.; Airoboman, A. A Review on Optimal Placement of Distributed Generators for Reliability Improvement on Distribution Network. *IEEE PES/IAS PowerAfr.* **2021**, 1–5. [CrossRef]
9. Jaber, A.S.; Satar, K.A.; Shalash, N.A. Short term load forecasting for electrical dispatcher of Baghdad city based on SVM-PSO method. In Proceedings of the 2018 2nd International Conference on Electrical Engineering and Informatics (ICon EEI), Batam, Indonesia, 16–17 October 2018; pp. 140–143.
10. Azrag, M.A.K.; Kadir, T.A.A.; Jaber, A.S. Segment Particle Swarm Optimization Adoption for Large-Scale Kinetic Parameter Identification of Escherichia Coli Metabolic Network Model. *IEEE Access* **2018**, *6*, 78622–78639. [CrossRef]
11. Jaber, A.S.; Satar, K.A.; Shalash, N.A. Short-term load forecasting for electrical dispatcher of Baghdad City based on SVM-FA. *Int. J. Adv. Comput. Sci. Appl.* **2018**, *9*, 300–304. [CrossRef]
12. Suresh, M.C.V.; Belwin, E.J. Optimal DG placement for benefit maximization in distribution networks by using Dragonfly algorithm. *Renew. Wind. Water Sol.* **2018**, *5*, 4. [CrossRef]
13. Ogunsina, A.A.; Petinrin, M.O.; Petinrin, O.O.; Offornedo, E.N.; Petinrin, J.O.; Asaolu, G.O. Optimal distributed generation location and sizing for loss minimization and voltage profile optimization using ant colony algorithm. *SN Appl. Sci.* **2021**, *3*, 248. [CrossRef]
14. Marimuthu, A.; Gnanambal, K.; Eswari, R.P.; Pavithra, T. Optimal Location and Sizing of DG Units to Improve The Voltage Stability in The Distribution System Using Particle Swarm Optimization Algorithm with Time Varying Acceleration Coefficients. In Proceedings of the International Conference on Innovations in Engineering and Technology (ICIET-2016), Bangkok, Thailand, 5–6 August 2016.
15. Gil-González, W.; Montoya, O.D.; Grisales-Noreña, L.F.; Vanegas, C.A.R.; Cabrera, A.M. Hybrid Optimization Strategy for Optimal Location and Sizing of DG in Distribution Networks. *Rev. Tecnura* **2020**, *24*, 47–61. [CrossRef]
16. Chandel, A.; Chauhan, D.S.; Singh, D. Enriched Technique for DG Placement and Sizing by GA Optimization. *Am. Eurasian J. Sci. Res.* **2017**, *12*, 260–270. [CrossRef]
17. Sayed, E.M.; Elamary, N.H.; Swief, R.A. Optimal Sizing and Placement of Distributed Generation (DG) Using Particle Swarm Optimization. *J. Phys. Conf. Ser.* **2021**, *2128*, 012023. [CrossRef]
18. Yuvaraj, T.; Devabalaji, K.R.; Ravi, K. Optimal Allocation of DG in the Radial Distribution Network Using Bat Optimization Algorithm. In *Advances in Power Systems and Energy Management*; Springer: Singapore, 2018; Volume 436, pp. 563–569. [CrossRef]
19. Suresh, M.C.V.; Edward, B.J. Optimal Placement of DG Units for Loss Reduction in Distribution Systems Using One Rank Cuckoo Search Algorithm. *Int. J. Grid Distrib. Comput.* **2018**, *11*, 37–44. [CrossRef]
20. Abedini, M.; Saremi, H. A Hybrid of GA and PSO for Optimal DG Location and Sizing in Distribution Systems with Load Uncertainty. *J. Basic. Appl. Sci. Res.* **2012**, *2*, 5103–5118.

21. Siddiqui, A.S.; Sarwar, M.; Althobaiti, A.; Ghoneim, S.S. Optimal Location and Sizing of Distributed Generators in Power System Network with Power Quality Enhancement Using Fuzzy Logic Controlled D-STATCOM. *Sustainability* **2022**, *14*, 3305. [CrossRef]
22. Nazir, M.S.; Abdalla, A.N.; Metwally, A.S.M.; Imran, M.; Bocchetta, P.; Javed, M.S. Cryogenic-Energy-Storage-Based Optimized Green Growth of an Integrated and Sustainable Energy System. *Sustainability* **2022**, *14*, 5301. [CrossRef]
23. Chen, W.; Liu, B.; Nazir, M.S.; Abdalla, A.N.; Mohamed, M.A.; Ding, Z.; Bhutta, M.S.; Gul, M. An Energy Storage Assessment: Using Frequency Modulation Approach to Capture Optimal Coordination. *Sustainability* **2022**, *14*, 8510. [CrossRef]
24. Qiming, Z.; Husheng, W.; Zhaowang, F. A review of intelligent optimization algorithm applied to unmanned aerial vehicle swarm search task. In Proceedings of the 11th International Conference on Information Science and Technology (ICIST), Chengdu, China, 21–23 May 2021; pp. 383–393.
25. Abdalla, A.N.; Ju, Y.; Nazir, M.S.; Tao, H. A Robust Economic Framework for Integrated Energy Systems Based on Hybrid Shuffled Frog-Leaping and Local Search Algorithm. *Sustainability* **2022**, *14*, 10660. [CrossRef]
26. Jaber, A.S.; Mohammed, K.S.; Shalash, N.A. Optimization of Electrical Power Systems Using Hybrid PSO-GA Computational Algorithm: A Review. *Int. Rev. Electr. Eng. (IREE)* **2020**, *15*, 502. [CrossRef]
27. An, H.K.; Javeed, M.A.; Bae, G.; Zubair, N.; Metwally, A.S.M.; Bocchetta, P.; Na, F.; Javed, M.S. Optimized Intersection Signal Timing: An Intelligent Approach-Based Study for Sustainable Models. *Sustainability* **2022**, *14*, 11422. [CrossRef]
28. Zamani, M.; Karimi-Ghartemani, M.; Sadati, N. FOPID controller design for robust performance using Particle Swarm Optimization. *Fract. Calc. Appl. Anal.* **2007**, *10*, 169–187.
29. Sadati, N.; Zamani, M.; Mahdavian, H. Hybrid particle swarmbased-simulated annealing optimization techniques. In Proceedings of the IECON 2006-32nd Annual Conference on IEEE Industrial Electronics, Paris, France, 7–10 November 2006; pp. 644–648.
30. Fogel, D.B. *Evolutionary Computation toward A New Philosophy of Machine Intelligence*; IEEE: Piscataway, NJ, USA, 1995.
31. Jaber, A.S.; Abdulbari, H.A.; Shalash, N.A.; Abdalla, A.N. Garra Rufa-inspired optimization technique. *Int. J. Intell. Syst.* **2020**, *35*, 1831–1856. [CrossRef]
32. Prakash, A.; Joseph, A.S.; Shanmugasundaram, R.; Ravichandran, C. A machine learning approach-based power theft detection using GRF optimization. *J. Eng. Des. Technol.* **2021**. [CrossRef]
33. Krishnan, V.A.; Kumar, N.S. Robust soft computing control algorithm for sustainable enhancement of renewable energy sources based microgrid: A hybrid Garra rufa fish optimization–Isolation forest approach. *Sustain. Comput. Inform. Syst.* **2022**, *35*, 100764. [CrossRef]
34. Kennedy, J.; Eberhart, R. Particle Swarm Optimization. In Proceedings of the ICNN'95-International Conference on Neural Networks, Perth, WA, Australia, 27 November–1 December 1995; pp. 1942–1948.
35. Modarresi, J.; Gholipour, E.; Khodabakhshian, A. A comprehensive review of the voltage stability indices. *Renew. Sustain. Energy Rev.* **2016**, *63*, 1–12. [CrossRef]
36. Venkatesan, C.; Kannadasan, R.; Alsharif, M.; Kim, M.-K.; Nebhen, J. Assessment and Integration of Renewable Energy Resources Installations with Reactive Power Compensator in Indian Utility Power System Network. *Electronics* **2021**, *10*, 912. [CrossRef]
37. Hung, D.Q.; Mithulananthan, N.; Bansal, R.C. Integration of PV and BES units in commercial distribution systems considering energy loss and voltage stability. *Appl. Energy* **2014**, *113*, 1162–1170. [CrossRef]
38. Singh, D.; Singh, D.; Verma, K. Multiobjective optimization for DG planning with load models. *IEEE Trans. Power Syst.* **2009**, *24*, 427–436. [CrossRef]

 *sustainability*

Article

Detecting Nontechnical Losses in Smart Meters Using a MLP-GRU Deep Model and Augmenting Data via Theft Attacks

Benish Kabir [1], Umar Qasim [2], Nadeem Javaid [1,*], Abdulaziz Aldegheishem [3], Nabil Alrajeh [4] and Emad A. Mohammed [5]

[1] Department of Computer Science, COMSATS University Islamabad, Islamabad 44000, Pakistan
[2] Department of Computer Science, University of Engineering and Technology at Lahore (New Campus), Lahore 54000, Pakistan
[3] Department of Urban Planning, College of Architecture and Planning, King Saud University, Riyadh 11574, Saudi Arabia
[4] Department of Biomedical Technology, College of Applied Medical Sciences, King Saud University, Riyadh 11633, Saudi Arabia
[5] Department of Engineering, Faculty of Science, Thompson Rivers University, 805 TRU Way, Kamloops, BC V2C 0C8, Canada
* Correspondence: nadeemjavaidqau@gmail.com

Abstract: The current study uses a data-driven method for Nontechnical Loss (NTL) detection using smart meter data. Data augmentation is performed using six distinct theft attacks on benign users' samples to balance the data from honest and theft samples. The theft attacks help to generate synthetic patterns that mimic real-world electricity theft patterns. Moreover, we propose a hybrid model including the Multi-Layer Perceptron and Gated Recurrent Unit (MLP-GRU) networks for detecting electricity theft. In the model, the MLP network examines the auxiliary data to analyze nonmalicious factors in daily consumption data, whereas the GRU network uses smart meter data acquired from the Pakistan Residential Electricity Consumption (PRECON) dataset as the input. Additionally, a random search algorithm is used for tuning the hyperparameters of the proposed deep learning model. In the simulations, the proposed model is compared with the MLP-Long Term Short Memory (LSTM) scheme and other traditional schemes. The results show that the proposed model has scores of 0.93 and 0.96 for the area under the precision–recall curve and the area under the receiver operating characteristic curve, respectively. The precision–recall curve and the area under the receiver operating characteristic curve scores for the MLP-LSTM are 0.93 and 0.89, respectively.

Keywords: deep learning; GRU; healthcare; MLP; non-technical losses; PRECON; smart cities; smart grids; smart meters

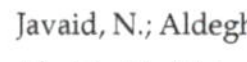 **check for updates**

Citation: Kabir, B.; Qasim, U.; Javaid, N.; Aldegheishem, A.; Alrajeh, N.; Mohammed, E.A. Detecting Nontechnical Losses in Smart Meters Using a MLP-GRU Deep Model and Augmenting Data via Theft Attacks. *Sustainability* **2022**, *14*, 15001. https://doi.org/10.3390/su142215001

Academic Editor: Nien-Che Yang

Received: 26 October 2022
Accepted: 10 November 2022
Published: 13 November 2022

Publisher's Note: MDPI stays neutral with regard to jurisdictional claims in published maps and institutional affiliations.

1. Background

One of the major achievements of smart grids was the development of the Advanced Metering Infrastructure (AMI) system [1]. This system reduces the danger associated with electricity theft by using its fine-grained computations and tracing ability [2]. However, an increase in the system's usage increases energy theft and consequently leads to a loss of electricity [3]. The loss of electricity is among the problems that reduce the performance of the power grids. There are two types of electricity losses. The first are known as Technical Losses (TLs) and the second are known as Non-Technical Losses (NTLs) [4]. The electric heating of resistive components in transformers, transmission lines, and other types of equipment causes TL, while electricity thefts, billing mistakes, and meter faults are the most common cause of NTL [3]. Electricity companies are particularly interested in reducing NTLs, since it accounts for a significant portion of the overall energy losses. Energy theft is the major type of NTL that involves bypassing meters, modifying the meter's readings, etc. The Electricity Consumption (EC) behavior of users may vary from customer to customer. Nonetheless, identifying NTL patterns among all of the usual patterns of EC is a crucial task. In order to capture different types of NTL behaviors, handcrafted feature engineering approaches have been used. However, these approaches are costly as well as time-consuming due to their reliance on expert knowledge [3].

On the one hand, energy theft has resulted in losses of more than 20% of India's total energy supply and 16% of China's accumulative energy supply [5]. On the other hand, financial losses due to energy theft are approximately 100 million and 6 billion dollars per year for Canada and USA, respectively [6], while Pakistan faces an annual loss of approximately 0.89 billion rupees as a result of NTLs [7]. Theft of energy has long been a severe problem in conventional power networks worldwide. Different users show different patterns of Electricity Consumption (EC). Nonetheless, distinguishing NTL patterns from regular EC patterns is challenging. To detect and address these NTLs, many approaches are employed [8,9]. These approaches are classified into three fundamental groups: hybrid-oriented, network-oriented, and data-driven-oriented detection systems. The data-driven methods have attracted the attention of academics and research scholars for performing Electricity Theft Detection (ETD) over the last few years.

The data-driven method is composed of machine learning-based classifiers that are used to detect NTLs [7]. These solutions are also used in various fields like healthcare, education, and transport. In [10], deep learning models were trained as binary classifiers to detect energy thefts. The authors investigated several deep learning models, such as the Convolutional Neural Network (CNN), Multi-Layer Perceptron (MLP), Long-Short Term Memory (LSTM), and Gated Recurrent Unit (GRU) networks. However, due to inefficient tuning of hyperparameters, these models exhibit poor generalization. To tackle the generalization issue, previous studies used the Grid Search Algorithm (GSA) to tune the hyperparameters of the models. However, the GSA requires high computational resources to find the optimal combination of parameters.

According to [11,12], ensemble models fail to identify diverse theft patterns of EC due to a significant imbalance in data, resulting in a high False Positive Rate (FPR). Therefore, we propose the use of a hybrid of neural networks referred as MLP-GRU to detect energy theft. Actual smart meter data and auxiliary information from the consumers are used for the data analysis.

The authors of [12] conducted a detailed analysis of ensemble models based upon boosting and bagging methods. They observed that the Random Forest (RF) model obtained the highest DR and the lowest FPR. Moreover, the authors implemented two data balancing techniques, i.e., the Synthetic Minority Oversampling Technique (SMOTE) and near-miss, to compare both oversampling and undersampling algorithms. However, there may be an increase in the chances of overlapping classes when using SMOTE, as it can increase the existence of noise. The problem of anomaly detection was addressed in [13]. In the proposed work, the authors used a deep learning approach this is capable of distinguishing between regular and anomalous consumption patterns. They also handled the drift concept by discriminating between nonmalicious and real anomalies. However, there is a substantial delay between the occurrence of an anomaly and its detection in the proposed approach.

Existing Machine Learning (ML) algorithms require an equal number of instances for each class during model training. For minority classes, these models have a poor predictive performance. For the detection of electricity thefts, there is a lack of theft data in the real world. Therefore, we synthetically generated the theft data using data balancing techniques [14]. Many studies have used different balancing techniques; however, such techniques have a high computational time and executional complexity. In [14], the authors proposed a hybrid technique, K-SMOTE, for data balancing. In the model, a k-means clustering algorithm is used to determine k clusters for abnormal samples. Afterwards, SMOTE is applied on the clusters of theft samples for interpolation to balance the complete data. Based on the balanced data, Random Forest (RF) classification is performed to detect electricity theft behavior. However, to determine optimal values of k and perform tuning of other hyperparameters for data balancing, an optimization algorithm is required.

With the emergence of smart meters, diverse types of energy theft cases have been introduced, and these are difficult to detect using the existing techniques. The authors of [15] presented a statistical and ML-based system designed to identify and alert customers about energy theft. In previous studies, several data-driven techniques for the NTL identification issue have been used. The majority of these studies have concentrated on boosting approaches while ignoring bagging methods, such as Extra Trees (ET) and RF. Furthermore, ML models, such as the Support Vector Machine (SVM) and neural networks, have high FPR values and low detection rates. Neural networks were used in [16] for the prediction of coalbed methane well production.

In [17], the authors employed an Extreme Gradient Boosting (XGBoost) technique to classify the malicious users. However, because of the imbalanced dataset, this technique has a high FPR and requires more onsite inspections. The authors of [18] introduced a boosting method called the Gradient Boosting Theft Detector (GBTD), which is based on three existing boosting models: XGBoost, light gradient boosting, and categorical boosting.

The data-driven methods can be broken down into nonsupervised and supervised learning. The nonsupervised learning techniques have acquired significant attention for their use in identifying energy theft nowadays. However, on big datasets, these techniques lack generalization and can also lead to high FPR values due to the fluctuations in load patterns. The authors of [19] exploited an unsupervised learning model called the Stack Sparse Denoising Auto-Encoder (SSDAE) detector, which extracts abstract features from large datasets. However, auto-encoders tune many hyperparameters, thereby consuming more processing time. Moreover, the SSDAE detector must be rectified regularly with incoming training samples. In [20], the authors introduced a novel solution to data augmentation and relevant feature extraction from high dimensional data using a Conditional Variational Auto-Encoder (CVAE) in conjunction with a CNN classifier.

Various experiments on energy theft identification in AMI have been carried out using ML techniques. The authors of [21] presented an unsupervised learning based anomalous pattern recognition technique to identify energy theft in data streams provided by smart meters. The technique only uses regular consumer usage data for model training. However, the classifier may recognize high energy usage patterns over weekdays and holidays. Furthermore, in [22], the authors proposed a Consumption Pattern-Based Energy Theft Detection (CPBETD) approach to leverage the predictability of consumers' benign and fraudulent class samples. However, the SVM misclassification rate limited the DR, resulting in a high FPR.

Most researchers have focused on EC nonmalicious patterns [23]. However, previous studies have shown poor detection rates and accuracy regarding NTL detection. In [23], the authors developed a hybrid K-means-DNN approach, which is a combination of the K-Nearest Neighbor (KNN) and Deep Neural Network (DNN). The approach detects electricity theft in power grids. However, its detection performance is low. The authors of [24] suggested a hybrid method that enhances the internal structure of the standard LSTM model combined with the Gaussian Mixture Model (GMM). However, the proposed method is applicable only for low dimensional space data and is not very robust. In [25], the authors proposed a hybrid technique based on the SVM and Decision Tree (DT) for detecting illegal consumers. However, no effective performance measures were used for the combined technique's evaluation.

Contribution List

The key contributions of this paper are as follows:

- A hybrid model, referred as MLP-GRU, that identifies NTLs using both metering data and auxiliary data is proposed.
- A data augmentation technique is used due to the scarcity of theft samples. This study uses six theft scenarios to create synthetic instances of EC by modifying the honest samples.

- Meanwhile, a Synthetic Minority Oversampling Technique (SMOTE) is employed to maintain a balance between synthetic and benign samples.
- An optimization algorithm, known as the Random Search Algorithm (RSA), is used to effectively tune the MLP-GRU model's hyperparameters.

The rest of the manuscript is structured as follows. A detailed discussion of the proposed model is provided in Section 2. Afterwards, performance evaluation metrics are described in Section 3. Section 4 discusses the simulation results, while the conclusion of the paper is given in Section 5.

2. Proposed System Model

The proposed work is an extended version of [26]. The model proposed for detecting electricity theft includes two stages: training and testing. These two stages are generally comprised of five major steps. Figure 1 depicts the complete methodology outline of this study.

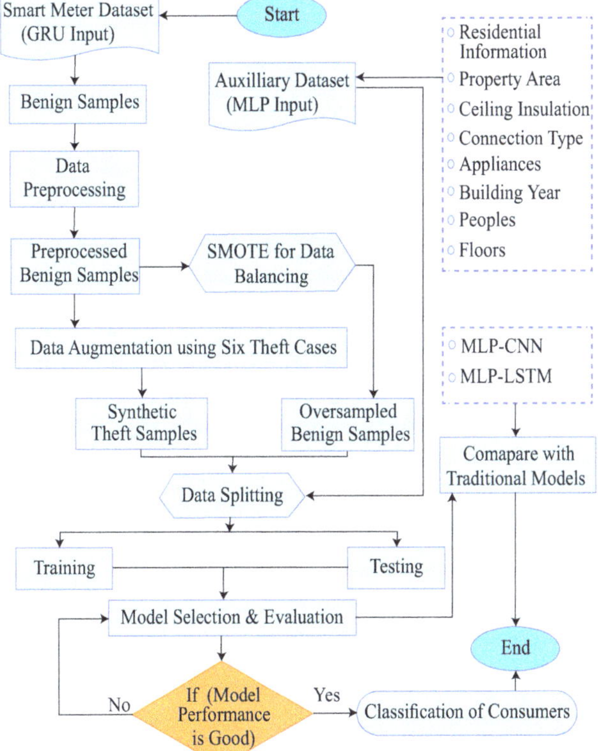

Figure 1. Methodology outline.

(1) The data preprocessing take place before the training step in the first stage. The data interpolation method is employed to fill in the dataset's missing values. Following that, a standard-scalar technique is used to normalize the data, which is a min-max procedure.

(2) Data augmentation is performed after the data have been standardized and cleaned. Different theft patterns are created by modifying the honest users' samples using six theft scenarios [18].

(3) Since the proportion of the theft class exceeds the benign class, SMOTE is applied on the benign class to balance the dataset.

(4)   Afterwards, the preprocessed data are used to train the model. The datasets from the smart meters and relative auxiliary information are sent to the GRU and MLP networks, respectively. The RSA is used to effectively tune the parameters of the classifiers.

(5)   In the last step, efficient performance metrics, such as the accuracy, F1-score, Area Under the Receiver Operating Characteristics Curve (AUC-ROC)m and Area Under the Precision–Recall Curve (PR-AUC) are used for evaluating the proposed model's performance.

During the second stage, we validated the model's performance by evaluating the unseen data to identify whether the new data belonged to the benign class or the theft class. These steps are shown in Figure 2 and are discussed in the following subsections.

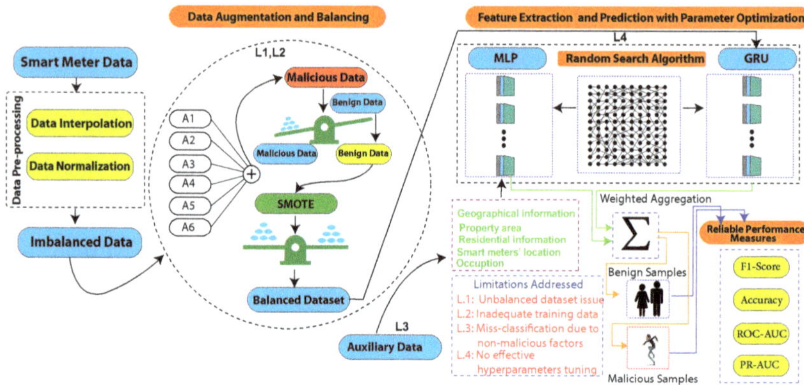

Figure 2. MLP-GRU model architecture.

2.1. Data Preprocessing

The EC data typically contain missing or incorrect numbers due to erroneous data transmission, short circuits in transmission equipment, smart meter failure, and storage problems. The classifier wrongly classifies fraudulent consumers due to missing data in the dataset. We applied an interpolation approach, the simple imputer, to fill in the missing values in the dataset [10]. It was used to impute missing data using the mean, median and so on.

Furthermore, data interpretation becomes complex when the data are spread across a vast scale as the execution time grows. Thus, we normalized the data through a standard-scalar technique, which was used to scale inconsistent data within 0 and 1 to improve the prediction models.

2.2. Data Balancing and Data Augmentation

In the real world, there a fewer nonhonest users' consumption samples as compared with the amount of benign users' samples. ML or deep learning models are biased towards majority class samples during training when the dataset is imbalanced. Moreover, they fail to recognize minority class instances that lead to performance degradation.

To address this issue, a variety of resampling techniques have been proposed in the literature [3,17,20]. Undersampling techniques result in the loss of critical data. In contrast, oversampling approaches replicate samples that are likely to be overfitted. The authors of [27] used the One-Dimensional-Wasserstein Generative Adversarial Network (WGAN), which takes a significant amount of time to generate synthetic patterns. Given the significant disparity between massive datasets of energy used and the shortcomings of previous methods, we created synthetic theft instances by altering benign samples in our proposed study. As shown in Figure 3, the Pakistan Residential Electricity Consumption (PRECON) dataset only contains normal users' samples. Electricity theft samples are also needed for training the deep learning classifiers to detect electricity theft. Thus, we performed data augmentations through synthetic theft attacks to get nonhonest users' patterns. The samples from fraudulent users were created by modifying the samples of normal users using the six theft attacks. The distribution of augmented data samples is depicted in Figure 4. The six existing theft cases were used to produce distinct malicious patterns using normal ones to train the deep learning classifiers with various theft patterns [10]. The generation of distinct theft patterns to provide diversity in the dataset is an essential feature.

Subsequently, SMOTE was employed to balance the minority class (benign) and majority class (theft) samples. Figure 5 shows the distribution of balanced data using SMOTE. When we generate malicious samples in the dataset, the proportion in the theft class exceeds the benign class. Therefore, we applied SMOTE to the benign class to balance the dataset. In this case, training ML or deep learning models on imbalanced datasets biases the model towards the majority class and adversely affects the model's performance. Thus, oversampling was performed on the data points of the benign class using SMOTE to balance the generated theft instances for each day.

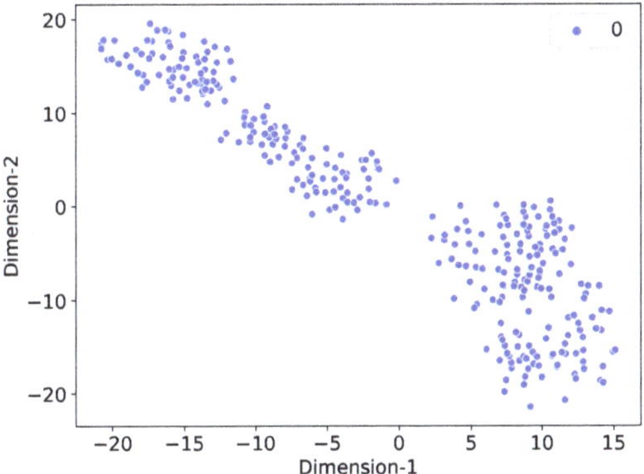

Figure 3. Imbalanced data distribution (Benign class).

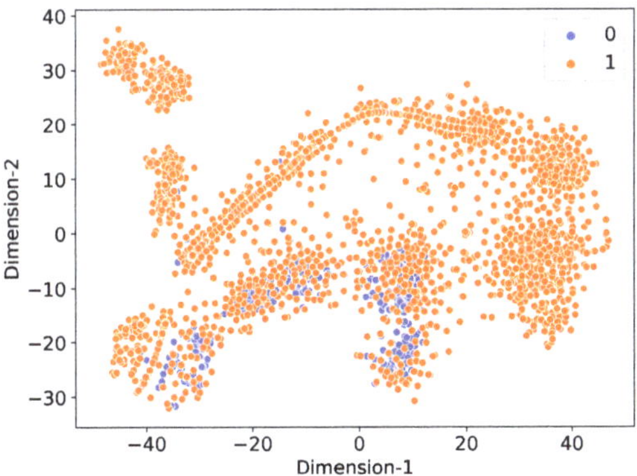

Figure 4. Augmented data using attacks.

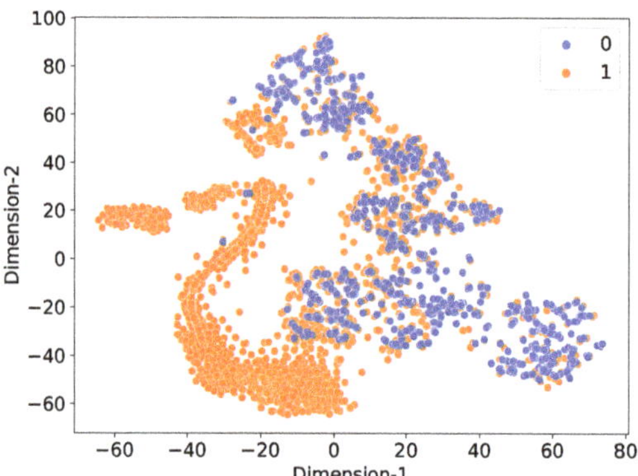

Figure 5. Balanced data distribution.

2.2.1. Six Theft Cases

Existing theft scenarios were used for generating theft data from different attacks by modifying the smart meters' data [18]. In the proposed model, we represent the real daily energy consumption of a home H as H_t, where $H_t = [H_1, H_2, H_3, \ldots, H_{48}]$ and $T = 48$ (total actual energy usage per day). We used these theft scenarios to modify the actual energy usage behavior, where t belongs to [1, 48]

(A1). $H_t = H_t * a$, where $a = $ rand $(0.1, 0.9)$,
(A2). $H_t = H_t * b_t$, where $b_t = $ rand $(0.1, 1.0)$,
(A3). $H_t = H_t * c_t$, where $c_t = $ rand $[0, 1]$,
(A4). $H_t = $ mean $(H) * d_t$, where $d_t = $ rand $(0.1, 1.0)$,
(A5). $H_t = $ mean (H),
(A6). $H_t = H_{T-t}$.

The first theft attack produces fraudulent patterns by multiplying the consumption of honest users with values randomly produced within the range of 0.1 to 0.9. In theft case 2, each consumer's meter reading is multiplied by a distinct random integer, ranging from (and including) 0.1 to 1.0. The generated values show a discontinuity in tracing the theft data and the manipulated values.

Theft case 3 is an on-off attack in which a consumer either submits the actual readings or a zero value is submitted as its EC. This means that the normal users' samples are multiplied by 1 during a random period t; otherwise, they are multiplied by zero. Furthermore, for theft attack 4, the average energy consumed for all users is multiplied with a randomly generated value in the range of 0.1 to 1.0 exclusively. As a result, the malicious users under-report the actual energy they consumed. For theft attack 5, the average energy consumed by all users is reported and is the same throughout the day. Theft case 6 changes the sequence of the real EC, for example, by shifting the order of consumption data from peak to off-peak hours [14].

The daily energy usage patterns and six distinct forms of theft cases are shown in Figures 6 and 7.

2.2.2. Hybrid MLP-GRU Network

The hybrid neural network, MLP-GRU, introduced in this work aims to integrate the metering data and auxiliary information. Table 1 shows the auxiliary dataset features with their descriptions. Our proposed method was influenced by the work undertaken in [4] to identify electricity theft, where the authors proposed a hybrid deep neural network, MLP-LSTM. In the proposed model, the GRU network receives the preprocessed EC data from the smart meters. It generalizes the embedding for a shorter processing time by employing few cells. Meanwhile, the auxiliary dataset is provided as an input for the MLP using 20 neurons. This design is highly efficient, since it allows simultaneous training on both forms of input data. Afterwards, the batch normalization layer is used to normalize the data until it is submitted to the final layer. In the model, the sigmoid activation function in the last layer only has one neuron. The subsections below provide a thorough description of each network.

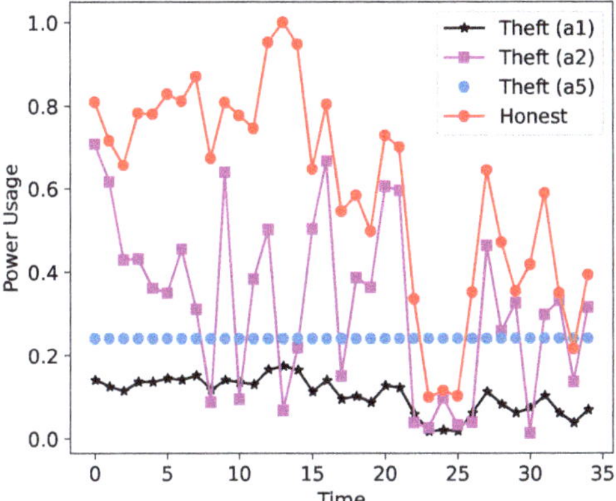

Figure 6. Attack patterns 1, 2, and 5.

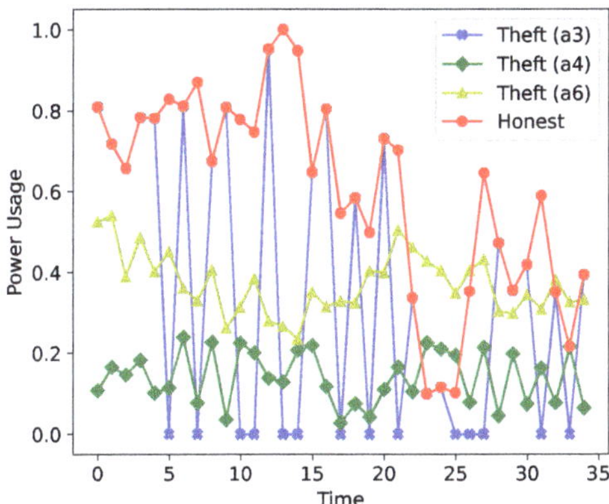

Figure 7. Attack patterns 3, 4, and 6.

Table 1. Representation of the shortcomings and the proposed solutions.

Shortcomings	Proposed Solutions	Evaluation
L1 and L2: imbalanced dataset issue and inadequate training data	S1: Employ six theft attacks on normal samples, then apply SMOTE to balance the dataset	V1: Comparison with oversampling techniques
L3: Misclassification as a result of non-malicious circumstances	S2: Integrate auxiliary data	V2: Performance comparison with traditional models
L4: Inappropriate tuning of model's hyperparameters	S3: RSA	V3: Compare the RSA with the existing GRA approach

2.2.3. Gated Recurrent Unit Network for Smart Meter Data

The GRU is a variant of the LSTM that overcomes the computational complexity of the LSTM by considering few gates, as it eliminates the output gate. The GRU includes an update gate (long-term memory) and a reset gate (short-term memory), as shown in Figure 8.

$$r_t = \sigma(X_t * V_r + H_{t-1} * W_r + B_r), \tag{1}$$

$$u_t = \sigma(X_t * V_u + H_{t-1} * W_u + B_u). \tag{2}$$

According to Equations (1) and (2) [15,28], r_t and u_t denote the number of times the reset gate and update gate have been enabled, respectively. V_r and V_u denote the weights of the input layer, while W_r and W_u indicate the recurrent weights of the GRU. The biases of the deep network are denoted by the variables B_r and B_u. X_t is the current input state and H_{t-1} is the previous layer input. All values of the reset and update gates are multiplied by the sigmoid activation function, denoted as σ [29].

$$D_n = c * d_1, c * d_2, c * d_3, \ldots c * d_i, (i = 1, 2, 3, \ldots, 365). \tag{3}$$

Equation (3) indicates the daily energy consumption data over the year. $c * d_i$ presents the 365 days of consumption records. The GRU network examines the whole EC history of smart meters on a daily basis and generates the final result. The final predicted outcome of the GRU network and the output of the MLP network are activated using a single activation function to generate a combined prediction.

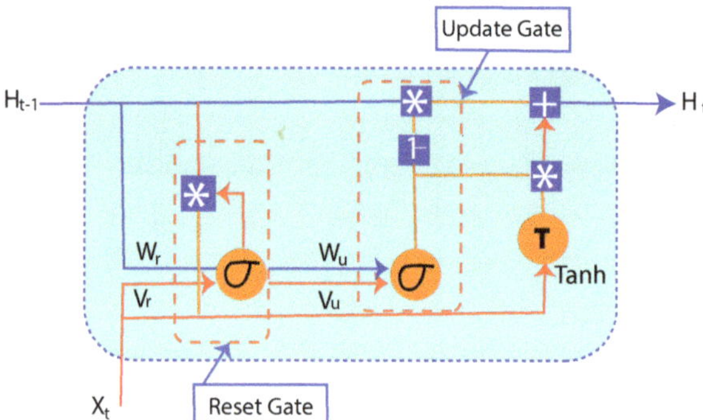

Figure 8. GRU model architecture.

2.2.4. Multi-Layered Perceptron Network with Auxiliary Data

The auxiliary dataset is analyzed using the MLP network. The MLP contains more than one hidden layer of neurons. The validation dataset is used to choose these hidden layers in the MLP network.

$$H_n = \sigma\left(\sum U_{i,n} * X_i + B_n\right), i = 1, 2, \ldots, N, \tag{4}$$

$$Y_n = \sigma(U_n * H_{n-1} + B_n). \tag{5}$$

According to Equations (4) and (5) [4], U_n denotes the weights of layer n, H_{n-1} indicates the previous hidden states of the input layer, and B_n represents the bias. After processing the input values, the activation function that activates the neuron is called and determines whether to send the values to the next layer or not. σ represents the sigmoid activation function. Equation (5) shows the output layer, which is denoted as Y_n. In this study, we used the Rectified Linear Unit (ReLU) activation function in the hidden layer, while for the final output layer, a sigmoid activation function was used for the binary classification [15]. To accelerate the network convergence, a batch normalization layer was added to standardize the input values. Afterwards, a dropout layer was added as a regularization technique to reduce overfitting.

2.2.5. Random-Search-Based Parameters' Optimization Algorithm

A critical challenge in the development of deep learning models is the appropriate setting of hyperparameters to achieve optimal results. Inappropriate selection of hyperparameters adversely affects the training activity as well as the time complexity of deep learning models. The fundamental goal of deep learning classifiers is to improve the accuracy of the classification results. Therefore, the selection of a suitable learning rate, number of neurons, number of hidden layers, batch size, activation function, epochs, and other hyperparameters of deep learning classifiers has significant impacts on the model's performance. In order to achieve optimal results, the important hyperparameters of the classifiers need to be optimized (tuned). In recent research, the GSA has been considered in many machine and deep learning algorithms for the tuning of hyperparameters [4,30]. For instance, we used hyperparameters hp1, hp2 and hp3 of an ML model M. The GSA specifies the range of values for each hyperparameter. Afterwards, it creates many different M versions using different combinations of hyperparameter values. This range of hyperparameter values is referred to as the grid.

Moreover, the manual selection of hyperparameters and GSA makes it somewhat easier for the user to define these essential parameters. However, both techniques take a long time to converge. On the other hand, GSA remains a computationally intensive method, particularly as the number of hyperparameters grows and the interval between discrete values shrinks [30].

In the case of high dimensionality, when several hyperparameters drastically grow, the GSA method suffers a lot and is computationally overburdened. It takes the maximum time during tuning, even in cases with a small number of hyperparameters. Since there is no guarantee of finding the best solution, in this study, a RSA method was employed to improve the classification accuracy of the models. The RSA is a stochastic optimization algorithm that is invaluable for finding the optimal solution globally with fast-running simulations. It performs searching using random combinations of hyperparameter values to train a model. Additionally, it is more effective in high-dimensional space.

The RSA consists of five major steps:

- The initial value is stored in a variable, denoted by x.
- If the values stored in x are target node values, the algorithm immediately stops with the success. Otherwise, it moves to the next step.
- The values of x are updated to get the optimal possible combination of x. We obtain the number of child nodes (values of x) and store them in another variable C.
- A value from all possible combinations of child node values is randomly selected.
- The values of x are replaced with the new values, and then the process returns to step 2 for validation, where the existing values are compared with the target values. The process continues until the final optimal solution is reached. Figure 9 shows the process of tuning hyperparameters with the RSA.

To optimize the hyperparameters, we used the following steps.

Step 1: The hyperparameters are initialized with their possible range. To train our hybrid MLP-GRU model, the hyperparameters, such as the activation function, epochs, the number of hidden layers, batch size, etc., are defined. The RSA samples a set of values for each of these hyperparameters and makes a grid of all available values from their respective distributions and uses it for training.

Step 2: During the evolution period, only one solution is retained. A random vector is added to the solution after each epoch. The process is repeated numerous times, resulting in the training of several models.

Step 3: The new solution is checked after it has been measured. If the new solution is superior to the old one, then it is acknowledged as the correct one; otherwise, the old one remains unchanged.

Step 4: The best combination of the values of the hyperparameters is eventually preserved.

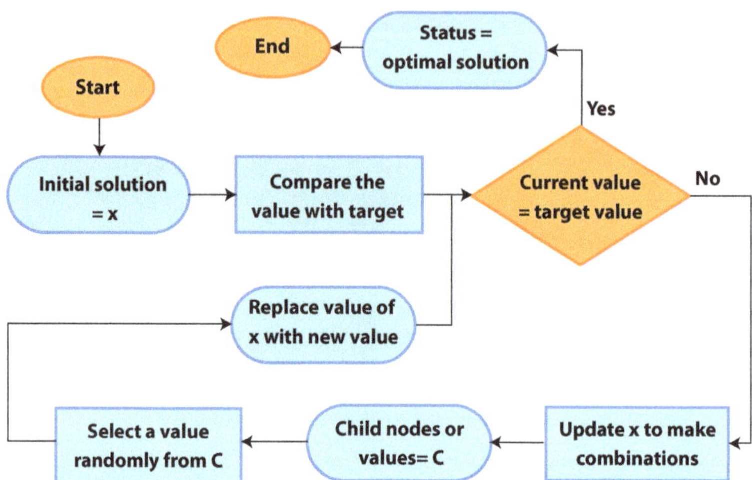

Figure 9. Flowchart of the Random Search Algorithm.

Table 2 displays the set of values that are searched and the best values that are discovered or revealed by the RSA during the tuning of the proposed MLP-GRU model. The best values are explored by tracking the results of the validation dataset. The RSA can monitor different random combinations of hyperparameters. To train the MLP-GRU model, the hyperparameters, like epochs, number of hidden layers, etc, are defined. The RSA can sample a set of values for epochs and hidden layers from their respective distributions that are used for training. The process is repeated several times until the desired results have been obtained. Table 3 shows the hyperparameters and their optimal values that were found using the GSA during the tuning of the existing MLP-LSTM model. However, we considered fewer hyperparameters due to their high computational time.

Table 2. Auxiliary Dataset Information.

Data Type	Description (MLP Input Data)	Size of Data
Residents' Information	Temporary residents and permanent residents	2
People	Total number of people including adults, children	3
Appliances	Number of appliances in a home including washing machine, fridge, iron, electronic devices, fans, AC, water-pump, UPS, water-dispenser, refrigerator and lightening devices	11
Connection Type	Single-phase and multi-phase	2
Rooms' Information	Number of rooms including bed room, living room, kitchen, washroom, dining room	6
Roof or Ceiling	The total height of ceiling, ceiling insulation used, ceiling insulation not used	2
Building Year	The year of building construction	1
Property Area	The area or location of house	1
Floors	The total number of floors in a building	1

Table 3. The Proposed MLP-GRU-Random Search Method.

Hyperparameter	Optimal Value	Values Range
Units	100	100, 10, 15, 50, 20, 35, 400, 25
Optimizer	Adam	Adam, Adamax and SGD
Dropout	0.01	0.3, 0.2, 0.5, 0.01, 0.1
Batch-size	32	10, 32, 25, 15
Activation function	relu	relu, elu, sigmoid, softmax, tanh and linear
Epochs	10	15, 25, 10, 20

3. Performance Measurement Indicators

In this section, we conduct a thorough examination of the proposed hybrid model's performance in comparison with the existing hybrid MLP-LSTM classifier. The accuracy, F1-score, PR-AUC, and ROC-AUC are effective performance indicators that are used to evaluate the performance of the techniques. These indicators are determined using the core confusion metrics, which are composed of four crucial error rates: False Positive (FP), False Negative (FN), True Positive (TP)m and False Negative (FN) [31]. These metrics indicate the total number of consumers wrongly classified as thieves, accurately labeled as fair consumers, erroneously identified as honest consumers, and correctly labeled as thieves [32]. Accuracy is a widely used performance measure that represents the percentage of correct model predictions. It offers the measures of predictability for TPs and TNs in the classifier. It quantifies how well the model predicts TPs and TNs. However, it frequently fails for imbalanced datasets.

$$\text{Accuracy} = (TP + TN)/(TP + TN + FP + FN). \tag{6}$$

Mathematically, Equation (6) exhibits accuracy [31]. Other metrics were employed due to the lack of a specific measure for FP and FN predictions. The harmonic mean of recall and precision is called the F1-score, which is calculated by Equation (7).

$$\text{F1-score} = 2 \times (\text{Recall} \times \text{Precision})/(\text{Recall} + \text{Precision}). \tag{7}$$

One of the main objectives of ETD is to enhance the TPR or Detection Rate (DR) while simultaneously reducing the FPR. Thus, the ROC-AUC is a useful metric for identifying NTLs in binary classification problems [33]. It demonstrates the relationship of TPR with FPR at different threshold values. A score is a number between 0 and 1 that represents how different the two classes are. An AUC score of 1 indicates a perfect detection method. In the case of an imbalanced dataset problem, it is more reliable in terms of evaluating the model's performance. FPR and TPR are beneficial for assessing the performance of a model. However, the precision of the model is not considered using these measures. Thus, the PR-AUC is a valuable performance measure that is used for evaluating the model's performance. It is more appropriate for imbalanced datasets as compared with balanced datasets.

4. Simulations and Findings

The simulation findings of the proposed model are discussed in this section. The PRECON dataset was used to test the proposed model. Python was used to carry out the simulations. The proposed model was implemented using an Intel Core i3 with 4 GB of RAM. Additionally, a Google Colaboratory application was used in conjunction with Python language packages such as NumPy, pandas, TensorFlow, Keras, etc. to simulate the data.

4.1. Data Acquisition

The proposed model was trained and tested on the actual smart meter dataset of PRECON [34], which is publicly available. It contains information about energy demand, including half-hourly electricity usage records from 42 residential properties. The PRECON smart meter dataset includes half-hourly energy consumption data with 48 features that belong to the normal users' consumption class, while the auxiliary dataset contains 28 features. Based on the dataset of honest users, we identified six types of theft attacks to generate malicious users' consumption patterns. We divided the consumption behavior of users on a half-hourly basis, while attacks were assessed on a yearly consumption basis. This helped us to analyze the daily consumption behavior of a user and identify nonmalicious users' patterns, as was done in [35,36] (to identify the periodicity in consumption). After applying attacks, we oversampled the minority class (benign) samples using SMOTE to balance the malicious attack class. The auxiliary information was provided to identify nonmalicious factors, which can cause a high misclassification rate. This included information on high energy consumption equipment and load profiles for the entire home. The utility provided the labeled dataset by inspecting it at least once. As a result, it is reasonable to assume that all samples belonged to trustworthy consumers. In addition, the dataset was split into a ratio of 80% training and 20% testing samples in a stratified manner, where 3494 instances were used in the training set and 874 were used in the testing set. Moreover, being motivated by [4], the proposed model was compared with the deep learning models and not the traditional machine learning models.

4.2. Evaluation Results

Figure 10 demonstrates the ROC curve of the proposed model after data balancing. The comprehensive scores of measurement are shown in Table 4. The proposed MLP-GRU model bet the single GRU classifier on the test data with an AUC score of 0.93. This indicates that the incorporation of auxiliary information such as permanent occupants, property area, and contracted power improve the performance by lowering the FPR. We also noticed that the performance of the hybrid MLP-LSTM was quite similar to our model, obtaining an AUC score of 0.89 due to the use of auxiliary information, except that the F1-score was relatively low with 0.89 compared with our model which obtained a score of 0.92. In contrast to our proposed model, the existing MLP-CNN classifier obtained an AUC score of 0.84 due to the lack of generalization in the CNN model.

Table 4. Hybrid MLP-LSTM-Grid Search.

Hyperparameter	Optimal Value	Values Range
Dropout	0.2	0.2, 0.5
Units	10	100, 10, 50
Optimizer	Adam	Adam and SGD
Activation function	sigmoid	relu and sigmoid

A comparative analysis of the proposed and existing models is depicted in Figure 10. On the x-axis and y-axis, the TPR and FPR are shown, respectively. The TPR indicates the proportion of correctly classified positive samples among all available data, whereas the FPR denotes the proportion of negative samples incorrectly classified as positive. The proposed model accurately categorized samples with high DR and low FPR values at the initial level. With a rising FPR, a small change was noticed after attaining a high TPR of 0.8. Hence, our proposed model's FPR was significantly lower than that of the existing model.

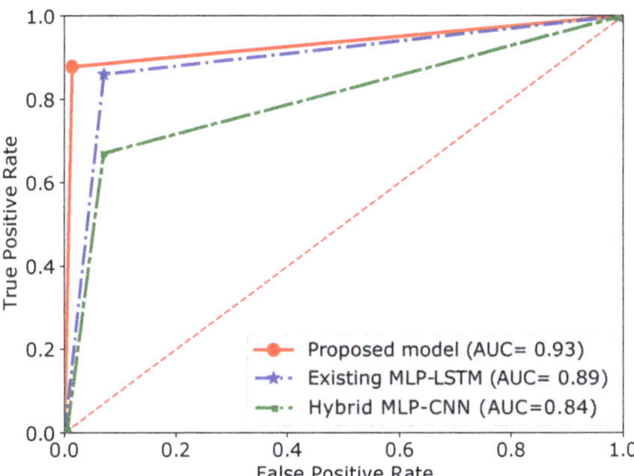

Figure 10. ROC-based performance comparison.

Subsequently, we were able to see an exponential periodic regain in the ROC curve of the proposed hybrid model. The models' stability and precision were enhanced by increasing the TPR and thereby reducing the FPR. A decrease in FPR minimizes the need for onsite inspections, which is a costly process as it involves the reliance on experts. Similarly, the PR-AUC curve is depicted in Figure 11. It indicates that, on the test datasets, our proposed model achieved a PR-AUC score of 0.91, which is substantially higher than those of existing models. Table 5 shows the accuracy, F1-score, and AUC values. The results indicate that the proposed MLP-GRU model surpasses the other state-of-the-art models. The computational complexity of the GRU classifier is minimal, since fewer gates are employed in GRU as compared with the LSTM classifier. In this regard, the GRU model requires a limited amount of hyperparameters for tuning, leading to a fast convergence rate. Moreover, we applied the RSA instead of the GSA, which is a computationally demanding process. The use of a small dataset also makes it better. In addition, the loss of the proposed model is shown in Figure 12. We ran 25 iterations. The loss declined with each move, settling at a 0-point minimum during training and testing. Training and testing data losses were the same. The proposed model works well regarding training and testing data, as seen in Figure 13.

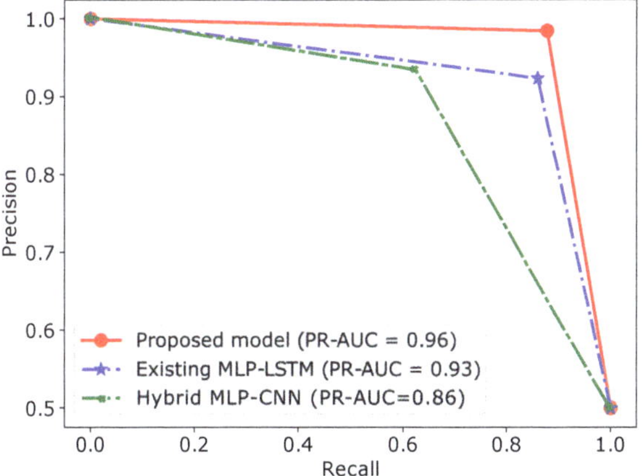

Figure 11. PR-AUC-based performance comparison.

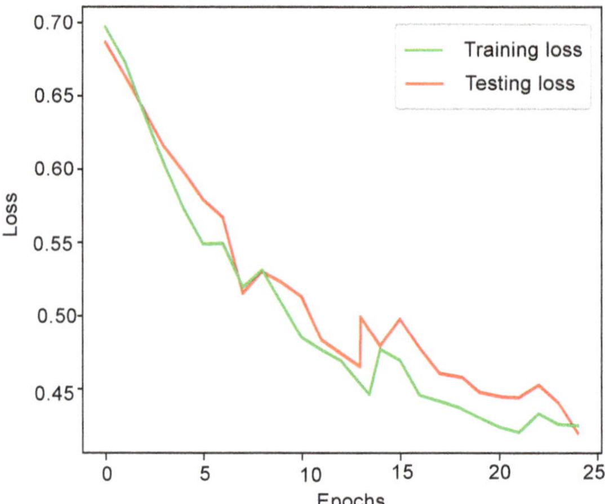

Figure 12. Accuracy-based analysis of the proposed solution.

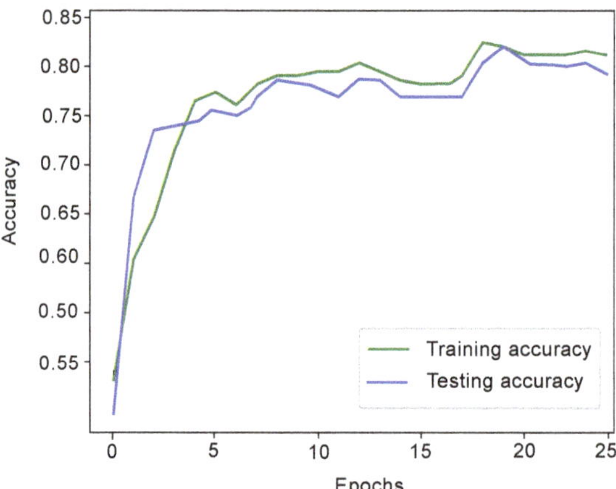

Figure 13. Loss-based analysis of the proposed solution.

It is critical to consider the execution time in a hyperparameter optimization process with real-world circumstances. Many studies have stated that finding acceptable hyperparameter values for a model can take a significant amount of time. As a result, many researchers do not consider parameter tuning due to the waste of time.

Figure 14 depicts the average execution times of the RSA and GSA, while Figure 15 shows the accuracy levels of the proposed and existing models. The results demonstrate that our proposed RSA approach takes less time than the existing MLP-LSTM classifier using the GSA method. The RSA creates a grid using a range of different hyperparameter values and picks random combinations from it to train the model. In contrast to the RSA, the GSA method makes a grid of hyperparameter values for each combination, which is computationally very expensive in terms of processing power and time. Besides, Table 5 compares the proposed model's performance with that of the existing models.

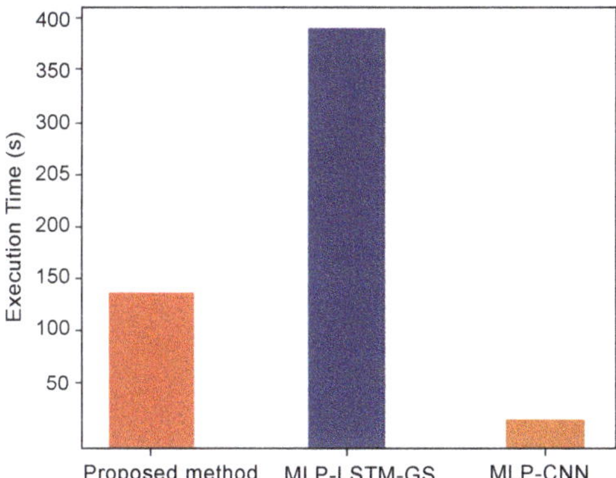

Figure 14. Execution time.

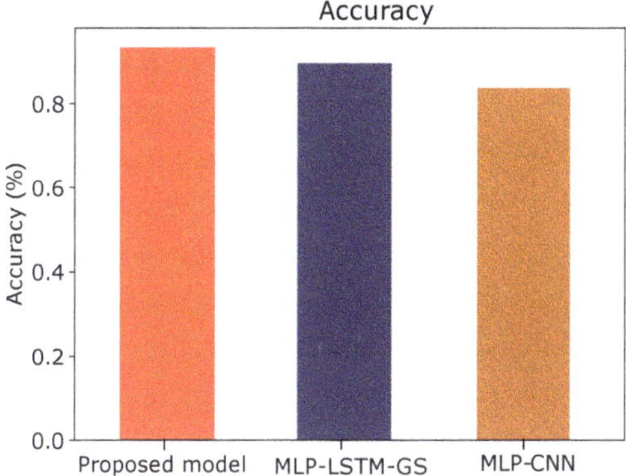

Figure 15. Accuracy analysis.

Table 5. Comparison of the proposed model's performance with that of existing models.

Models	Accuracy	AUC	F1-Score	Time Required (s)
Proposed model	0.93	0.93	0.92	144
MLP-LSTM-GS	0.89	0.89	0.89	391
MLP-CNN	0.67	0.84	0.71	26

5. Conclusions

In the proposed study, a hybrid deep learning model, MLP-GRU, was developed using metering data and auxiliary information. The MLP network receives auxiliary data, whereas the GRU network takes nonsequential or metering data for detecting electricity theft. Additionally, the EC datasets contain a small number of malicious samples that skew the model in favor of the majority class. The issue of the biased dataset is tackled through data augmentation in which synthetic theft instances are created via six different theft attacks on benign samples. Afterwards, the imbalanced dataset problem is resolved using SMOTE. The effectiveness of our proposed hybrid model was assessed against changes in EC usage patterns and various attack types. The PRECON dataset was used to run the simulations. According to the findings, the proposed model outperforms the existing hybrid MLP-LSTM and other conventional models. The results demonstrate that the proposed model performs much better with a ROC-AUC value of 0.93 and a PR-AUC value of 0.96 when auxiliary data are included with the metering data. In the future, there is a need to exploit more sophisticated optimization algorithms for adjusting hyperparameters of deep learning classifiers to find the optimal ETD results. We will also use other residential areas of the PRECON dataset to conduct a detailed analysis of consumers' consumption behaviors.

Author Contributions: Writing—original draft preparation, B.K.; writing—review and editing, N.J. and E.A.M.; visualization, U.Q.; supervision, N.J.; project administration, A.A.; funding acquisition, project management, evaluating the prior works/literature review, and conducting a critical study, A.A. and N.A. All authors have read and agreed to the published version of the manuscript.

Funding: The authors thank King Saud University, Riyadh, Saudi Arabia for providing the funding for Project number (RSP-2021/295).

Data Availability Statement: The data is available at http://web.lums.edu.pk/~eig/CXyzsMgyXGpW1 sBo, accessed on 15 September 2022.

Acknowledgments: The authors would like to acknowledge the support of Researchers Supporting Project number (RSP-2021/295), King Saud University, Riyadh, Saudi Arabia.

Conflicts of Interest: The authors declare no conflict of interest.

Abbreviations

The following abbreviations are used in this manuscript:

Abbreviation	Full Form
AMI	Advanced Metering Infrastructure
CNN	Convolutional Neural Network
CPBETD	Consumption Pattern Based Electricity Theft Detector
CVAE	Conditional Variational Auto Encoder
DNN	Deep Neural Network
DR	Detection Rate
EC	Electricity Consumption
ETD	Electricity Theft Detection
ETs	Extra Trees
FPR	False Positive Rate
GBTD	Gradient Boosting Theft Detector
GRU	Gated Recurrent Unit
GMM	Gaussian Mixture Model
GSA	Grid Search Algorithm
KNNs	K-Nearest Neighbors
LR	Logistic Regression
LSTM	Long-Short Term Memory
ML	Machine Learning
MLP	Multi-Layer Perceptron
NTL	Nontechnical Loss
PRECON	Pakistan Residential Electricity Consumption
RF	Random Forest
RSA	Random Search Algorithm
SVM	Support Vector Machine
SMOTE	Synthetic Minority Oversampling Technique
SSDAE	Stacked Sparse Denoising Auto-Encoder
SETS	Smart Energy Theft System
TL	Technical Loss
TPR	True Positive Rate
WGAN	Wasserstein Generative Adversarial Network
XGBoost	Extreme Gradient Boosting
H_{t-1}	Previous Layer Input
r	Reset Gate
t	Time Period
u	Update Gate
X_t	Current Input State

References

1. Praju, T.K.S.; Samal, S.; Saravanakumar, R.; Yaseen, S.M.; Nandal, R.; Dhabliya, D. Advanced metering infrastructure for low voltage distribution system in smart grid based monitoring applications. *Sustain. Comput. Inform. Syst.* **2022**, *35*, 100691.
2. Otuoze, A.O.; Mustafa, M.W.; Abioye, A.E.; Sultana, U.; Usman, A.M.; Ibrahim, O.; Abu-Saeed, A. A rule-based model for electricity theft prevention in advanced metering infrastructure. *J. Electr. Syst. Inf. Technol.* **2022**, *9*, 2. [CrossRef]
3. Fei, K.; Li, Q.; Zhu, C. Non-technical losses detection using missing values' pattern and neural architecture search. *Int. J. Electr. Power Energy Syst.* **2022**, *134*, 107410. [CrossRef]
4. Buzau, M.M.; Tejedor-Aguilera, J.; Cruz-Romero, P.; Gómez-Expósito, A. Hybrid deep neural networks for detection of non-technical losses in electricity smart meters. *IEEE Trans. Power Syst.* **2019**, *35*, 1254–1263. [CrossRef]
5. Kocaman, B.; Tümen, V. Detection of electricity theft using data processing and LSTM method in distribution systems. *Sādhanā* **2020**, *45*, 286. [CrossRef]
6. Aslam, Z.; Javaid, N.; Ahmad, A.; Ahmed, A.; Gulfam, S.M. A combined deep learning and ensemble learning methodology to avoid electricity theft in smart grids. *Energies* **2020**, *13*, 5599.
7. Ghori, K.M.; Abbasi, R.A.; Awais, M.; Imran, M.; Ullah, A.; Szathmary, L. Performance analysis of different types of machine learning classifiers for non-technical loss detection. *IEEE Access* **2019**, *8*, 16033–16048. [CrossRef]
8. Shehzad, F.; Javaid, N.; Aslam, S.; Javaid, M.U. Electricity theft detection using big data and genetic algorithm in electric power systems. *Electr. Power Syst. Res.* **2022**, *209*, 107975. [CrossRef]
9. Pamir Javaid, N.; Javaid, S.; Asif, M.; Javed, M.U.; Yahaya, A.S.; Aslam, S. Synthetic Theft Attacks and Long Short Term Memory-Based Preprocessing for Electricity Theft Detection Using Gated Recurrent Unit. *Energies* **2022**, *15*, 2778. [CrossRef]
10. Li, S.; Han, Y.; Yao, X.; Yingchen, S.; Wang, J.; Zhao, Q. Electricity theft detection in power grids with deep learning and random forests. *J. Electr. Comput. Eng.* **2019**, *2019*, 4136874. [CrossRef]
11. Saeed, M.S.; Mustafa, M.W.; Sheikh, U.U.; Jumani, T.A.; Mirjat, N.H. Ensemble bagged tree based classification for reducing non-technical losses in multan electric power company of Pakistan. *Electronics* **2019**, *8*, 860. [CrossRef]
12. Gunturi, S.K.; Sarkar, D. Ensemble machine learning models for the detection of energy theft. *Electr. Power Syst. Res.* **2020**, *192*, 106904. [CrossRef]
13. Fenza, G.; Gallo, M.; Loia, V. Drift-aware methodology for anomaly detection in smart grid. *IEEE Access* **2019**, *7*, 9645–9657. [CrossRef]
14. Qu, Z.; Li, H.; Wang, Y.; Zhang, J.; Abu-Siada, A.; Yao, Y. Detection of electricity theft behavior based on improved synthetic minority oversampling technique and random forest classifier. *Energies* **2020**, *13*, 2039. [CrossRef]
15. Li, W.; Logenthiran, T.; Phan, V.T.; Woo, W.L. A novel smart energy theft system (SETS) for IoT-based smart home. *IEEE Internet Things J.* **2019**, *6*, 5531–5539. [CrossRef]
16. Yang, R.; Qin, X.; Liu, W.; Huang, Z.; Shi, Y.; Pang, Z.; Zhang, Y.; Li, J.; Wang, T. A Physics-Constrained Data-Driven Workflow for Predicting Coalbed Methane Well Production Using Artificial Neural Network. *SPE J.* **2022**, *27*, 1531–1552. [CrossRef]
17. Yan, Z.; Wen, H. Electricity theft detection base on extreme gradient boosting in AMI. *IEEE Trans. Instrum. Meas.* **2021**, *70*, 2504909. [CrossRef]
18. Punmiya, R.; Choe, S. Energy theft detection using gradient boosting theft detector with feature engineering-based preprocessing. *IEEE Trans. Smart Grid* **2019**, *10*, 2326–2329. [CrossRef]
19. Huang, Y.; Xu, Q. Electricity theft detection based on stacked sparse denoising autoencoder. *Int. J. Electr. Power Energy Syst.* **2021**, *125*, 106448. [CrossRef]
20. Gong, X.; Tang, B.; Zhu, R.; Liao, W.; Song, L. Data Augmentation for Electricity Theft Detection Using Conditional Variational Auto-Encoder. *Energies* **2020**, *13*, 4291. [CrossRef]
21. Park, C.H.; Kim, T. Energy Theft Detection in Advanced Metering Infrastructure Based on Anomaly Pattern Detection. *Energies* **2020**, *13*, 3832. [CrossRef]
22. Jokar, P.; Arianpoo, N.; Leung, V.C. Electricity theft detection in AMI using customers' consumption patterns. *IEEE Trans. Smart Grid* **2015**, *7*, 216–226. [CrossRef]
23. Maamar, A.; Benahmed, K. A hybrid model for anomalies detection in AMI system combining K-means clustering and deep neural network. *Comput. Mater. Continua* **2019**, *60*, 15–39. [CrossRef]
24. Ding, N.; Ma, H.; Gao, H.; Ma, Y.; Tan, G. Real-time anomaly detection based on long short-Term memory and Gaussian Mixture Model. *Comput. Electr. Eng.* **2019**, *79*, 106458. [CrossRef]
25. Jindal, A.; Dua, A.; Kaur, K.; Singh, M.; Kumar, N.; Mishra, S. Decision tree and SVM-based data analytics for theft detection in smart grid. *IEEE Trans. Ind. Inform.* **2016**, *12*, 1005–1016. [CrossRef]
26. Kabir, B.; Ullah, A.; Munawar, S.; Asif, M.; Javaid, N. Detection of non-technical losses using MLPGRU based neural network to secure smart grids. In Proceedings of the 15th International Conference on Complex, Intelligent and Software Intensive System (CISIS), Asan, Republic of Korea, 1–3 July 2021; ISBN 978-3-030-50454-0.
27. Kong, X.; Zhao, X.; Liu, C.; Li, Q.; Dong, D.; Li, Y. Electricity theft detection in low-voltage stations based on similarity measure and DT-KSVM. *Int. J. Electr. Power Energy Syst.* **2021**, *125*, 106544.
28. Gul, H.; Javaid, N.; Ullah, I.; Qamar, A.M.; Afzal, M.K.; Joshi, G.P. Detection of non-technical losses using SOSTLink and bidirectional gated recurrent unit to secure smart meters. *Appl. Sci.* **2020**, *10*, 3151.

29.	Kuo, P. H.; Huang, C.J. An electricity price forecasting model by hybrid structured deep neural networks. *Sustainability* **2018**, *10*, 1280. [CrossRef]

30.	George, S.; Sumathi, B. Grid search tuning of hyperparameters in random forest classifier for customer feedback sentiment prediction. *Int. J. Adv. Comput. Sci. Appl.* **2020**, *11*, 173–178.

31.	Ghori, K.M.; Imran, M.; Nawaz, A.; Abbasi, R.A.; Ullah, A.; Szathmary, L. Performance analysis of machine learning classifiers for non-technical loss detection. *J. Ambient. Intell. Humaniz. Comput.* **2020**, *18*, 1–16. [CrossRef]

32.	Razavi, R.; Gharipour, A.; Fleury, M.; Akpan, I.J. A practical feature-engineering framework for electricity theft detection in smart grids. *Appl. Energy* **2019**, *238*, 481–494. [CrossRef]

33.	Aslam, Z.; Ahmed, F.; Almogren, A.; Shafiq, M.; Zuair, M.; Javaid, N. An attention guided semi-supervised learning mechanism to detect electricity frauds in the distribution systems. *IEEE Access* **2020**, *8*, 221767–221782. [CrossRef]

34.	Nadeem, A.; Arshad, N. PRECON: Pakistan residential electricity consumption dataset. In Proceedings of the Tenth ACM International Conference on Future Energy Systems, Phoenix, AZ, USA, 25–28 June 2019; pp. 52–57. Available online: http://web.lums.edu.pk/~eig/CXyzsMgyXGpW1sBo (accessed on 15 September 2022).

35.	Zheng, Z.; Yang, Y.; Niu, X.; Dai, H.N.; Zhou, Y. Wide and deep convolutional neural networks for electricity-theft detection to secure smart grids. *IEEE Trans. Ind. Inform.* **2017**, *14*, 1606–1615. [CrossRef]

36.	Arif, A.; Javaid, N.; Aldegheishem, A.; Alrajeh, N. Big data analytics for identifying electricity theft using machine learning approaches in micro grids for smart communities. *Concurr. Comput. Pract. Exp.* **2021**, *33*, e6316. [CrossRef]

sustainability

Article

A Hybrid Deep Learning-Based Model for Detection of Electricity Losses Using Big Data in Power Systems

Adnan Khattak [1], Rasool Bukhsh [2,*], Sheraz Aslam [3,*], Ayman Yafoz [4], Omar Alghushairy [5] and Raed Alsini [4]

1 Department of Computer Science, Abasyn University, Islamabad 44000, Pakistan
2 Department of Computer Science, COMSATS University Islamabad, Islamabad 44000, Pakistan
3 Department of Electrical Engineering, Computer Engineering, and Informatics, Cyprus University of Technology, Limassol 3036, Cyprus
4 Department of Information Systems, Faculty of Computing and Information Technology, King Abdulaziz University, Jeddah 21589, Saudi Arabia
5 Department of Information Systems and Technology, College of Computer Science and Engineering, University of Jeddah, Jeddah 21589, Saudi Arabia
* Correspondence: rasoolbax.rb@gmail.com (R.B.); sheraz.aslam@cut.ac.cy (S.A.)

Citation: Khattak, A.; Bukhsh, R.; Aslam, S.; Yafoz, A.; Alghushairy, O.; Alsini, R. A Hybrid Deep Learning-Based Model for Detection of Electricity Losses Using Big Data in Power Systems. *Sustainability* 2022, 14, 13627. https://doi.org/10.3390/su142013627

Academic Editor: Andreas Kanavos

Received: 8 August 2022
Accepted: 5 October 2022
Published: 21 October 2022

Abstract: Electricity theft harms smart grids and results in huge revenue losses for electric companies. Deep learning (DL), machine learning (ML), and statistical methods have been used in recent research studies to detect anomalies and illegal patterns in electricity consumption (EC) data collected by smart meters. In this paper, we propose a hybrid DL model for detecting theft activity in EC data. The model combines both a gated recurrent unit (GRU) and a convolutional neural network (CNN). The model distinguishes between legitimate and malicious EC patterns. GRU layers are used to extract temporal patterns, while the CNN is used to retrieve optimal abstract or latent patterns from EC data. Moreover, imbalance of data classes negatively affects the consistency of ML and DL. In this paper, an adaptive synthetic (ADASYN) method and TomekLinks are used to deal with the imbalance of data classes. In addition, the performance of the hybrid model is evaluated using a real-time EC dataset from the State Grid Corporation of China (SGCC). The proposed algorithm is computationally expensive, but on the other hand, it provides higher accuracy than the other algorithms used for comparison. With more and more computational resources available nowadays, researchers are focusing on algorithms that provide better efficiency in the face of widespread data. Various performance metrics such as F1-score, precision, recall, accuracy, and false positive rate are used to investigate the effectiveness of the hybrid DL model. The proposed model outperforms its counterparts with 0.985 Precision–Recall Area Under Curve (PR-AUC) and 0.987 Receiver Operating Characteristic Area Under Curve (ROC-AUC) for the data of EC.

Keywords: class imbalance; gated recurrent units; convolutional neural network; electricity theft detection; non-technical losses; smart grids

1. Introduction

Electricity has become a basic need in the modern world, as it is used in homes, businesses, and industry. To distribute electricity to these sectors, a network is formed, which is called the power grid. Technically, the power grid consists of a production side and a demand side. Electricity generation is increased or decreased depending on the demand side's needs. Unfortunately, some of the electricity produced is lost during generation, transmission, and distribution. Energy losses are divided into two main classes: non-technical losses (NTL) and technical losses. Various methods, techniques, and tools are in practice or are proposed to address technical losses.

On the demand side, one of the NTLs is electricity theft. Electricity loss is a major issue for power utility companies, as it causes major disruption to their operations, which leads to loss of revenue, increased generation load, and excessive electricity bills for legitimate consumers. Moreover, electricity loss also causes issues related to economic growth and power infrastructure stability. NTL, also known as commercial losses, happen mostly due to electricity theft and fraud. Power utility companies still lose large amounts of revenue due to unlawful electricity theft and fraud by electricity consumers. This theft places a heavy burden on the power grid infrastructure and results in fires that threaten public safety. They also cause loss of revenue for electrical generation companies [1–3]. It is a challenge to address power caused by theft. Theft can be done by tampering with electricity meters, double-tapping attacks, changing meter readings through communication links, and using shunt devices. It is an open secret that power utilization is strongly connected with the development of a country and is hence a vital measure that shapes the foundation of industrialization. With the consistently increasing need for power usage, electricity theft is at a peak. Fossil fuel combustion from electricity generation causes 70% of greenhouse gas (GHG) emissions [4]. In spite of endeavors to reduce GHG outflows, electricity theft overshadows these endeavors in developing countries. The capacity to create electric power is diminished as a result of resources lost to energy theft. Due to electricity theft, unnecessary blackouts/load-shedding occur, which encourages users to opt alternative energy resources to fulfill their requirements, including using petrol and diesel generators that cause GHG emissions.

The majority of climate talks have focused on how to lower GHG emissions; very few have examined the consequences of energy theft. By continuously monitoring the electrical system and isolating energy-theft hotspots from a distance, Smart Meters (SM) are suggested as a strategy to prevent energy theft. All transformers, distribution poles, and customer houses should have SMs. The measurements are subsequently transmitted over a communication network to the distribution company's database for examination, and if trouble areas are found, power is cut off remotely. This technology would enhance performance, which would immediately result in a decrease in GHG emissions while also increasing total returns to the distribution firm. It would also promote transparency in the metering process.

Moreover, NTLs cause USD 75 billion in lost revenue in the United States. This amount is enough to power 77,000 households for a year [5]. A World Bank report shows that China, Brazil, and India suffer 16%, 25%, and 6% losses in electricity supply, respectively [6]. According to Joker et al. [7] such losses are not only limited to developing countries; developed countries such as the U.S. and the U.K. bear losses of USD 6 billion and GBP 173 million, respectively, each year. The above discussion shows that an efficient electricity theft detection (ETD) model is required to detect NTLs. In the literature, hardware devices, and data-driven and game-theoretic approaches are used to detect NTLs. Hardware-based approaches use sensors and radio identification tags to distinguish between honest and malicious samples. However, these approaches are expensive, require huge maintenance costs, and do not provide optimal results under extreme weather conditions [3,8–10]. Methods based on game theory design a utility function among electric utilities, stakeholders, and customers. However, it is difficult to implement an accurate utility function. Moreover, these approaches are less accurate and have a high false-positive rate (FPR) [11–14].

The introduction of smart power grids opens new opportunities for ETD. A smart grid is an upgraded version of a conventional power grid and consists of smart meters, sensors, and computing devices that have self-healing mechanisms and communication technologies. The smart meters and sensors obtain data on consumers' electricity consumption (EC), electricity prices, and the status of the grid infrastructure [15,16]. The data-driven approaches are trained on the collected EC data to distinguish between honest and malicious samples. These approaches have received a lot of focus from the research community, but they have the following limitations: curse of dimensionality, class imbalance problems, and low detection rates for standalone ML and DL models. Moreover, conventional ML models such as k-nearest neighbors and naïve Bayes have high FPRs. As mentioned in the literature, electric utilities cannot tolerate low detection rates and high FPRs because for on-site inspection they have limited resources.

This paper presents a hybrid DL model (named HGC) that is a combination of a gated recurrent unit (GRU) and a convolutional neural network (CNN). GRU extracts temporal features, while CNN retrieves abstract patterns from EC data. The advantages of the models are summarized in the HGC model. It also outperforms existing models. The uneven distribution of class patterns leads to poor performance. This problem leads to majority class bias, which leads to incorrect results. In this paper, a hybrid approach consisting of undersampling and oversampling methods is presented to deal with the uneven distribution of class samples. The main contributions of the paper are listed below.

- We present an HGC model that combines the advantages of GRU and CNN. It is the first study that combines the advantages of sequential and non-sequential models.
- A CNN model extracts latent or abstract patterns, while a GRU retrieves temporal patterns from EC data. The curse of dimensionality is addressed with both DL models.
- The adaptive approach of synthetic minority oversampling and TomekLinks are used to discuss the problem of class imbalance.
- The performance of the HGC model is evaluated using a real EC dataset obtained from the State Grid Corporation of China (SGCC).
- To verify the real efficiency of the proposed model, extensive experimentation is performed based on recall, accuracy, precision, F1 score and FPR.

The rest of the paper is organized as follows. Section 2 presents an overview of related literature. We present the Problem Statement in Section 3, followed by Materials and Methods in Section 4. The Proposed Model is outlined in Section 5. Section 6 contains the Experimental Analysis and Discussion. The Experimental Outcome and Arguments are discussed in Section 7. Finally, we come to an end in Section 8.

2. Related Literature

The tools and techniques proposed in the literature to detect NTLs are studied in this part of the document. In [5], a model combining CNN and multilayer perceptron (MLP) is used. It integrates the advantages of both DL models, which is why it gives better results than standalone models. The first model is employed to extract hidden, abstract patterns, while the latter one is used for extracting meaningful information. The class imbalance problem, however, is not addressed, which makes the ML and DL models biased towards majority class samples and ignore minority ones. Moreover, MLP does not give results on sequential datasets. Joker et al. [7] propose an electricity theft detector that is developed using an SVM classifier to differentiate between malicious and honest customers. It is the first study that integrates a ML model and hardware devices to capture drift changes in data that can happen due to many reasons: e.g., a different number of members in a household or weather changes. Some authors utilize random undersampling to solve the uneven distribution of class samples. However, this technique creates underfitting. Moreover, they utilize hardware devices that make the proposed solution expensive. In [17], the authors propose a theft detector that contains gradient boosting classifiers. The authors introduce the concept of stochastic features, which enhance the detection rate and reduce the FPR. Moreover, they conduct a comparative study and prove that boosting classifiers perform better than SVM on an Irish dataset. Moreover, electricity theft cases are updated by arguing that existing theft cases' resemblance to real-time samples is the least. Random oversampling is employed to handle the uneven distribution of class samples, which creates an overfitting problem. The curse of dimensionality is a big nuisance and reduces the detection-rate of ML and DL models. In [18], the authors use heuristic techniques to select optimal combination of features from EC data, which solves overfitting, memory constraints, and computational overhead issues. However, they use accuracy as a fitness function to evaluate the efficacy of meta-heuristic techniques, which is not a good practice.

In [19], a long short-term memory (LSTM)-dependent framework is suggested. It is proposed for differentiating between malicious and normal patterns as well as changes due to drift. Based on our knowledge, this is the first study that considers drift changes with malicious patterns and reduces FPR. The Power utilities are unable to bear high FPR due to their limited resources to inspect on the site. Fenza et al. [20] propose a model that integrates the benefits of both CNN and random forest. The former is used to obtain abstract features, while the latter is used to differentiate malicious and normal patterns in EC data. The class imbalance problem is handled using SMOTE, which creates overfitting. In [21], a DL model is proposed that integrates the benefits of both LSTM and MLP. This is the first article that has leveraged the benefits of both sequential and non-sequential data. The class imbalance problem is not considered, which is why ML and DL models give biased results. In [22], an ensemble deep CNN is used for detection of atypical behaviors in EC data. Imbalanced data are a severe issue in ETD and is handled through random bagging. Finally, a well-known voting ensemble strategy is utilized to decide between malicious and normal patterns. Ghori et al. [23] conduct a comparison study between different conventional ML classifiers using a real EC dataset. The ANN and boosting classifiers such as LightBoost, CatBoost, and XGBoost give better performance than other models. Moreover, the curse of dimensionality is dealt with by selecting optimal combination features.

In [24], the authors put forward a fascinating technique for NTL detection using smart meter data. Moreover, auxiliary information is utilized to enhance the accuracy of ML models. Different features are built using distance and density outlier-detection methods. The proposed model is employed in smart grids to distinguish illegitimate patterns from legitimate patterns. In [25], Hasan et al. put forward the idea of identifying low-voltage stations and comparing the performance of supervised and unsupervised learning methods. The suggested method gives better results in contrast to SVM and DT-SVM.

Ismail et al. [26], merge the integrated model of CNN and LSTM. This is the first study that integrates the benefits of both DL learning models. Moreover, the uneven distribution of class samples is another severe issue. SMOTE is utilized to handle this issue. The proposed hybrid model achieves 89% accuracy, which is more than conventional ML and DL models.

The poisoning attack problem in smart grids is proposed by Maamar et al. [27]. They introduce a sequential and parallel DL-based autoencoder based on GRU and LSTM models. The deep neural network performs better than a shallow neural network. In [28], it is revealed that existing studies mostly monitor attacks on the consumer side. No one focuses on the distribution side, where hackers hack utility meters and create higher electricity bills. In their study, they introduce a hybrid C-RNN-based model and prove that it performs well compared to other DL models. The proposed model is evaluated on SCADA meter readings.

In [29], a new hybrid approach is introduced that integrates the benefits of k-mean clustering and a deep neural network. Irish Smart Energy Trials data are used for model evaluation. However, if the authors utilize other advanced clustering algorithms, then proposed model increases the performance. Shehzad et al. [30] introduce a smart system for ETD. The system integrates the benefits of statistical methods and different DL models such as MLP, LSTM, RNN, and GRU. The proposed technique is evaluated on real data from Singaporean homes. However, the performance of the suggested technique is not checked using other performance measures such as F1-score, recall, precision, FPR, ROC-AUC, and PR-AUC.

3. Problem Statement

In [3], the authors propose a theft detector consisting of an SVM to discriminate between malicious and normal samples. However, they do not use a feature selection or extraction approach to deal with the curse of dimensionality. Overfitting leads to high accuracy when using training data compared to test data when ML and DL models are used. Moreover, in [17], the black-hole algorithm (BHA) is used to handle the high dimensional data. BHA is a meta-heuristic method that requires high and complex computations to find an optimal feature combination with which ML models achieve better results. For this reason, it is not suitable for real-time smart-grid applications. Moreover, the problem of class imbalance is another serious problem in ETD. There are more samples of normal classes than malicious classes. Zheng et al. [1] do not use any approach to solve this problem. In [8], SMOTE is used to improve the minority class samples. However, with this approach, there is a tendency for the ML or DL models to run into an overfitting problem as sample size increases. In [3], a random undersampling approach is employed to compensate for the unequal distribution of normal and malicious samples. However, this approach removes important information and creates the problem of underfitting. In the literature, authors usually use conventional ML models such as SVM, DT, and NB. These models have low detection rates and high FPRs. Therefore, an efficient framework with accurate identification of NTLs in EC needs to be proposed.

4. Materials and Methods

Section 4.1 is about acquiring the dataset; data preprocessing is covered in Section 4.2, which include handling missing values, removing outliers, normalizing data values, and class imbalance problems; and in Section 5 the proposed model is discussed.

4.1. Acquiring the Dataset

In this study, to appraise the performance of the suggested model, data from the State Grid Corporation of China (SGCC) are used, as it is the only publicly available dataset; it includes 42,372 records of consumers, 3615 of which are thieves, while the rest are ordinary consumers (https://github.com/henryRDlab/ElectricityTheftDetection (accessed on 2 March 2022)). Each consumer has a label, either 1 or 0, where 0 represents a normal consumer, and 1 represents a malicious consumer. SGCC assigns the labels after conducting on-site inspections. The dataset is in tabular form, with rows representing consumers, and columns indicating the daily EC of each consumer from 1 January 2014 to 31 October 2016. Facts and figures regarding the SGCC dataset are mentioned in Table 1. Here, it is important to mention that the dataset contains some incorrect and missing values. Therefore, to handle this issue, data preprocessing is used, as described in Section 4.2.

Table 1. Details about the data.

Description	EC Time Window	Class of Customer	Power Source	Data Resolution	Total Customers	Honest Customers	Thief Customers
Values	1 January 2014 to 31 October 2016	Residential	Utility	Daily data	42,372	38,757	3615

4.2. Data Preprocessing

Data preprocessing is an important step and includes the following steps: removal of missing values and outliers, normalization of data values, feature extraction or selection, and handling the class imbalance problem.

4.2.1. Handling the Missing Values

The SGCC dataset contains missing values and non-numeric values, indicated by 'NAN'. These values occur for many reasons, such as improper operation of smart meters, human typos, data storage problems, and distribution line faults. If the data contain missing values, ML and DL methods do not produce good results. If the records with missing values are removed, it may also take away important information which creates the problem of underfitting. The missing values are tackled with linear imputation to avoid the problem of underfitting. The mathematical equations are given below.

$$f(z_i) = \begin{cases} \dfrac{z_{i,j-1} + z_{i,j+1}}{2}, & z_{i,j} = NaN, z_{i,j\pm1} \neq NaN, \\ 0, & z_{i,j-1} = NaN \ or \ z_{i,j+1} = NaN, \\ z_{i,j}, & z_{i,j} \neq NaN. \end{cases} \tag{1}$$

In Equation (1), z_i denotes the EC of consumer i on the current day, and z_{i-1} and z_{i+1} show the EC of the previous day and the next day, respectively.

4.2.2. Removing the Outliers

Some outliers are also found in the data. In the preprocessing of the data, one of the most important steps is to remove or treat the outliers. In the literature, experimental results show the sensitivity of the ML and DL models to splitting data and generating false results. To treat the outliers, the three-sigma rule (TSR) is used in this study. The mathematical equation of the TSR is given below.

$$f(z_i) = (z_i) * \sigma(z_i) \quad if\ z_{i,j} > \mu(z_i) + 3 * \sigma(z) \quad otherwise\ f(z_i) = z_i \tag{2}$$

In Equation (2), z_i shows the EC history of a consumer i, $\mu(z_i)$ represents the averaging of EC, and $\sigma(z_i)$ denotes the standard deviation.

4.2.3. Normalizing the Data Values

After performing the above steps, normalization of the data is done by a min–max method. The reason for this is that ML and DL do not work well on diverse data. The mathematical equation is given below.

$$z_{i,j} = \frac{z_{i,j} - min(Z_i)}{max(Z_i) - min(Z_i)} \tag{3}$$

In Equation (3) $min(Z_i)$, represents the minimum EC, while $max(Z_i)$ denotes the maximum EC of consumer i.

Algorithm 1 shows the data pre-processing phase, which contains following steps: handling the missing values, removing the outliers, and normalizing the data values.

Algorithm 1: Data pre-processing phase.

---

1 **Data:** EC data: Z
2 $X = (z_{i,j}, y_i), (z_{i+1,j}, y_{i+1}), ..., (z_{m,n}, y_m)$
3 m = number of records, n = number of features
4 **Variables:** min_i = minimum consumption, max_i = maximum consumption, $\bar{z}_i$ = mean consumption, σ_i = standard deviation,
5 **for** $i \leftarrow m$ **do**
6   **for** $j \leftarrow n$ **do**
7   **Handle the missing data:**
8   **if** $z_{i,j-1}\ \&\&\ z_{i,j+1} \neq NaN\ \&\&\ z_{i,j} == NaN$ **then**
9             $z_{i,j} = (z_{i,j-1} + z_{i,j+1})/2$
10   **end**
11   **if** $z_{i,j-1}\ ||\ z_{i,j+1} == NaN$ **then**
12             $z_{i,j} = 0$
13   **end**
14   **Remove anomalies:**
15   **if** $z_{i,j} > \bar{z}_i + 3\sigma_i$ **then**
16             $z_{i,j} = \bar{z}_i + 3\sigma_i$
17   **end**
18   **Data normalization through min–max method:**
19         $z_{i,j} = \frac{z_{i,j} - min_i}{max_i - min_i}$
20   **end**
21 **end**
22 **Result:** $Z_{normalized} = Z$

4.2.4. Class Imbalance Problem

The problem of class imbalance or uneven distribution of class samples is a severe issue in ETD, where there are more samples of one class than other classes. When ML or DL are trained on an imbalanced dataset, they provide biased results with high FPRs. As mentioned in the literature, power generation companies cannot tolerate high FPRs because they have limited resources for on-site inspections. Two approaches are generally used in the literature to deal with class imbalance problems: undersampling and oversampling. In the former, replicates of the minority class are generated, while in the latter, samples are eliminated to balance the classes. However, both techniques have the following drawbacks: overfitting, duplication of existing data, and loss of information. In this paper, a hybrid sampling approach based on adaptive synthetic sampling (ADASYN) and TomekLinks is proposed. The former uses oversampling while the latter uses undersampling to solve the problem of class imbalance. The proposed hybrid approach solves the problems of undersampling, oversampling, and duplication of data. A detailed description of ADASYN and TomekLinks can be found below.

ADASYN (Adaptive Synthetic):

To solve underfitting, ADASYN is employed to generate minority class samples, which are harder to learn. The overall working mechanism of that sampling approach is elaborated below.

- The ratio of the minority to the majority class is calculated using the below equation:

$$d = \frac{m_{min}}{m_{maj}} \tag{4}$$

where m_{min} is the total number of minority class samples, and m_{maj} is the number of majority class samples in the dataset.

- The ratio of how many samples will be generated is decided using the following equation:

$$G = (m_{maj} - m_{min})\beta \tag{5}$$

where G is the total number of minority class samples that will be generated to handle undersampling; β is a random number whose value is between 1 and 0, with 0 indicating that no samples of the minority class will be generated, while 1 shows that minority samples will be generated until both classes have an equal number of samples, $\beta = (0, 1)$.

- In this step, the number of majority class samples near minority class samples are calculated using k-nearest neighbors. After that, each minority class sample is associated with a different number of neighbors that belong to the majority class.

$$r_j = \frac{majority}{k} \tag{6}$$

Here, r_j shows the dominance of the majority class samples over each minority class sample. A higher r_j shows that it is difficult for ML and DL models to learn/remember the patterns of minority class samples. Thus, a greater number of samples are created for minority class samples that are surrounded by large/maximum numbers of majority class samples. This phenomena gives an adaptive nature to ADASYN.

- To normalize the r_j values, we use

$$r_j = \frac{r_j}{\sum r_j} \qquad \sum r_j = 1 \tag{7}$$

- For minority class samples, we compute the amount of synthetic samples with

$$G_j = G r_j \tag{8}$$

- In the last step, Algorithm 1 selects the minority class samples from training data and generates new samples. If training data contain m number of minority class samples, then new samples are created using the following equation.

$$s_j = x_j + (x_j - x_{random}) * \lambda, j = 1 \ldots m. \tag{9}$$

In the above equation, λ is a random number between 1 and 0, $_j$ is the newly generated sample, x_j is a first sample of training data, and x_{random} is a randomly selected sample from the training data.

TomekLink:

TomekLink is used for undersampling class imbalance problems. It is a modification of Condensed Nearest Neighbor ((CNN), not to be confused with Convolutional Neural Network). It uses the following rules to select pairs of observations (e.g., X and Y) that satisfy the properties listed below:

- The observation that X's nearest neighbor is Y (and vice versa);
- The observation that X and Y belong to different classes: either the minority class or the majority class.

Mathematically, this is expressed as (X_{min} and X_{maj}), representing the Euclidean distance between X_{min} and X_{maj}, where X_{min} and X_{maj} belong to the minority and majority classes, respectively. If there is no sample X_k that satisfies the following conditions:

$$d(X_{min}, X_k) < d(X_{min}, X_{maj}) \tag{10}$$

$$d(X_{maj}, X_k) < d(X_{min}, X_{maj}) \tag{11}$$

The pair (X_{min}, X_{maj}) are TomekLink samples, which removes noise and duplicated values from data. Consequently, ML and DL models learn diverse patterns from data and do not get stuck in underfitting.

5. Proposed Model

In [5], a combined MLP and CNN model is proposed, which proves that the hybrid model outperforms standalone models of ML and DL. In [22], the authors present CNNs with LSTMs. GRUs and LSTMs utilize different approaches toward gating information to prevent the vanishing gradient problem. RNNs have two variants: GRU and LSTM. The vanishing gradient problem is solved by the author of [31] by comparing the performance of GRU and LSTM with an RNN model using different sequential datasets. Extensive experimentation are performed by Ding et al. on 10,000 LSTM and RNN architectures [32]. The final results advocate that GRU outperform as compared to all contemporary models. For the above reasons, in this research paper a hybrid DL model is presented that combines the advantages of both GRU and CNN models. The GRU extracts the time-related patterns, while the CNN retrieves abstract or latent pattern data. The HGC model consists of the following parts/modules: GRU, CNN, and Hybrid. One-dimensional data are fed as input to the GRU module, while 2D data are fed as input to the CNN to learn abstract features. The hybrid module takes the extracted features from both modules as input and combines them to discriminate between malicious and normal patterns. From the literature, hybrid models work well because they allow combined training and testing of both DL models. In the following, the individual modules are explained in more detail.

5.1. Gated Recurrent Unit (GRU)

GRU is an enhanced form of a recurrent neural network (RNN). One of the main problems in RNNs is the vanishing gradient problem, which stops the learning process and pushes the sequential DL models into local optima. To solve the prior problem, GRU model was introduced. GRU structure consists of an update gate and a reset gate that affect the learning of temporal patterns from EC data. Basically, the information to be passed to the next layers or units is determined by the update gate. Otherwise, the amount of information from the past that should be forgotten is determined by the reset gate. This information is not important for future decisions. The GRU layers are trained on past data, learn and remember the important information, and remove the redundant values that are not important for distinguishing between malicious and normal patterns. These GRU layers are able to retrieve time-related patterns from EC data. The equations of the update and reset gates are given below.

$$UG_t = \sigma(U_{ug}, [hdn_{t-1}, Z_t]), \tag{12}$$

$$RG_t = \sigma(U_{rg}, [hdn_{t-1}, Z_t]), \tag{13}$$

$$\hat{hdn}_t = \tanh(U, [r_t * hdn_{t-1}, Z_t]), \tag{14}$$

$$hdn_t = (1 - UG_t) * hdn_{t-1} + UG_t * \hat{h}_t. \tag{15}$$

$$Dense_{GRU} = Flaten(hdn_t * W_{GRU} + b_{GRU}) \tag{16}$$

where Z_t and hdn_{t-1} show the input value and hidden layer value of the previous time step, respectively, UG_t indicates the update gate, RG_t shows the reset gate, U_{ug} and $U_{rg}r$ are weights of the update and reset gates, respectively. $Dense_{GRU}$ layers are used to merge extracted features of GRU and CNN models to enhance the prediction accuracy. The hyperparameter settings for GRU are mention in Table 2.

Algorithm 2 describes the working mechanism of the proposed hybrid DL model containing a GRU, a CNN, and fully connected layers.

Algorithm 2: Working of HGC model.

---

1   **Data:** EC data: $Z_{Balance}$
2   Data in 1D format:
3   $Z_{1D} = z_{i,j}, z_{i,j+1}, z_{i,j+2}, \ldots, z_{l,m}$
4   $l = 42372, \ m = 1034$
5   Convert data to 2D format
6   $Z_{2D} = \begin{bmatrix} x_{1,1} & \cdots & x_{1,k} \\ \vdots & \ddots & \vdots \\ x_{j,1} & \cdots & x_{m,k} \end{bmatrix}$
7   Pass Z_{1D} data to GRU model
8   **for** $i < Epoch$ **do**
9      $r_t = \sigma(U_{rg}, [hdn_{t-1}, x_t])$
10     $\hat{hdn}_t = \tanh(U, [r_t * hdn_{t-1}, x_t])$
11     $hdn_t = (1 - z_t) * hdn_{t-1} + z_t * \hat{hdn}_t$
12     $Dense_{GRU} = relu(U \cdot hdn_t, b])$
13     $Fl_{GRU} = flatten(Dense_{GRU})$
14     $Z_{2D}[u,v] = (Z_{2D})[m,v] = \sum_j \sum_k f[j,k] Z_{2D}[m-j, v-k]$
15     $u,v \Rightarrow$ dimension of output matrix
16     $Fl_{CNN} = flatten(Z_{2D})$
17     $h_{HGC} = (W_{HGC} \cdot [Fl_{CNN}, Fl_{GRU}] + b)$
18     $Dense_{layer} = [U \cdot h_{HGC} + b]$
19     $b \Rightarrow$ bias term, $U \Rightarrow$ weight
20     $Y_{NTL} = \sigma(Dense_{layer})$
21     $Loss(Y_{NTL}, Y) = -\sum_{i=i}^{42372}(Y_i \cdot \log Y_{NTL_i})$
22     $Y_{NTL} = Predicted, Y = Actual$
23   Reduce the loss value
24   **end**
25   **Result:** Y_{NTL}

Table 2. GRU hyperparameter settings.

Model	GRU Layers	Activation Function	Dropout Rate	Kernel Initializer	Hyperparameter	Epochs
GRU	40	Sigmoid	0.4	he_{normal}	Optimal values	15

5.2. Convolutional Neural Network (CNN)

The CNN algorithm belongs to the group of DL models. It is mainly used in the recognition of images and videos. It is an extended version of the MLP. It takes images as input, learns important features using a weight-learning mechanism, and develops a relationship between learned features and labels. Technically, CNN design consists of a number of convolution layers with filters (kernels), pooling layers, then one or more fully connected (FC) layers; it applies a softmax function to classify an object with probabilistic values between 0 and 1. Each layer has its own functionality that extracts abstract or latent features that cannot be detected by the human eye. In this study, a CNN model is used to extract latent patterns from data provided by electric utilities. The extracted features are fed into the hybrid layer to make final decisions about malicious and normal consumers. The final hidden layer of the CNN model is shown below.

$$Dense_{CNN} = Flaten(X * W_{CNN} + b_{CNN}) \tag{17}$$

where W_{CNN} and b_{CNN} represent the weight and bias values, respectively, of hidden CNN layers and the feature matrix by X. The hyperparameter settings for CNN are explained in Table 3.

Table 3. CNN hyperparameter settings.

Model	Filters	Strides	Padding	Activation	Batch Size	Epochs	Time
CNN	32	1	Same	ReLu	64	15	202 s

5.3. Hybrid Module

The GRU model learns temporal patterns from 1D data, while CNN extracts the patterns, which are viewed through the human eye from 2D data. The extracted features of both models are concatenated using Keras API and then passed to a hybrid layer that decides whether there is an anomaly in the EC data; h_{HGC} is the last hidden layer of the hybrid module. Its output is passed to the sigmoid function to give a final decision about malicious and normal consumers.

$$h_{HGC} = (W_{HGC}[Dense_{CNN} + Dense_{GRU}] + b_{HGC}), \quad Y_{NTL} = \sigma(h_{HGC}) \tag{18}$$

where W_{HGC} and b_{HGC} represent the weight and bias values of the hybrid layer, and σ denotes a sigmoid function. The settings of hyperparameter for HGC are mention in Table 4 and the pictorial representation of the proposed framework is given in Figure 1.

Table 4. HGC parameter settings.

Model	Layers	Dense	Batch Size	Epochs	Optimize	Time (s)	Dropout	Activation	Kernal Initializer	Pool Size
GRU	40	20	64	10	ADAM	1704	0.4	-	he_{normal}	-
CNN	32	20	64	10	ADAM	1704	-	ReLu	-	2×2

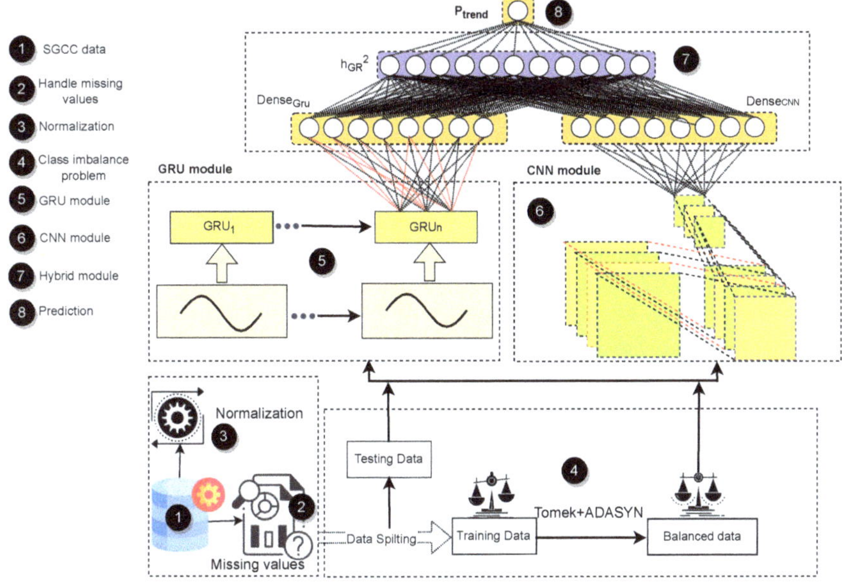

Figure 1. Proposed system model.

6. Experimental Setting and Analysis

In this section, we analyze the performance of the proposed model on the SGCC dataset using various performance measures. We also compare the results obtained with the proposed model to those of benchmark models.

6.1. Performance Measures

Uneven distribution of class samples is a critical problem in ETD, where the number of samples of the normal class is higher than that of the malignant class. When an ML or DL model is trained on this type of data, it attracts majority class samples and ignores minority class samples, producing false results/alarms. The literature indicates that electric utilities cannot tolerate false alarms due to limited resources for on-site testing. Although the training dataset is balanced with the proposed sampling technique, the test data are unbalanced. Therefore, appropriate performance measures are needed to evaluate the performance of the benchmark and proposed models. In this paper, the performance measures used are accuracy, F1 score, recall, ROC-AUC, and PR-AUC. To calculate the above measures, we use a confusion matrix: a confusion table that contains true negative (TN), true positive (TP), false negative (FN), and false positive (FP) results.

6.1.1. Accuracy

Accuracy is the ratio between the number of correct predictions and the total number of records in the dataset.

$$Accuracy = \frac{TN + TP}{TN + TP + FN + FP} \tag{19}$$

where TN and TP are the sums of total number of true negatives and true positives, respectively, and TN, TP, FN, and FP are the sums of true negatives, true positives, false negatives, and false positives, respectively.

6.1.2. Recall

Recall is determined by dividing the correctly predicted positive records by the total number of positive records. The equation of recall is given below, as described in [33]:

$$Recall = \frac{TP}{FN + TP} \tag{20}$$

where *FN* is the number of dishonest consumers predicted by the model as honest consumers.

6.1.3. F1-Score

The F1-score is also a good performance measure for imbalanced datasets. When ML/DL models have a high F1-score, they are considered good for predictions in real-world scenarios. The equation for the F1-score is given below, as described in [34,35]

$$F1 - Score = \frac{2 * precision * recall}{precision + recall} \tag{21}$$

To calculate the precision, the number of true positives divided by the sum of false positives and true positives, as mentioned in [33].

The ROC curve is obtained by plotting recall and FPR on the y-axis and x-axis, respectively. It is a good measure for imbalanced datasets because it is not skewed toward the majority class. Its value ranges from 0 to 1. However, ROC only considers the recall/true positive rate, so it focuses on positive records and ignores the negative ones. The PR curve is another important measure that considers recall and precision simultaneously and gives equal importance to twain classes.

6.2. Implementation Environment

The proposed and benchmark models are implemented using Google Colaboratory [36], which provides distributed computing power. Their performance is studied using the SGCC dataset collected from the largest electric utility in China. DL models are implemented using TensorFlow (v2.8.2), while ML models are trained and evaluated using the Scikit library (v1.0.2), and the Keras API is used to develop the hybrid model.

6.3. Proposed Deep Learning Model Performance Analysis

In this section, we analyze the performance of the proposed model using accuracy and loss curves for training and testing data. Figure 2 shows the performance of the model on training and test data using accuracy curves. Both curves move side-by-side with a small difference, indicating that the proposed model does not suffer from overfitting. However, after the fourth epoch, the test accuracy starts to decrease, which means that the model suffers from overfitting. Thus, if more than four epochs are trained, the performance of the model decreases. To improve the model's performance in the future, meta-heuristic algorithms will be used to help select the optimal parameters for deep and machine learning to avoid overfitting. It is very complex and time-consuming to select these parameters manually.

Figure 3 also shows the same phenomena using loss curves on training and testing data. The value of loss can be decreased with more epochs.

However, there is a high probability that the model encounters overfitting, which affects generalization. In addition, the proposed model consists of GRU, CNN, and dense layers. The gates like, update and reset in the GRU layer control the information flow through network. These gates remember valuable information and ignore redundant and noisy patterns from the data. CNN layers help the proposed hybrid model learn global/abstract patterns from EC data and reduce the curse of dimensionality, which directly increases the convergence speed. The literature shows that dropout layers simplify the model and prevent overfitting. Finally, the dense layer takes inputs from the GRU and CNN models and passes them to a sigmoid function to distinguish between normal and malicious samples. For all these reasons, a hybrid model performs better than the individual models.

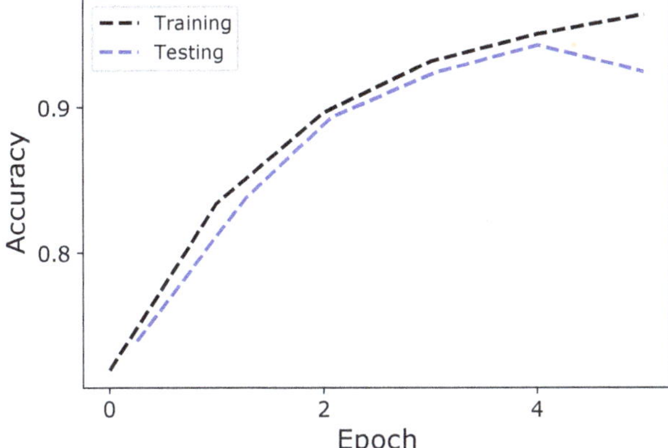

Figure 2. Accuracy curves on training and testing data.

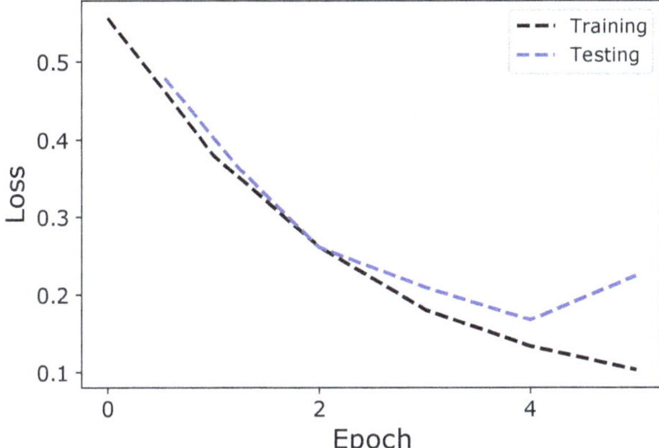

Figure 3. Loss curves on training and testing data.

6.4. Benchmark Models

This section implements various DL and ML models that have previously been proposed in the literature and compares their performance with that of the proposed hybrid model.

6.4.1. Wide and Deep Convolutional Neural Network

In [5], Zheng et al. propose a DL model that is a fusion of CNN and ANN. This is the first study to combine the advantages of both models. The authors feed 2D data to a CNN, while 1D data are fed into an ANN to learn local and global patterns from the SGCC dataset. However, the ANN model does not give good results on 1D data because it is designed for tabular data. In this work, we use the same hyperparameter settings and the same dataset for a fair comparison.

6.4.2. Logistic Regression (LR)

This is a basic supervised learning model used for binary classification. It is also known as a single-layer neural network. It simply contains an input layer whose values are multiplied by weights, and the resulting value is fed into a sigmoid function that produces either 0 or 1 as input. LR consists of various solvers such as Newton's method and stochastic gradient descent that are used to tune the hyperparameters.

6.4.3. Decision Tree (DT)

DTs are used in both regression and classification tasks. They consist of a root node, edges, and leaf nodes that are used to predict the result. A DT works like the human mind and creates a tree-like structure in which the dataset is divided into many branches based on features. The best attributes/features are selected based on the information gain and Gini index criteria as root nodes. DTs are easy to implement and give good results on smaller datasets. However, for larger datasets there is a risk of overfitting. In addition, a small change in the data leads to poor generalization.

6.4.4. Support Vector Machine (SVM)

SVMs are a supervised learning model used for both regression and classification purposes. They are able to classify linear and nonlinear data by using the power of kernel functions. These kernel functions draw a decision boundary to classify between normal and malicious samples after converting non-linear data into linear patterns. In [7], the authors develop a current theft detector based on consumption patterns using an SVM classifier to draw a decision boundary between benign and stolen samples. From the literature, SVM is well-suited for smaller datasets, as it requires a lot of computational time to draw a decision boundary between normal and malicious patterns for larger datasets. In this work, the RBF kernel is used for the SGCC dataset due to the nonlinearity of the data.

6.4.5. Random Forest (RF)

An ensemble technique called RF is used to solve complex problems by training multiple decision trees on datasets. It has applications in banking, e-commerce, and other fields. RFs control the problem of DF overfitting and increase precision. They give good results with little adjustment of hyperparameters. They also minimize overfitting and increase the precision when the number of DTs is increased during the training period. However, they require a lot of computation time for larger datasets, since multiple DTs are trained on a single dataset, which reduces their effectiveness in real-world problems.

6.4.6. Naive Bayes Classifier

This is a classification method derived from Bayes' theorem. The Naive Bayes (NB) does not consider the linkage between inputted features and targeted column, and uses the probability distribution to distinguish between normal and malicious samples. There are many versions developed depending on the type of dataset. In today's world, there are many applications in various fields such as sentiment analysis, email filtering, recommender systems, spam, and natural language processing. In this work, we use Gaussian NB since the SGCC dataset has continuous features.

7. Experimental Results and Discussions

The performance of the proposed HGC model is compared with the state-of-the-art classifiers. The same datasets with different ratios for training and testing are used for DT, NB, LR, CNN, GRU, RF, SVM, and WDCNN. As discussed earlier, the CNN design consists of a number of convolution layers with filters (kernels) and pooling layers, followed by one or more fully connected (FC) layers, and applies a softmax function to classify an object with probabilistic values between 0 and 1. Each layer has its own functionality and extracts abstract or latent features that cannot be detected by the human eye.

The GRU layers have two important gates; update and reset. These are used to learn necessary patterns and remove unnecessary values. As discussed earlier, the flow of information is controlled by GRU gates to improve the performance of the model. The GRU-extracted features are then combined with the latent or abstract patterns. The proposed HGC model extracts abstract and periodic patterns from EC data using GRU and CNN hence HGC outperforms as compared to counterparts of it. The combination of optimal features helps the HGC to attain 0.96 PR-AUC and 0.97 ROC-AUC values, which are higher than those of all the above-mentioned classifiers. The performance of proposed model is compared with conventional models using PR and ROC curves in Figures 4 and 5. The proposed hybrid model achieves better results than its counterparts. SVM achieves 0.88 ROC-AUC and 0.85 PR-AUC. We use a linear kernel instead of an RBF kernel to train the SVM model on EC data because the dataset contains a large number of records and features, which increases the model computation time, so it is not suitable for larger datasets.

LR is a conventional ML model that distinguishes between normal and malignant samples using a sigmoid function. It achieves 0.86 and 0.88 for PR-AUC and ROC-AUC, respectively, which is better than SVM, but has lower performance than other models. It has a large number of applications in various fields because it is easy to implement and is suitable for linearly separable datasets, but in the SGCC dataset, malicious and normal samples are not linearly separable. Therefore, LR gives lower performance compared to other models [30].

RF gets 0.76 PR-AUC and 0.75 ROC-AUC, while DT gets 0.80 ROC-AUC and 0.85 PR-AUC on the EC dataset. DT gives better results than RF. DT provides good performance on smaller datasets but has overfitting on larger datasets, and small changes in the data reduce its generalization ability. RF is an ensemble method designed to overcome the overfitting/low generalization of DT. It controls overfitting but has low PR-AUC and ROC-AUC, as seen in Figures 4 and 5, because RF takes the average of all DT prediction results.

In addition, NB is a conventional classifier that classifies between normal and malignant samples using Bayes theorem. It obtains 0.71 and 0.65 PR-AUC and ROC-AUC values, respectively. Unlike other conventional ML and ensemble models, it gives poor results. It assumes that there is an independent relationship between the attributes and the target features.

Moreover, CNN gains 0.96 ROC-AUC and 0.94 PR-AUC values, while GRU gains 0.96 and 0.96 ROC-AUC and PR-AUC values on the EC dataset, which are higher than the PR-AUC and ROC-AUC values of conventional ML models. Technically, a CNN consists of a number of convolution layers with filters (kernels) and pooling layers, followed by one or more fully connected (FC) layers. In addition, the convolutional layer is used to remove redundant, overlapping, and noisy values from the EC data. GRU also gives good results that are in the acceptable range, as it has update and reset gates to help remember periodic patterns. In [5], the authors combine the merits of the ANN and CNN models to develop a hybrid model. Their proposed model achieves a value of 0.96 PR-AUC and 0.97 ROC-AUC. In the literature, the authors demonstrate that the hybrid model performs better than the DL models and the standalone ML model. Therefore, in this research, the Keras API is used to develop a hybrid model. It integrates the advantages of both GRU and CNN models. The former learns the temporal patterns, while the latter derives global and abstract patterns from EC data. The extracted features of both models are merged and passed to a fully linked layer for the classification of theft and normal patterns. The proposed model achieves better results than the standalone DL and the previously proposed hybrid DL models for the above reasons. It achieves 0.987 ROC-AUC values and 0.985 PR-AUC values on EC data, as observed in Tables 5 and 6.

Tables 5 and 6 show the performance analysis of the ML and DL models at 70% and 60% training ratios, respectively. It can be seen that the proposed model maintains its superiority and gives better results at both training ratios. For the DL models, performance increases as the size of the training data increases because DL models are inherently sensitive to the size of the training data. On the other side, the increased or decreased performance of conventional ML models follow the power law [37]. This law states that beyond a certain point, the performance of ML models increases with the increase of the amount of data. After this point, the models face the problem of overfitting, which affects their generalizability. In this work, RF and NB give poor results compared to other conventional ML models. Although both models perform well on balanced datasets, they show poor performance due to the following limitations.

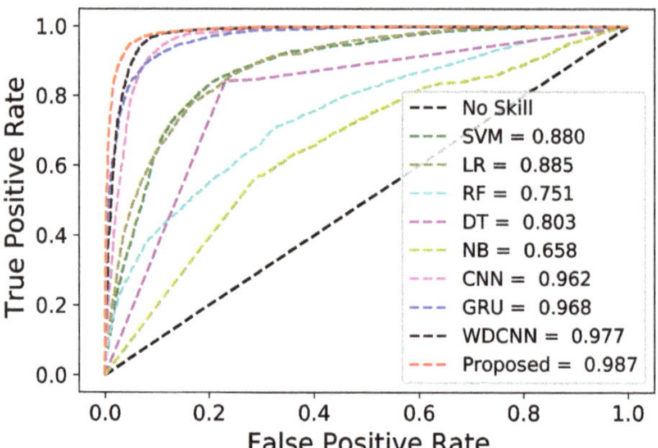

Figure 4. ROC curves of proposed and benchmark models.

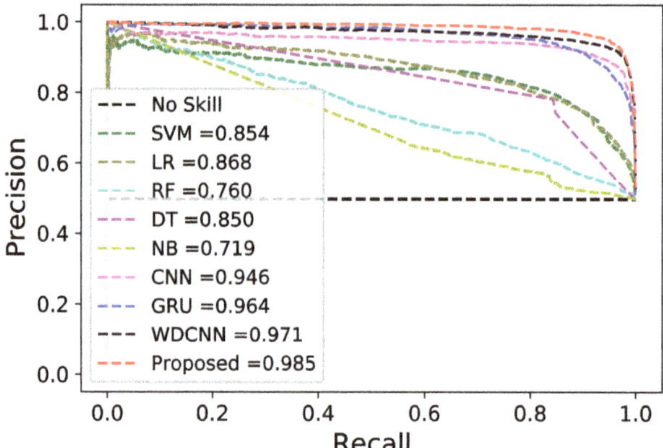

Figure 5. PR curves of proposed and benchmark models.

Table 5. Performance analysis of DL and ML using 70% training data.

ML/DL Models	Accuracy	F1-Score	Recall Score	PR_{AUC}	ROC_{AUC}
LR	0.8040	0.8068	0.714	0.868	0.885
SVM	0.8165	0.8200	0.800	0.854	0.880
RF	0.6912	0.696	0.6128	0.756	0.748
DT	0.8056	0.8118	0.7826	0.850	0.803
NB	0.6261	0.649	0.608	0.719	0.658
CNN	0.914	0.918	0.877	0.946	0.962
GRU	0.9074	0.9080	0.919	0.964	0.968
WDCNN	0.9397	0.9408	0.919	0.971	0.977
HGC	0.9438	0.9452	0.91709	0.985	0.987

Table 6. Performance analysis of DL and ML using 60% training data.

ML/DL Models	Accuracy	F1-Score	Recall Score	PR_{AUC}	ROC_{AUC}
LR	0.804	0.807	0.796	0.868	0.883
SVM	0.811	0.815	0.797	0.855	0.877
RF	0.677	0.680	0.672	0.756	0.748
DT	0.801	0.808	0.781	0.848	0.799
NB	0.619	0.645	0.604	0.715	0.650
CNN	0.916	0.916	0.916	0.955	0.966
GRU	0.925	0.926	0.910	0.968	0.973
WDCNN	0.936	0.938	0.906	0.971	0.779
HGC	0.947	0.948	0.921	0.985	0.987

NB accounts for the independent relationship between features and target variables that does not exist in real EC data, while RF controls for overfitting by the average performance of all DTs. The literature shows that the performance of DL models depends on the size of the training data. Large datasets yield high values for performance measures. ROC analysis of different hybrid models is given in Table 7. In [38], CNN-LSTM and LSTM RUSBoost achieve 0.817 and 0.879 ROC values, respectively, while in [30], MLP–LSTM achieves 0.92 ROC, and HG^2 achieves 0.93 ROC. In our case, our proposed model maintains its superiority and performs better than the above-mentioned hybrid models by achieving 0.98 ROC.

The computation time of the ML and DL models is given in Table 8. NB and LR have a lower computation time in contrast to other ML models because the former only computes the probability distribution of all features and provides the final results, whereas LR is a single-layer neural network that multiplies the inputs with weights and distinguishes between malignant and normal samples. For the above reasons, they require little computational time compared to other ML models.

In ETD, SVM is a well-known classifier. RF requires more training time than DT because it trains multiple DTs on the SGCC dataset and computes the average of multiple estimators. Moreover, the training time of DL models depends on the number of hidden layers, the size of the dataset, the stack size, and the number of neurons in each layer. GRU and CNN are DL models that take 2364 and 202 seconds to train, respectively. GRU requires more training time because it has update and reset gates that extract temporal patterns from SGCC data and save the important information in memory networks, while CNN only retrieves abstract/latent patterns by using convolution functions and max-pooling layers, which is why they have low computation time. Moreover, HGC takes 1704 seconds to train with the SGCC dataset. It has a lower computation time than GRU because it converges in 5 epochs, whereas GRU converges in 15 epochs. In addition, HGC requires more training time than the CNN model because it integrates the benefits of both models. Moreover, at the present time, meta-heuristic techniques are receiving attention from the research community for feature selection and hyperparameter optimization in ML and DL models. Therefore, in this study, BHA, a meta-heuristic technique, is used for feature selection. The literature demonstrates that these techniques have high computational complexity. For this reason, a small portion of the dataset is used to evaluate the ability of BHA for feature selection. The selected data consist of 10,000 records and 30 days of EC values from 42,372 records. BHA takes 3000 seconds to select the optimal combination of features/attributes from the selected EC data, which is more than the time required by all DL models: GRU, CNN, WDCNN, and HGC. The above results show that the computational time of BHA increases as the amount of data increases. Therefore, these types of real-time applications are not suitable for the smart grid. Moreover, the increased dataset size enhances the performance of DL models. Hence, the performance of these models depend on the size of training dataset. In canse of convolution ML models, the performance is enhanced by following the power law. Their performance stop improving after certain point of training [37].

From the literature, hybrid models work well because they combine training and testing of both DL models and have better generalization capabilities than many other machine and deep learning models. However, HGC maintains dominance over the state-of-the-art DL models and shows better performance on varieties of training ratios over SGCC dataset. Nexus to the above, there is no free lunch. The cost benefit analysis is a trade-off between computational time and accuracy. The proposed algorithm is computationally expensive, but on the other hand, it provides higher accuracy than the other algorithms used for comparison. With more and more computational resources available these days, researchers are focusing on algorithms that provide better efficiency in the face of widespread data.

Table 7. ROC performance analysis of hybrid models.

Hybrid Models	CNN-LSTM	LSTM-RUSBoost	MLP-LSTM	HG^2	Proposed Model
ROC	0.817	0.879	0.92	0.93	0.98

Table 8. Computation time of ML and DL models.

ML/DL Models	SVM	LR	DT	RF	NB	SVM + BHA	CNN	GRU	WDCNN	HGC
Time (s)	1618	4	52	281	1	3000	202	2364	304	1704
Epoch	-	-	-	-	-	-	15	15	15	5

8. Conclusions and Future Work

Electricity theft is an unavoidable issue that causes power losses in both; developed and developing countries. As a result, power utility companies have major disruptions in their operations, leading to loss of revenue. Moreover, electricity loss also causes issues with economic growth and power infrastructure stability. In this study, a combined DL model for NTL detection is presented that incorporates a GRU and a CNN. To remove null and undefined values, EC data are pre-processed by normalization. In addition, uneven distribution of class samples is another problem in ETD that affects the effectiveness of the ML and DL models. In this paper, a hybrid approach is used to address these problems. The performance of the proposed model is evaluated on the SGCC dataset in real-time using various performance metrics and compared with SVM, LR, CNN, GRU, RF, DT, NB, and WDCNN. The model achieves 0.987, 0.985, 0.94, 0.94, and 0.91 ROC-AUC, PR-AUC, accuracy, F1-score, and recall score on the SGCC dataset, respectively. The obtained results are better than those of other ML and DL models. However, despite the proposed model outperforming substitute techniques, it is too sensitive to changes in input data. The presented model will help many industrial applications to identify normal and abnormal samples or records. To improve the model's performance and avoid overfitting, meta-heuristic algorithms help select the optimal parameters for deep and machine learning. It is very complex and time consuming to select these parameters manually.

In the future, meta-heuristic techniques will be used to achieve optimal hyperparameter tuning in DL models.

Author Contributions: Conceptualization, A.K., R.B. and R.A.; Data curation, A.K.; Methodology, R.B., O.A. and R.A.; Project administration, S.A. and A.Y.; Resources, A.Y. and O.A.; Software, A.Y. and O.A.; Supervision, R.B. and S.A.; Validation, S.A.; Visualization, R.A.; Writing—original draft, A.K. and R.B.; Writing—review & editing, R.B., S.A., A.Y., O.A. and R.A. All authors have contributed equally and have read and agreed to the published version of the manuscript.

Funding: This research received no external funding.

Institutional Review Board Statement: Not applicable.

Informed Consent Statement: Not applicable.

Data Availability Statement: Not applicable.

Acknowledgments: The authors of this study would like to thank the anonymous reviewers and the editor for their insightful comments and suggestions to improve our work.

Conflicts of Interest: The authors declare no conflict of interrest.

Nomenclature

ANN	Artificial Neural Network (NN)	z_i	EC of consumer i at current day
ADASYN	Adaptive Synthetic	z_{i-1}	EC of consumer i at previous day
AUC	Area Under the Curve	z_{i+1}	EC of consumer i at next day
BHA	Black Hole Algorithm		
CNN	Convolutional Neural Network	$\mu(z_i)$	Represents average E
DT-SVM	Decision Tree-SVM	$min(Z_i)$	Minimum EC
DL	Deep Learning	$max(Z_i)$	Maximum EC
DE	Differential Evolution	m_{min}	Total number of minority class
DT	Decision Tree	m_{max}	Total number of majority class
DNN	Deep Neural Network	G	Total number of minority data to be generated
EC	Electricity Consumption	ETD	Electricity Theft Detection
FP	False Positive	FN	False Negative
FPR	False Positive Rate	GRU	Gated Recurrent Unit
HGC	Hybrid GRU–CNN	LSTM	Long Short-Term Memory
KNN	K-Nearest Neighbor	β	Ratio of minority: majority data desired after ADASYN
LR	Linear Regression	MLP	Multi-Layer Perceptron
NTL	Non-Technical Loss	RNN	Recurrent Neural Network
PR-AUC	Precision–Recall Area Under Curve	ROC-AUC	Receiver Operating Characteristic Area Under Curve
RBF	Radial Basis Function	RF	Random Forest
SGCC	State Grid Corporation of China	SVM	Support Vector Machine
TP	True Positive	TN	True Negative
TL	Technical Loss	TSR	Three Sigma Rule
WADCNN	Wide And Deep Convolution NN	λ	Number between 0–1

References

1. Leon, C.; Biscarri, F.; Monedero, I.; Guerrero, J.; Biscarri, J.; Millan, R. Variability and Trend-Based Generalized Rule Induction Model to NTL Detection in Power Companies. *IEEE Trans. Power Syst.* **2011**, *26*, 1798–1807. [CrossRef]
2. Glauner, P.; Meira, J.; Valtchev, P.; State, R.; Bettinger, F. The Challenge of Non-Technical Loss Detection Using Artificial Intelligence: A Survey. *Int. J. Comput. Intell. Syst.* **2017**, *10*, 760. [CrossRef]
3. McLaughlin, S.; Holbert, B.; Fawaz, A.; Berthier, R.; Zonouz, S. A Multi-Sensor Energy Theft Detection Framework for Advanced Metering Infrastructures. *IEEE J. Sel. Areas Commun.* **2013**, *31*, 1319–1330. [CrossRef]
4. David, N. The Effects of Energy Theft on Climate Change and Its Possible Prevention Using Smart Meters: Case Study Nigeria. *Int. J. Sci. Eng. Res.* **2018**, *9*. 1775.
5. Zheng, Z.; Yang, Y.; Niu, X.; Dai, H.; Zhou, Y. Wide and Deep Convolutional Neural Networks for Electricity-Theft Detection to Secure Smart Grids. *IEEE Trans. Ind. Inform.* **2018**, *14*, 1606–1615. [CrossRef]
6. Aryanezhad, M. A novel approach to detection and prevention of electricity pilferage over power distribution network. *Int. J. Electr. Power Energy Syst.* **2019**, *111*, 191–200. [CrossRef]
7. Jokar, P.; Arianpoo, N.; Leung, V. Electricity Theft Detection in AMI Using Customers' Consumption Patterns. *IEEE Trans. Smart Grid* **2016**, *7*, 216–226. [CrossRef]
8. Lo C.; Ansari, N. Consumer: A Novel Hybrid Intrusion Detection System for Distribution Networks in Smart Grid. *IEEE Trans. Emerg. Top. Comput.* **2013**, *1*, 33–44. [CrossRef]
9. Xiao, Z.; Xiao, Y.; Du, D. Non-repudiation in neighborhood area networks for smart grid. *IEEE Commun. Mag.* **2013**, *51*, 18–26. [CrossRef]
10. Khoo, B.; Cheng, Y. Using RFID for anti-theft in a Chinese electrical supply company: A cost-benefit analysis. In Proceedings of the IEEE Wireless Telecommunications Symposium, New York, NY, USA, 13–15 April 2011; pp. 1–6.
11. Angelos, E.; Saavedra, O.; Cortés, O.; de Souza, A. Detection and Identification of Abnormalities in Customer Consumptions in Power Distribution Systems. *IEEE Trans. Power Deliv.* **2011**, *26*, 2436–2442. [CrossRef]
12. Depuru, S.; Wang, L.; Devabhaktuni, V.; Nelapati, P. A hybrid neural network model and encoding technique for enhanced classification of energy consumption data. In Proceedings of the IEEE Power and Energy Society General Meeting, San Diego, CA, USA, 24–29 July 2011; pp. 1–8.

13. Depuru, S.; Wang, L.; Devabhaktuni, V.; Green, R. High performance computing for detection of electricity theft. *Int. J. Electr. Power Energy Syst.* **2013**, *47*, 21–30.

14. Jiang, H.; Wang, K.; Wang, Y.; Gao, M.; Zhang, Y. Energy big data: A survey. *IEEE Access* **2016**, *4*, 3844–3861. [CrossRef]

15. Batalla-Bejerano, J.; Trujillo-Baute, E.; Villa-Arrieta, M. Smart meters and consumer behaviour: Insights from the empirical literature. *Energy Policy* **2020**, *144*, 111610. [CrossRef]

16. Punmiya, R.; Choe, S. Energy theft detection using gradient boosting theft detector with feature engineering-based preprocessing. *IEEE Trans. Smart Grid* **2019**, *10*, 2326–2329. [CrossRef]

17. Ramos, C.C.O.; Rodrigues, D.; de Souza, A.N.; Papa, J.P. On the study of commercial losses in brazil: A binary black hole algorithm for theft characterization. *IEEE Trans. Smart Grid* **2016**, *9*, 676–683. [CrossRef]

18. Fenza, G.; Gallo, M.; Loia, V. Drift-aware methodology for anomaly detection in smart grid. *IEEE Access* **2019**, *7*, 9645–9657. [CrossRef]

19. Li, S.; Han, Y.; Yao, X.; Yingchen, S.; Wang, J.; Zhao, Q. Electricity theft detection in power grids with deep learning and random forests. *J. Electr. Comput. Eng.* **2019**, *2019*, 4136874. [CrossRef]

20. Buzau, M.M.; Tejedor-Aguilera, J.; Cruz-Romero, P.; Gómez-Expósito, A. Hybrid deep neural networks for detection of non-technical losses in electricity smart meters. *IEEE Trans. Power Syst.* **2019**, *35*, 1254–1263.

21. Rouzbahani, H.M.; Karimipour, H.; Lei, L. An Ensemble Deep Convolutional Neural Network Model for Electricity Theft Detection in Smart Grids. In Proceedings of the 2020 IEEE International Conference on Systems, Man, and Cybernetics (SMC), Toronto, ON, Canada, 11–14 October 2020; pp. 3637–3642.

22. Ghori, K.M.; Abbasi, R.A.; Awais, M.; Imran, M.; Ullah, A.; Szathmary, L. Performance analysis of different types of machine learning classifiers for non-technical loss detection. *IEEE Access* **2019**, *8*, 16033–16048. [CrossRef]

23. Buzau, M.; Tejedor-Aguilera, J.; Cruz-Romero, P.; Gomez-Exposito, A. Detection of Non-Technical Losses Using Smart Meter Data and Supervised Learning. *IEEE Trans. Smart Grid* **2019**, *10*, 2661–2670.

24. Kong, X.; Zhao, X.; Liu, C.; Li, Q.; Dong, D.; Li, Y. Electricity theft detection in low-voltage stations based on similarity measure and DT-KSVM. *Int. J. Electr. Power Energy Syst.* **2021**, *125*, 106544. [CrossRef]

25. Hasan, M.; Toma, R.; Nahid, A.; Islam, M.; Kim, J. Electricity Theft Detection in Smart Grid Systems: A CNN-LSTM Based Approach. *Energies* **2019**, *12*, 3310. [CrossRef]

26. Ismail, M.; Shaaban, M.; Naidu, M.; Serpedin, E. Deep Learning Detection of Electricity Theft Cyber-Attacks in Renewable Distributed Generation. *IEEE Trans. Smart Grid* **2020**, *11*, 3428–3437. [CrossRef]

27. Maamar, A.; Benahmed, K. A Hybrid Model for Anomalies Detection in AMI System Combining K-means Clustering and Deep Neural Network. *Comput. Mater. Contin.* **2019**, *60*, 15–39. [CrossRef]

28. Li, W.; Logenthiran, T.; Phan, V.; Woo, W. A Novel Smart Energy Theft System (SETS) for IoT-Based Smart Home. *IEEE Internet Things J.* **2019**, *6*, 5531–5539. [CrossRef]

29. Manoharan, H.; Teekaraman, Y.; Kirpichnikova, I.; Kuppusamy, R.; Nikolovski, S.; Baghaee, H.R. Smart grid monitoring by wireless sensors using binary logistic regression. *Energies* **2020**, *15*, 3974. [CrossRef]

30. Shehzad, F.; Javaid, N.; Almogren, A.; Ahmed, A.; Gulfam, S.M.; Radwan, A. A Robust Hybrid Deep Learning Model for Detection of Non-Technical Losses to Secure Smart Grids. *IEEE Access* **2021**, *9*, 128663–128678. [CrossRef]

31. Chung, J.; Gulcehre, C.; Cho, K.; Bengio, Y. Empirical evaluation of gated recurrent neural networks on sequence modeling. *arXiv* **2014**, arXiv:1412.3555.

32. Ding, N.; Ma, H.; Gao, H.; Ma, Y.; Tan, G. Real-time anomaly detection based on long short-Term memory and Gaussian Mixture Model. *Comput. Electr. Eng.* **2019**, *79*, 106458. [CrossRef]

33. Hand, D.; Christen, P. A note on using the F-measure for evaluating record linkage algorithms. *Stat. Comput.* **2018**, *28*, 539–547. [CrossRef]

34. Gu, Y.; Cheng, L.; Chang, Z. Classification of Imbalanced Data Based on MTS-CBPSO Method: A Case Study of Financial Distress Prediction. *J. Inf. Process. Syst.* **2019**, *15*, 682–693.

35. Douzas, G.; Bacao, F.; Fonseca, J.; Khudinyan, M. Imbalanced Learning in Land Cover Classification: Improving Minority Classes' Prediction Accuracy Using the Geometric SMOTE Algorithm. *Remote Sens.* **2019**, *11*, 3040. [CrossRef]

36. Bisong, E. *Building Machine Learning and Deep Learning Models on Google Cloud Platform*; Springer: Berkeley, CA, USA, 2019.

37. Sun, C.; Shrivastava, A.; Singh, S.; Gupta, A. Revisiting unreasonable effectiveness of data in deep learning era. In Proceedings of the IEEE International Conference on Computer Vision, Venice, Italy, 22–29 October 2017.

38. Adil, M.; Javaid, N.; Qasim, U.; Ullah, I.; Shafiq, M.; Choi, J. LSTM and Bat-Based RUSBoost Approach for Electricity Theft Detection. *Appl. Sci.* **2020**, *10*, 4378. [CrossRef]

 *sustainability*

Article

Load Frequency Control and Automatic Voltage Regulation in a Multi-Area Interconnected Power System Using Nature-Inspired Computation-Based Control Methodology

Tayyab Ali [1], Suheel Abdullah Malik [1], Ibrahim A. Hameed [2,*], Amil Daraz [3], Hana Mujlid [4] and Ahmad Taher Azar [5,6,7,*]

1 Department of Electrical Engineering, FET, International Islamic University, Islamabad 44000, Pakistan
2 Department of ICT and Natural Sciences, Norwegian University of Science and Technology, Larsgårdsve-gen, 2, 6009 Ålesund, Norway
3 School of Information Science and Engineering, Ningbotech University, Ningbo 315100, China
4 Department of Computer Engineering, Taif University, Taif 21944, Saudi Arabia
5 College of Computer and Information Sciences, Prince Sultan University, Riyadh 11586, Saudi Arabia
6 Automated Systems and Soft Computing Lab (ASSCL), Prince Sultan University, Riyadh 12435, Saudi Arabia
7 Faculty of Computers and Artificial Intelligence, Benha University, Benha 13518, Egypt
* Correspondence: ibib@ntnu.no (I.A.H.); aazar@psu.edu.sa or ahmad.azar@fci.bu.edu.eg or ahmad_t_azar@ieee.org (A.T.A.)

check for
updates

Citation: Ali, T.; Malik, S.A.; Hameed, I.A.; Daraz, A.; Mujlid, H.; Azar, A.T. Load Frequency Control and Automatic Voltage Regulation in a Multi-Area Interconnected Power System Using Nature-Inspired Computation-Based Control Methodology. *Sustainability* **2022**, *14*, 12162. https://doi.org/10.3390/su141912162

Academic Editors: Herodotos Herodotou, Sheraz Aslam and Nouman Ashraf

Received: 19 August 2022
Accepted: 20 September 2022
Published: 26 September 2022

Publisher's Note: MDPI stays neutral with regard to jurisdictional claims in published maps and institutional affiliations.

Abstract: The stability control of nominal frequency and terminal voltage in an interconnected power system (IPS) is always a challenging task for researchers. The load variation or any disturbance affects the active and reactive power demands, which badly influence the normal working of IPS. In order to maintain frequency and terminal voltage at rated values, controllers are installed at generating stations to keep these parameters within the prescribed limits by varying the active and reactive power demands. This is accomplished by load frequency control (LFC) and automatic voltage regulator (AVR) loops, which are coupled to each other. Due to the complexity of the combined AVR-LFC model, the simultaneous control of frequency and terminal voltage in an IPS requires an intelligent control strategy. The performance of IPS solely depends upon the working of the controllers. This work presents the exploration of control methodology based on a proportional integral–proportional derivative (PI-PD) controller with combined LFC-AVR in a multi-area IPS. The PI-PD controller was tuned with recently developed nature-inspired computation algorithms including the Archimedes optimization algorithm (AOA), learner performance-based behavior optimization (LPBO), and modified particle swarm optimization (MPSO). In the earlier part of this work, the proposed methodology was applied to a two-area IPS, and the output responses of LPBO-PI-PD, AOA-PI-PD, and MPSO-PI-PD control schemes were compared with an existing nonlinear threshold-accepting algorithm-based PID (NLTA-PID) controller. After achieving satisfactory results in the two-area IPS, the proposed scheme was examined in a three-area IPS with combined AVR and LFC. Finally, the reliability and efficacy of the proposed methodology was investigated on a three-area IPS with LFC-AVR with variations in the system parameters over a range of Â ± 50%. The simulation results and a comprehensive comparison between the controllers clearly demonstrates that the proposed control schemes including LPBO-PI-PD, AOA-PI-PD, and MPSO-PI-PD are very reliable, and they can effectively stabilize the frequency and terminal voltage in a multi-area IPS with combined LFC and AVR.

Keywords: PI-PD controller; load frequency control; automatic voltage regulator; nature-inspired optimization; multi-area interconnected power system

1. Introduction

Research efforts and specializations in power systems are increasing day by day to acquire reliable power with nominal voltage and frequency. In a power system, the main goal is to provide nominal voltage and frequency to all consumers without any interruption. The simultaneous control of load frequency and terminal voltage in an interconnected electrical power system is the fundamental area of research for all practicing engineers. The mutilation of frequency or voltage can spoil the performance and life expectancy of equipment associated with IPS [1]. The active and reactive powers can change with load demands in IPS. The active power can be adjusted by a speed governor in an LFC loop, whereas reactive power can be controlled by an exciter in an AVR loop. In order to fulfill the active power demand, a turbine input is continuously regulated in LFC, or else the changing frequency will vary the machine's speed. In AVR, terminal voltage remains within the prescribed limit if the excitation of generators is regulated properly to match the reactive power demand. A lot of literature is available on individual AVR or LFC systems; however, relatively less research work has been carried out on combined LFC-AVR due to its complex design. The PID controller was extensively used in multi-area IPS due to its simple design and easier installation. For instance, the artificial electric field algorithm-based hybridized approach to tune the fuzzy PID controller was suggested for combined LFC and AVR with the incorporation of different energy storage devices [1]. A particle swarm-optimized Ziegler–Nicholas (PSO-ZN)-based PID controller was examined for AVR-LFC control in PV integration and a conventional power system [2]. PI and PID with filter (PIDF) controllers based on the sine cosine algorithm were also inquired for a two-area, two-source IPS. The redox flow batteries were assimilated for further improvements in the system dynamics [3]. The doctor and patient optimization (DPO)-based accelerating PID controller (PIDA) was proposed for the LFC-AVR problem in a multi-area IPS with renewable energy sources [4]. The PID controller was employed for collective AVR-LFC in a two-area IPS. A nonlinear threshold-accepting algorithm was explored to find the optimum parameters of the PID controller [5]. PI and I controllers for AVR and an LFC loop were also investigated for a single-area IPS [6]. In [7], due to the inclusion of deregulated environments in IPS, a fuzzy logic controller (FLC) was recommended for a two-area LFC-AVR problem. A fractional order controller ($PID^{\mu}F$) based on the lightning search algorithm (LSA) was also proposed for LFC-AVR with wind and a reheat thermal plant as the generating companies (GENCOs) of area-1,and with diesel and a nonlinear reheat thermal plant as the GENCOs of area-2 under deregulated environments [8].The PID controller was optimized with the hybridization of the artificial electric field algorithm and differential evolution for a two-area IPS with a joint LFC-AVR [9,10]. In [11], PID with the firefly algorithm was employed for a two-area IPS with AVR-LFC. The moth flame optimization (MFO)-based fractional order PID controller was proposed for both LFC and AVR loops [12]. For a single-area synchronous generator, the combined LFC-AVR was explored using a hardware environment [13]. In [14,15], the authors inspected the firefly algorithm, particle swarm optimization, and the genetic algorithm-based PID controller for AVR-LFC loops. The novel state-observer (SO)-based integral double-derivative controller based on magneto-tactic-bacteria optimization (MBO) was presented for voltage–frequency control in a hybrid IPS [16]. The model predictive controller (MPC) was also used to improve AVR-LFC responses [17]. In [18], the heuristic computation-based two degrees of freedom state-feedback PI controller was exploited for the AVR loop in synchronous generators. A combination of the bacterial foraging optimization algorithm and particle swarm optimization was utilized to tune the PI controller for the AVR system with a static synchronous compensator [19]. In [20], a sliding mode controller with the addition of a gene ralized extended state observer was successfully explored to optimize the LFC loop in a multi-area IPS. The PID controller tuned with the many optimizing liaisons (MOL) algorithm was applied to a two-area IPS with non-reheat thermal sources in the presence of GDB [21]. Moreover, a comprehensive research work was presented for individual LFC loops as presented in [22–37]. A brief literature summary of AVR-LFC is provided in Table 1. It can be seen that much less attention has been given to the combined LFC-AVR problem in multi-area IPS due to its complex structure. The literature survey also depicts that modified forms of the PID controller were explored very rarely for combined AVR-LFC. Different modified forms such as PI, PIDF, PID$^\mu$F, and FO-IDF have been explored due to their excellent time response characteristics with fast convergence, but the PI-PD controller has not been employed for combined LFC-AVR multi-area IPS. Due to its modified structure having a control branch in the feedback path, complex systems can be well optimized with PI-PD as compared to classical control schemes such PI and PID, etc. To obtain optimal controller parameters, an intelligent tuning algorithm is needed, which can optimize the controller with minimum error/fitness. In the past, nature-inspired optimization algorithms have received a lot of attention from researchers because of their strengths and abilities to tackle a variety of complex optimization issues in engineering. These strategies have also been used successfully to obtain optimal controller parameters. The classical nature-inspired computing techniques have shown very satisfactory performances for both individual and combined LFC-AVR. Moreover, researchers have also presented some novel nature-computing algorithms such as dandelion optimizer [38], modified particle swarm optimization (MPSO), bald eagle search (BES) [39], the transient search algorithm (TSO) [40], learner performance-based behavior optimization (LPBO) [41], the Archimedes optimization algorithm (AOA) [42], etc. These recently introduced techniques such as MPSO, LPBO, and AOA have not been considered for the optimal tuning of the PI-PD control scheme. It will be worth choosing these nature-inspired techniques for the optimization of multi-area IPS with combined LFC-AVR. Keeping in mind the existing research gap, the nature-inspired computation-based PI-PD control scheme is proposed in this research for multi-area IPS with combined AVR-LFC. The main contributions of this work are:

1. The modeling of combined AVR-LFC for two-area and three-area IPS;
2. The modeling of the PI-PD control scheme and its optimization using the Archimedes optimization algorithm (AOA), learner performance-based behavior optimization (LPBO), and modified particle swarm optimization (MPSO);
3. The formulation of fitness functions for the optimization of proposed controller;
4. Further, a comprehensive performance comparison is carried out between LPBO-PI-PD, AOA-PI-PD, and MPSO-PI-PD in two-area IPS. Moreover, the efficacy of the proposed control schemes has been tested in a three-area IPS with a combined LFC-AVR problem;
5. The reliability of the proposed control methodology has been illustrated by altering the system parameters of three-area IPS over a range of $\hat{A} \pm 50\%$.

Table 1. Literature on ALR-LFC.

Reference	Year	Research Area	Controller	Tuning Schemes	Area/System	Nonlinearities	Additional Incorporation
[2]	2021	AVR-LFC	PID	PSO-ZN	Two Area	-	-
[3]	2020	AVR-LFC	PI, PIDF	CSA	Two Area	-	RFBs, UPFC
[4]	2022	AVR-LFC	PIDA	DPO	Two Area		-
[5]	2019	AVR-LFC	PID	NLTA	Single Area	-	-
[6]	2014	AVR-LFC	PI	Not given	Single Area	-	Damper Winding
[7]	2019	AVR-LFC	PID, FLC	Fuzzy Logic	Two Area	-	DC Link, Deregulated Environment
[8]	2018	AVR-LFC	PIDF, PID$^\mu$F	LSA	Two Area	GDB, GRC	SMES, IPFC, Deregulated Environment
[9]	2020	AVR-LFC	PID	DE-AEFA	Two Area	GRC	IPFC and RFBs
[10]	2020	AVR-LFC	PID	DE-AEFA	Two Area	GRC	HVDC link with the existing AC tie-line
[11]	2019	AVR-LFC	PID	FO	Two Area	-	-
[12]	2019	AVR-LFC	FO-PID	MFO	Two Area	GDB, BD	-
[13]	2020	AVR-LFC	PI	HIL Strategy	Single Area	-	-
[14]	2019	AVR-LFC	PID	FA, GA, PSO	Single Area	-	-

Table 1. *Cont.*

Reference	Year	Research Area	Controller	Tuning Schemes	Area/System	Nonlinearities	Additional Incorporation
[15]	2015	AVR-LFC	PID	PSO	Two Area	-	-
[16]	2018	AVR-LFC	SO-IDD	MBO	Two Area	GRC, GDB	-
[17]	2020	AVR-LFC	MPC	MPC	Two Area	-	-
[18]	2020	AVR	2DOF state-feedback PI control	VSA, WOA, SCA GWO, SSA, WCA	AVR for Synchronous Generator	-	-
[19]	2021	AVR	PI	Hybrid BFOA-PSO	Standalone Wind–Diesel Power System	-	STATCOM
[20]	2019	LFC	Observer-based nonlinear sliding mode control	LMI	Two Area	GRC, GDB	-
[21]	2021	LFC	PID	MOL	Two Area	GDB	-
Proposed Work	2022	AVR-LFC	PI-PD	AOA, LPBO, MPSO	Two Area, Three Area	-	-

Table 2 demonstrates the nomenclature used in this study. This research paper is organized in following way: The power system model is described in Section 2. The proposed control methodology is presented in Section 3. Section 4 contains a description and flow charts of nature-inspired computation algorithms including LPBO, AOA, and MPSO. The implementation and results of the proposed techniques are summarized in Section 5. Lastly, conclusions and future guidelines are given in Section 6.

Table 2. Nomenclature.

Acronym	Definition	Acronym	Definition
AOA	Archimedes optimization algorithm	IPS	Interconnected power system
NLTA	Nonlinear threshold-accepting algorithm	LPBO	Learner performance-based behavior optimization
AVR	Automatic voltage regulator	ΔP_{tie}	Tie-line power deviation
PI-PD	Proportional integral–proportional derivative	V_t	Terminal voltage
MPSO	Modified particle swarm optimization	LFC	Load frequency control
R_i	Speed regulation	Δf	Frequency deviation
K_G	Governor gain	B	Area bias factor
T_G	Time constant of governor	ΔP_D	Load deviation
K_a	Amplifier gain	K_t	Turbine gain
T_a	Time constant of amplifier	T_t	Time constant of turbine
K_g	Generator gain	K_e	Exciter gain
T_g	Time constant of generator	T_e	Time constant of exciter
K_p	Power system gain	ΔX_G	Valve position of governor
T_p	Time constant of power system	ΔP_G	Deviation in the output of generator
T_{12}, T_{21}	Tie-line synchronizing time constants	K_i	Coupling coefficient of AVR-LFC loops

2. Power System Model

The multi-area IPS model under study is shown in Figure 1. The terminal voltage was maintained at nominal value by stabilizing the generator fields, while the load frequency was regulated by controlling real power. Figure 1a represents the AVR-LFC model of a power system for a single area, where *i* and *j* represent area-1 and area-2, respectively [5].

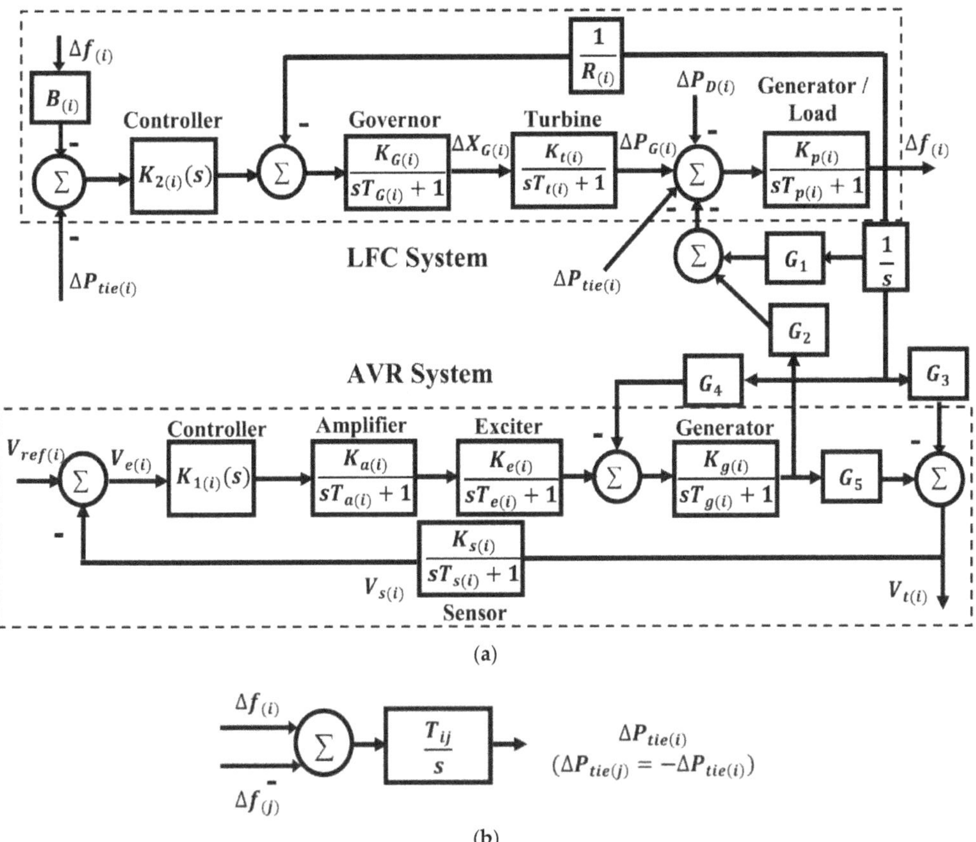

(a)

(b)

Figure 1. (**a**) Single-area AVR–LFC [5]. (**b**) Connections of two-area tie–line.

$V_{t(i)}$, $V_{e(i)}$, $V_{ref(i)}$, and $V_{s(i)}$ refer to the terminal output, error, reference, and sensor voltage in area-1, respectively. The AVR system of area-i consists of a controller ($K_{1(i)}(s)$), amplifier ($\frac{K_{a(i)}}{sT_{a(i)}+1}$), generator ($\frac{K_{g(i)}}{sT_{g(i)}+1}$), exciter ($\frac{K_{e(i)}}{sT_{e(i)}+1}$), and sensor ($\frac{K_{s(i)}}{sT_{s(i)}+1}$). Area-1's LFC system has a controller $K_{2(i)}(s)$, turbine ($\frac{K_{t(i)}}{sT_{t(i)}+1}$), governor ($\frac{K_{G(i)}}{sT_{G(i)}+1}$), speed regulation ($R_i$), and generator/load ($\frac{K_{p(i)}}{sT_{p(i)}+1}$). $\Delta f_{(i)}$ denotes frequency deviation (Hz), $\Delta X_{G(i)}$ shows the valve position of the governor (p.u.MW), $\Delta P_{G(i)}$ represents the deviation in the output of the generator (p.u.MW), $\Delta P_{D(i)}$ (p.u.MW) denotes the deviation in load, speed regulation is represented by $R_{(i)}$(Hz p.u.MW-1), and $\Delta P_{tie(i)}$ is the tie–line power. The purpose of tie–line is to interconnect multiple areas in IPS. Figure 1b shows the tie–line connections. The synchronization coefficient between area-i and area-j is represented by T_{ij}.

3. Proposed Control Methodology

The proportional integral derivative (PID) controller is commonly utilized in industrial applications owing to its easier implementation and simpler structure. The PID controller provides a satisfactory performance in most of the systems; however, the modified forms of the PID control structure have shown improved performance in many control systems, such as the AVR-LFC interconnected power system. The proportional integral–proportional derivative controller (PI-PD) is a modified version of PID, which is designed in such a way to eliminate system errors with optimum transient and steady state response [43]. The PI part of PI-PD exists in a feed forward path and directly responds to the error signal coming from the summing junction. The PD part is located in the feedback path, and it is unaffected by sudden changes in the set point specification. The closed-loop response can be improved significantly with the addition of a controller part in the feedback path. The PI-PD controller has been successfully employed in the recent past in different applications [44–50]. The proposed control methodology with the combined LFC-AVR system is given in Figure 2. The transfer function of PI-PD controllers is represented as:

$$U(s) = (K_{p1} + \frac{K_i}{s})E(s) - (K_{p2} + K_d s)Y(s) \tag{1}$$

$$E(s) = Y(s) - R(s) \tag{2}$$

where $U(s)$, $Y(s)$, $R(s)$, and $E(s)$ denote the control, output, reference, and error signals, respectively. The cost function (J) is minimized to obtain the best possible parameters of the controllers. J depends upon $E(s)$, which is basically the difference between the output and reference signal.

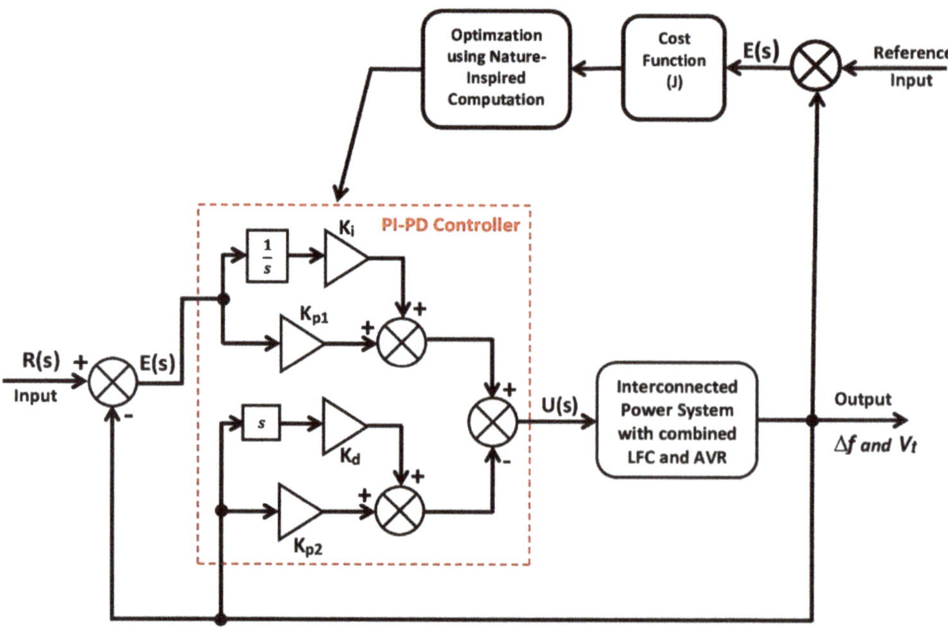

Figure 2. Proposed control methodology with combined LFC−AVR system.

In order to minimize the error signal, different types of performance indices can be used such as the integral of the squared value of the error signal (ISE), the integral of the time multiplied with the absolute value of the error signal (ITAE), the integral of the time multiplied with the squared value of the error signal (ITSE), and the integral of the absolute value of error (IAE) represented by the following equations:

$$J_{\text{ISE,two−area}} = \int_0^T [\Delta f_1^2 + \Delta f_2^2 + \Delta V_{t1}^2 + \Delta V_{t2}^2 + \Delta P_{tie12}^2]dt \tag{3}$$

$$J_{\text{ITAE,two−area}} = \int_0^T t[|\Delta f_1| + |\Delta f_2| + |\Delta V_{t1}| + |\Delta V_{t2}| + |\Delta P_{ptie12}|]dt \tag{4}$$

$$J_{\text{ITSE,two-area}} = \int_0^T t[\Delta f_1^2 + \Delta f_2^2 + \Delta V_{t1}^2 + \Delta V_{t2}^2 + \Delta P_{tie12}^2]dt \tag{5}$$

$$J_{\text{IAE,two-area}} = \int_0^T [|\Delta f_1| + |\Delta f_2| + |\Delta V_{t1}| + |\Delta V_{t2}| + |\Delta P_{ptie12}|]dt \tag{6}$$

For three-area IPS, we can write:

$$J_{\text{ISE,three-area}} = \int_0^T [\Delta f_1^2 + \Delta f_2^2 + \Delta f_3^2 + \Delta V_{t1}^2 + \Delta V_{t2}^2 + \Delta V_{t3}^2 + \Delta P_{tie1}^2 + \Delta P_{tie2}^2 + \Delta P_{tie3}^2]dt \tag{7}$$

$$J_{\text{IAE,three-area}} =$$
$$\int_0^T [|\Delta f_1| + |\Delta f_2| + |\Delta f_3| + |\Delta V_{t1}| + |\Delta V_{t2}| + |\Delta V_{t3}| + |\Delta P_{ptie1}| + |\Delta P_{ptie2}| + |\Delta P_{ptie3}|]dt \tag{8}$$

$$J_{\text{ITSE,three-area}} = \int_0^T t[\Delta f_1^2 + \Delta f_2^2 + \Delta f_3^2 + \Delta V_{t1}^2 + \Delta V_{t2}^2 + \Delta V_{t3}^2 + \Delta P_{tie1}^2 + \Delta P_{tie2}^2 + \Delta P_{tie3}^2]dt \tag{9}$$

$$J_{\text{ITAE,three-area}} = \int_0^T t[|\Delta f_1| + |\Delta f_2| + |\Delta f_3| + |\Delta V_{t1}| + |\Delta V_{t2}| + |\Delta V_{t3}| + |\Delta P_{ptie1}| + |\Delta P_{ptie2}| + |\Delta P_{ptie3}|]dt \tag{10}$$

where,

$$\Delta V_{t1} = V_{ref} - V_{t1} \quad \Delta V_{t2} = V_{ref} - V_{t2} \quad \Delta V_{t3} = V_{ref} - V_{t3} \tag{11}$$

$$\Delta P_{ptie1} = \Delta P_{ptie12} + \Delta P_{ptie13} \quad \Delta P_{ptie2} = \Delta P_{ptie21} + \Delta P_{ptie23} \quad \Delta P_{ptie3} = \Delta P_{ptie31} + \Delta P_{ptie32} \tag{12}$$

When the cost function is minimized, the algorithm returns the best optimum parameters of the controller. To optimize the cost function (J), nature-inspired computation algorithms including LPBO, AOA, and MPSO were adapted.

4. Nature-Inspired Computation Algorithms

Due to their ability to solve complex valued problems, nature-inspired computation algorithms have gained brilliant attention in IPS. Keeping in view their remarkable contribution, an effort was made in this research to optimize the combined LFC and AVR-based IPS using nature-inspired computation techniques.

4.1. Learner Performance-Based Behavior Optimization

Rashid and Rahman presented a novel nature-inspired learner performance-based behavior optimization (LPBO) technique in 2020. The basic concept behind this algorithm is based on the fact that how students are admitted to different departments of a university is based on their high school performance. After admission, students must be able to improve their intellectual level to improve their skills. In this way, both exploitation and exploration phases are preserved. In this algorithm, a random population is generated with various ranges of grade point average (GPA). The applications of some of these learners will be rejected or accepted based on their fitness. After that, the population is divided in to subpopulation. Fitness is calculated and is then sorted into separate groups. The new population's structure is changed using crossover and mutation operators. A specified number of learners is acquired by different departments based on the minimum GPA criteria. This rejection and acceptance process is continued until all departments have their vacancies filled. Population fitness is improved in each iteration based on group learning, intellectual level, and teaching level [41]. Figure 3 presents the flow chart of the LPBO algorithm. Note that the LPBO population represents the PI-PD controller's parameters in this case.

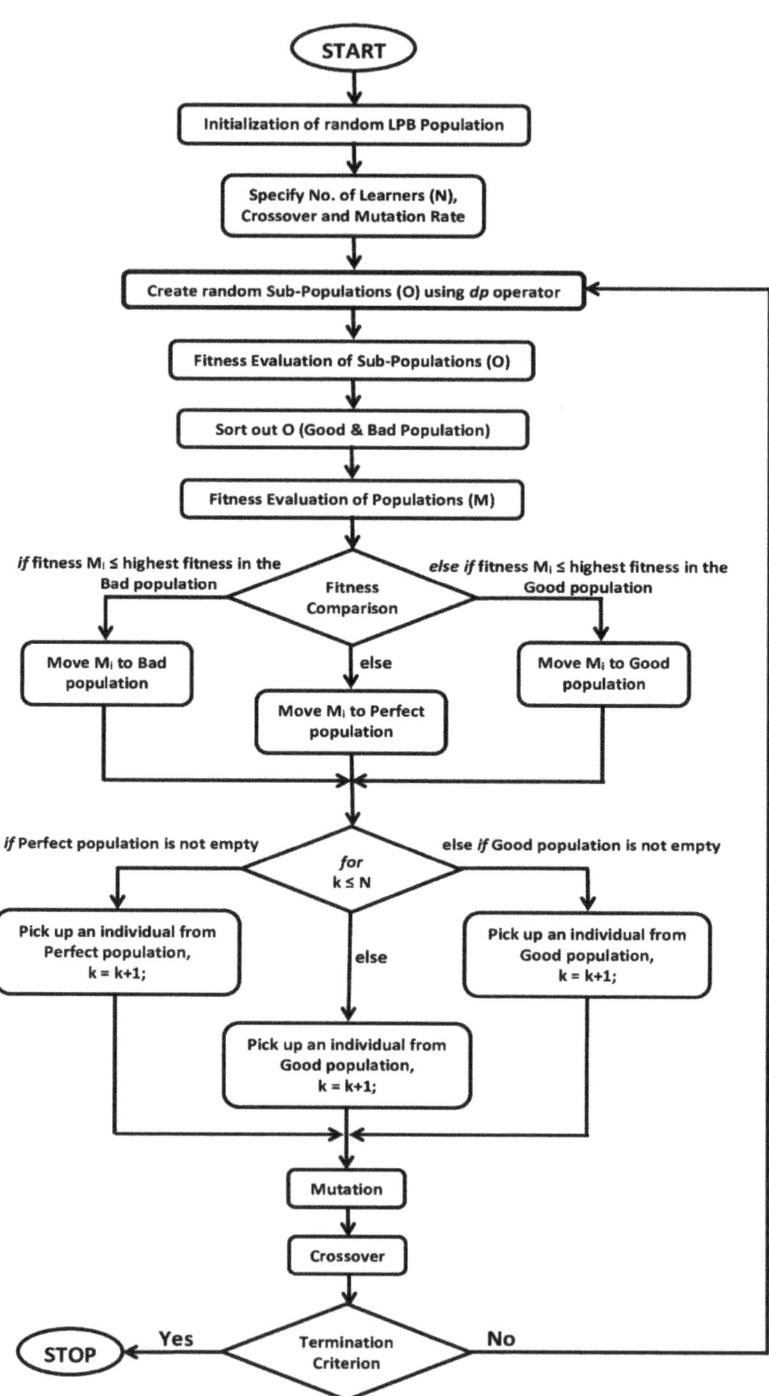

Figure 3. Flow chart of LPBO algorithm.

4.2. Archimedes Optimization Algorithm

The Archimedes optimization algorithm (AOA) is a new state-of-the-art algorithm based on the Archimedes principle. It deals with both convex and non-convex problems. It was invented in 2021 by Fatmaand Houssein. It defines the relationship between a buoyant force and an object submerged in water. The object will sink if the displaced fluid weight is less than the weight of the object. Similarly, if the displaced fluid and object weight are equal, the object floats on the fluid. An object has volume, acceleration, and density that results in the buoyancy force, as a result fluid's net force is always zero. AOA is a very effective nature-inspired algorithm in a way that it analyzes a problem with a global optimum solution. AOA fences in both exploitation and exploration phases since it is a global optimization algorithm. A comprehensive area must be examined to identify the global optimum solution of a given problem. Firstly, the fluid's random position is initialized, and then AOA evaluates the initial population fitness to discover the best possible solution until the selection criteria are met. The density and volume of each object changes at each AOA iteration. The new density, volume, and acceleration are obtained using the object's fitness. The AOA population represents the PI-PD controller's parameters [42]. Figure 4 presents a flow chart diagram of AOA.

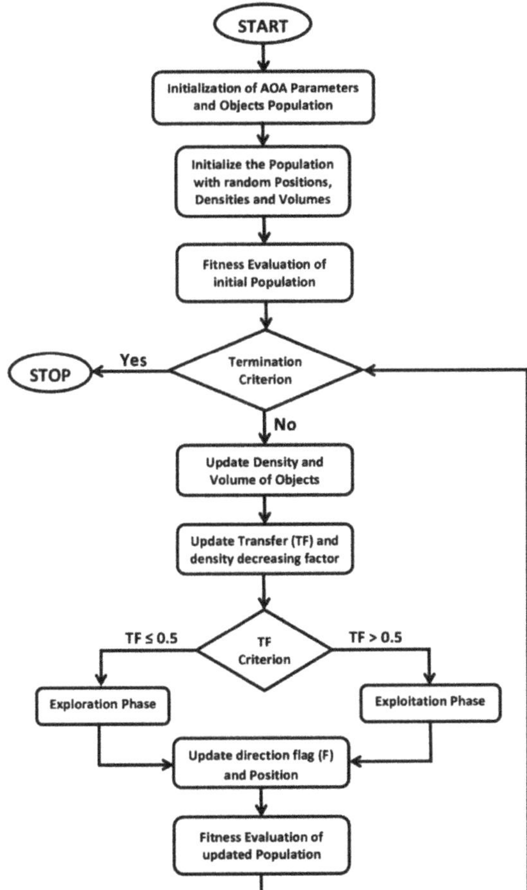

Figure 4. Flow chart of AOA.

4.3. Modified Particle Swarm Optimization (PSO)

Drawing inspiration from swarm intelligence, Eberhart and Kennedy proposed the particle swarm optimization (PSO) algorithm in 1995. In PSO, the movement of particles (candidate solutions) over a defined search space depends upon their velocity and position. The movement of particles is incited by the best possible positions known as local bests. These local bests lead particles toward the best possible position [51]. In modified particle swarm optimization (MPSO), the global learning coefficient is updated using a combination of existing local and global learning coefficients. The modification in the PSO algorithm is being made to improve the convergence characteristics of the controller. Figure 5 depicts the flow chart of the MPSO algorithm. Remember that in this research work, the particles represent the PI-PD controller's parameters.

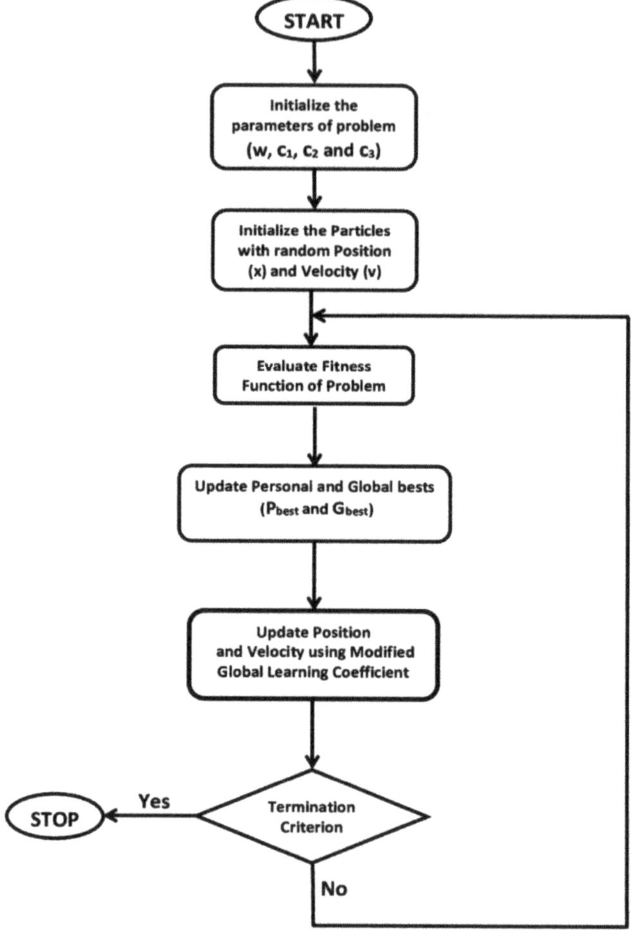

Figure 5. Flow chart of MPSO algorithm.

5. Implementation and Results Discussion

Multiple simulations were carried out in MATLAB/Simulink to express the validation of the proposed control methodology. Firstly, a two-area, two-source IPS with combined LFC and AVR was optimized using LPBO, AOA, and MPSO. ITSE was chosen as the error criterion, due to efficient error convergence characteristics. After achieving successful results, the proposed methodology was applied to a three-area, three-source IPS with LFC-AVR loops.

5.1. Optimization of Two-Area Interconnected Power System

The two-area IPS model under investigation with a collective LFC-AVR system is shown in Figure 6. The system parameters of the two-area IPS are specified in Appendix A. The system parameters of area-1 and area-2 were chosen from [5] for a direct comparison of the proposed methodology with the NLTA-PID controller. The parameters of optimization algorithms such as MPSO, LPBO, and AOA used in simulations are given in Table 3. The tie–line connection between area-1 and are-2 can be established using Figure 7. The optimal parameters of MPSO-PI-PD, LPBO-PI-PD, and AOA-PI-PD control schemes are given in Table 4. For the sake of the assessment of the proposed control schemes, the evaluation of the time response of each schemes was carried out and comparisons were made with the results of NLTA-PID [5]. Further, a comparison between the proposed control schemes such as MPSO-PI-PD, LPBO-PI-PD, and AOA-PI-PD is also presented in detail in this section.

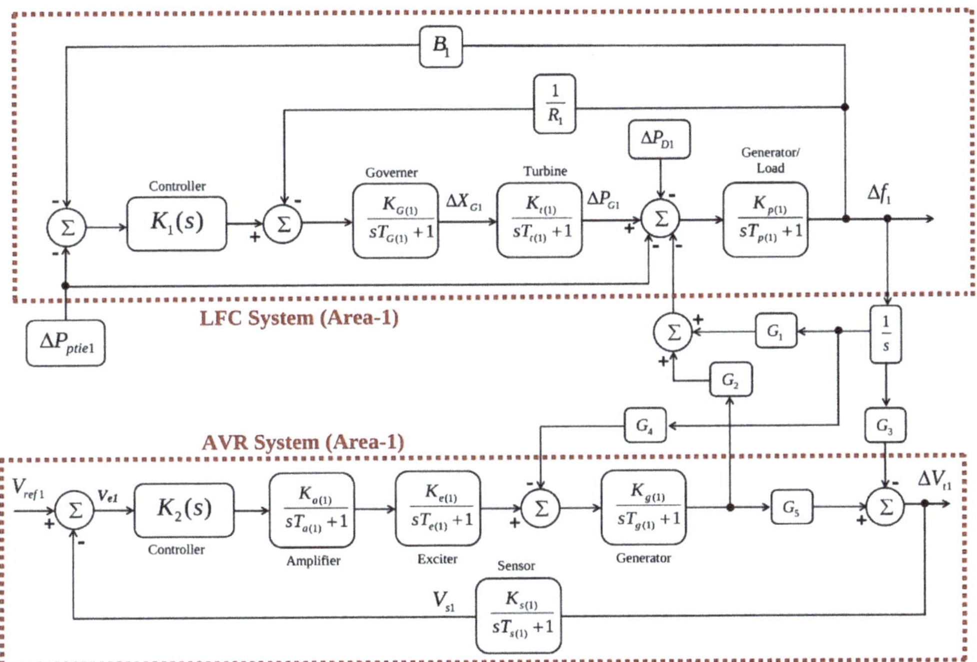

Figure 6. *Cont.*

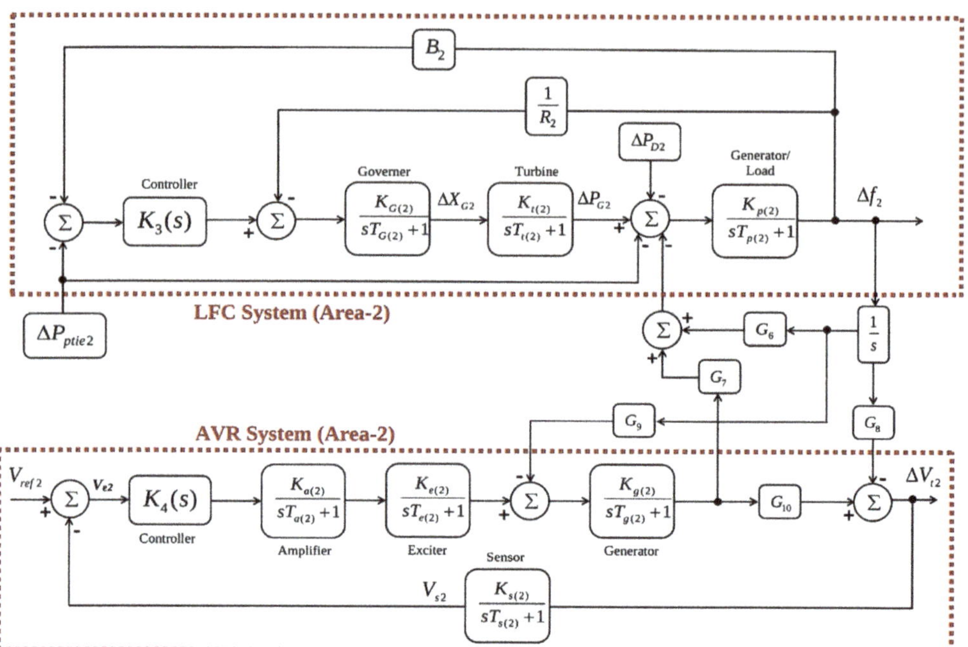

Figure 6. Two-area IPS with combined LFC−AVR.

Table 3. Parameters of optimization techniques.

MPSO		LPBO		AOA	
Parameter	Value	Parameter	Value	Parameter	Value
Population size	20	Population size	20	Population size	20
Iterations	10	Iterations	10	Iterations	10
Inertia Weight Damping Ratio	1	Crossover Percentage	0.7	C_1 (constant)	2
Personal Learning Coefficient	2.74	Mutation Percentage	0.3	C_2 (constant)	6
Global Learning Coefficient	2.88	Mutation Rate	0.03	C_3 (constant)	2
Max. Velocity Limit	0.2	Number of Mutants	6	C_4 (constant)	0.5
Min. Velocity Limit	−0.2	Number of Offspring	14	Range of Normalization (u,l)	0.9, 0.1

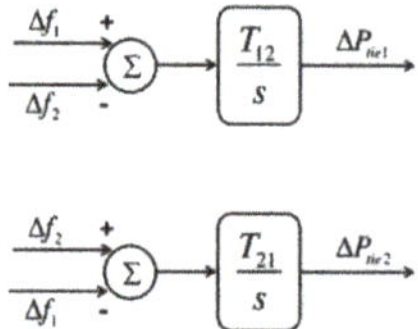

Figure 7. Tie–line connection.

Table 4. Optimal values of controller parameters (area-2).

Area	Controller Parameters	NLTA-PID [5]	Controller Parameters	Proposed Control Schemes		
				MPSO-PI-PD	**LPBO-PI-PD**	**AOA-PI-PD**
Area-1	K_{p1}	1.995	K_{p1}	1.061	1.064	1.61
	K_{i1}	1.943	K_{i1}	0.630	1.396	1.512
	K_{d1}	1.079	K_{p2}	1.162	1.071	1.88
	K_{p2}	1.994	K_{d1}	1.621	1.795	1.263
	K_{i2}	1.295	K_{p3}	1.063	1.850	1.01
	K_{d2}	1.107	K_{i2}	1.419	0.772	1.68
	-	-	K_{p4}	0.812	0.140	0.68
	-	-	K_{d2}	0.283	0.483	0.37
Area-2	K_{p3}	1.956	K_{p5}	0.564	0.965	0.90
	K_{i3}	1.919	K_{i3}	0.792	0.667	0.67
	K_{d3}	0.655	K_{p6}	0.775	0.670	1.44
	K_{p4}	1.283	K_{d3}	1.106	0.616	1.60
	K_{i4}	0.586	K_{p7}	1.903	1.522	1.50
	K_{d4}	0.819	K_{i4}	1.376	1.325	1.85
	-	-	K_{p8}	0.799	0.507	0.74
	-	-	K_{d4}	0.822	0.526	0.52
	ITSE	2.84	ITSE	0.250	0.164	0.1892

Figure 8 shows the frequency deviation curves of area-1 and area-2 using NLTA-PID [5], MPSO-PI-PD, LPBO-PI-PD, and AOA-PI-PD control techniques in a two-area IPS, respectively. It can be seen that the proposed control schemes provided a very satisfactory frequency deviation response. For the area-1 LFC, the settling time of NLTA-PID [5] was lower than the proposed schemes but at the cost of a high undershoot. NLTA-PID provided an undershoot of -0.285, whereas the proposed MPSO-PI-PD, LPBO-PI-PD, and AOA-PI-PD provided -0.130, -0.135, and -0.115, respectively. It can be noticed that the proposed MPSO-PI-PD, LPBO-PI-PD, and AOA-PI-PD provided 54%, 52.6%, and 60%, respectively, better undershoot responses as compared to the NLTA-PID controller in area-1. For area-2, NLTA-PD provided a quick settling, but it provided an undershoot of -0.275, whereas the proposed MPSO-PI-PD, LPBO-PI-PD, and AOA-PI-PD provided -0.135, -0.170, and -0.120, respectively. It was verified that the proposed MPSO-PI-PD, LPBO-PI-PD, and AOA-PI-PD provided 51%, 38%, and 56%, respectively, better undershoot responses as compared to the NLTA-PID controller. The percentages of overshoots and steady state (s-s) errors were almost zero with each proposed technique.

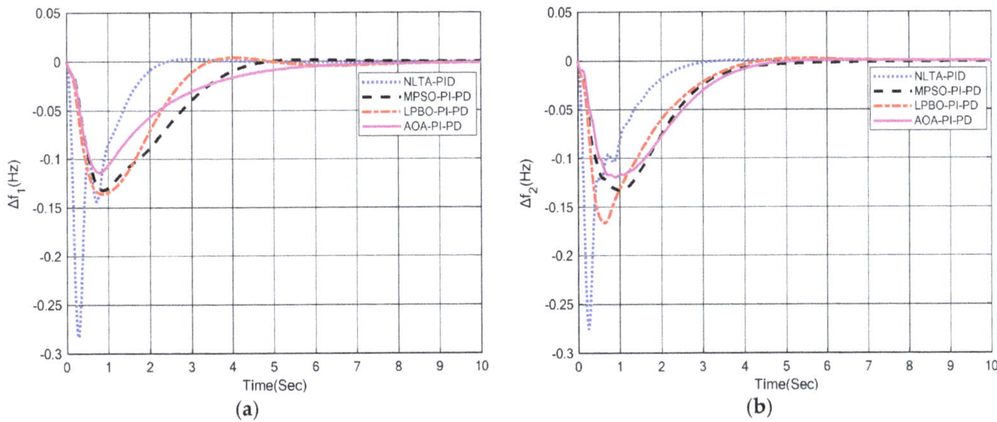

Figure 8. LFC response with PI−PD control scheme. (**a**) Δf_1; (**b**) Δf_2.

Figure 9 shows the terminal voltage of area-1 and area-2 using the NLTA-PID, MPSO-PI-PD, LPBO-PI-PD, and AOA-PI-PD control techniques in a two-area IPS, respectively. It is clear that the proposed control schemes provided a very satisfactory transient response in both area-1 and area-2. It is identified that NLTA-PID provided 18% and 17% overshoot in area-1 and area-2, respectively, but the proposed technique provided a negligible overshoot percentage at the cost of the settling time with all tuning techniques. It can be observed that the proposed LPBO-PI-PD and AOA-PI-PD control schemes produced settling times approximately the same as those achieved by NLTA-PID.

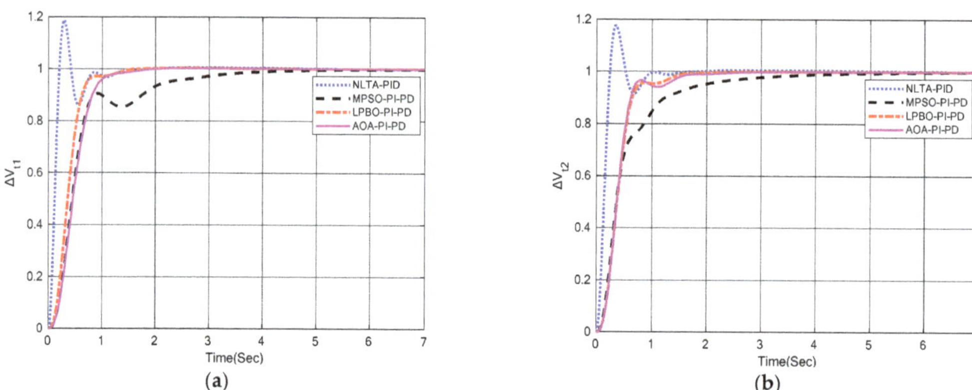

Figure 9. AVR response with PI-PD control scheme. (**a**) V_{t1}; (**b**) V_{t2}.

Figure 10 shows the tie–line power response using NLTA-PID, MPSO-PI-PD, LPBO-PI-PD, and AOA-PI-PD control techniques in a two-area IPS, respectively. It can be observed from the results that LPBO-PI-PD and AOA-PI-PD provided tie–line power responses with no undershoot; however, this was at the cost of a slightly small overshoot. In addition, the tie–line power responses yielded by MPSO-PI-PD, LPBO-PI-PD, and AOA-PI-PD were satisfactory.

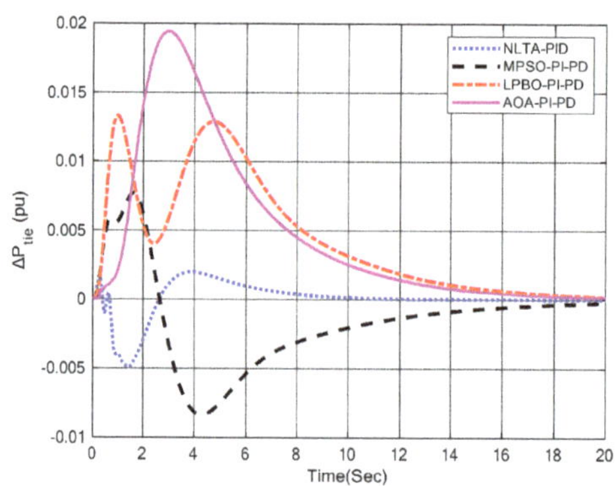

Figure 10. Tie–line power response in two−area IPS with combined LFC−AVR.

Tables 5 and 6 show the summary of LFC and AVR responses using NLTA-PID, MPSO-PI-PD, LPBO-PI-PD and AOA-PI-PD control schemes in a two-area IPS, respectively.

Table 5. LFC responses (area-2).

Control Scheme	Area-1				Area-2			
	Settling Time	% Overshoot	Undershoot	s-s Error	Settling Time	% Overshoot	Undershoot	s-s Error
NLTA-PID [5]	2.1204	0.0005	−0.285	0	2.592	0	−0.275	0
MPSO-PI-PD	4.5407	0	−0.13	0	4.92	0	−0.135	0
LPBO-PI-PD	6.9478	0.005	−0.135	0	4.043	0	−0.17	0
AOA-PI-PD	6.6752	0	−0.115	0	4.69	0	−0.12	0

Table 6. AVR responses (area-2).

Control Scheme	Area-1				Area-2			
	Rise Time	Settling Time	% Overshoot	s-s Error	Rise Time	Settling Time	% Overshoot	s-s Error
NLTA-PID [5]	0.1287	1.24	18.80	0	0.154	0.887	17.75	0
MPSO-PI-PD	0.6532	3.30	0	0	1.077	3.17	3.2971×10^{-4}	0
LPBO-PI-PD	0.4546	1.22	0.28	0	0.464	1.381	0	0
AOA-PI-PD	0.610	1.23	0.27	0	0.435	1.499	0	0

Figure 11 shows the graphical comparison of the performance parameters of NLTA-PID, MPSO-PI-PD, LPBO-PI-PD, and AOA-PI-PD control techniques in a two-area IPS, respectively. It is very clear that the proposed PI-PD control schemes provided relatively better responses in terms of the undershoot in LFC and overshoot percentage in AVR as compared to the NLTA-PID controller. From Tables 5 and 6 and Figure 11, it is concluded that the proposed MPSO-PI-PD, LPBO-PI-PD, and AOA-PI-PD were effective for maintaining the frequency and voltage within the prescribed values with a satisfactory performance in a two-area IPS.

5.2. Three-Area, Three-Source System

In this section, the proposed methodology is applied to a three-area IPS model with combined LFC-AVR. The model under study is presented in Figure 12, while the model parameters are provided in Appendix B.

The optimal values of MPSO-PI-PD, LPBO-PI-PD, and AOA-PI-PD for a three-area IPS with combined LFC and AVR are given in Table 7. Figure 13 shows the frequency deviation response using MPSO-PI-PD, LPBO-PI-PD, and AOA-PI-PD control techniques in a three-area IPS, respectively.

Table 7. Optimal values of controller parameters (area-3).

Area	Controller Parameters	Proposed Control Schemes		
		MPSO-PI-PD	LPBO-PI-PD	AOA-PI-PD
Area-1	K_{p1}	1.0995	0.66	1.51
	K_{i1}	1.1028	0.59	1.29
	K_{p2}	1.2737	0.96	−0.38
	K_{d1}	0.831	0.53	0.55
	K_{p5}	1.1106	0.77	0.86
	K_{i3}	0.9076	0.61	0.71
	K_{p6}	0.8639	1.48	1.55
Area-2	K_{d3}	1.3118	1.03	0.86
	K_{p7}	1.7917	1.68	1.91
	K_{i4}	1.8286	1.57	1.97
	K_{p8}	0.9068	0.83	1.074
	K_{d4}	0.6882	0.73	1.071

Table 7. *Cont.*

Area	Controller Parameters	Proposed Control Schemes		
		MPSO-PI-PD	**LPBO-PI-PD**	**AOA-PI-PD**
Area-3	K_{p3}	1.5371	1.56	0.88
	K_{i2}	1.965	1.62	1.91
	K_{p4}	1.2543	0.85	1.13
	K_{d2}	0.5936	0.56	0.5
	K_{p9}	0.7914	0.78	1.9
	K_{i5}	1.0795	1.12	1.26
	K_{p10}	1.2741	0.66	1.64
	K_{d5}	0.8581	1.56	0.42
	K_{p11}	1.2282	1.29	1.63
	K_{i6}	1.4326	1.3	1.69
	K_{p12}	0.9527	0.77	1.43
	K_{d6}	0.5874	0.45	1.33
	ITSE	0.3507	0.34485	0.4853

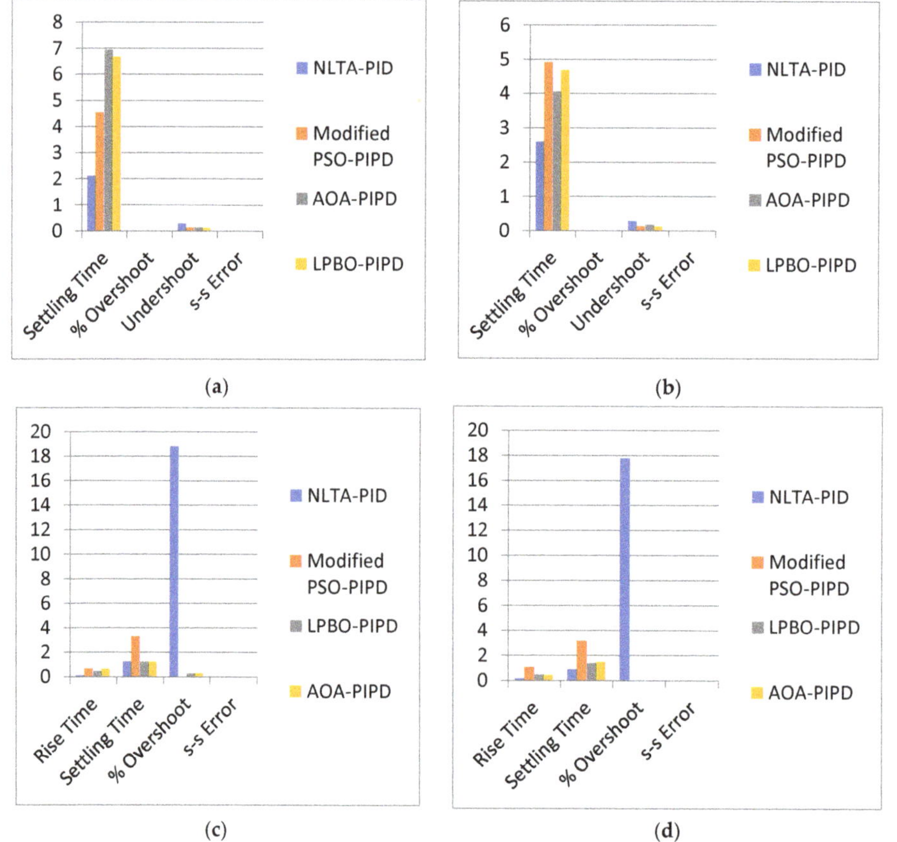

Figure 11. Graphical comparison of performance parameters. (**a**) Δf_1; (**b**) Δf_2; (**c**) V_{t1}; (**d**) V_{t2}.

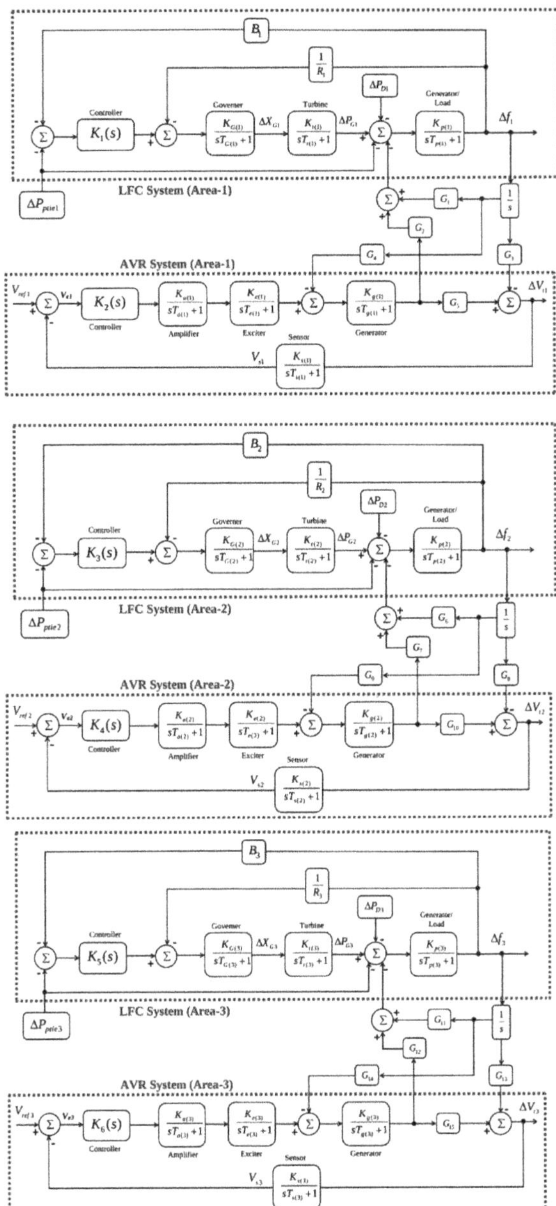

Figure 12. Three−area IPS with combined LFC−AVR.

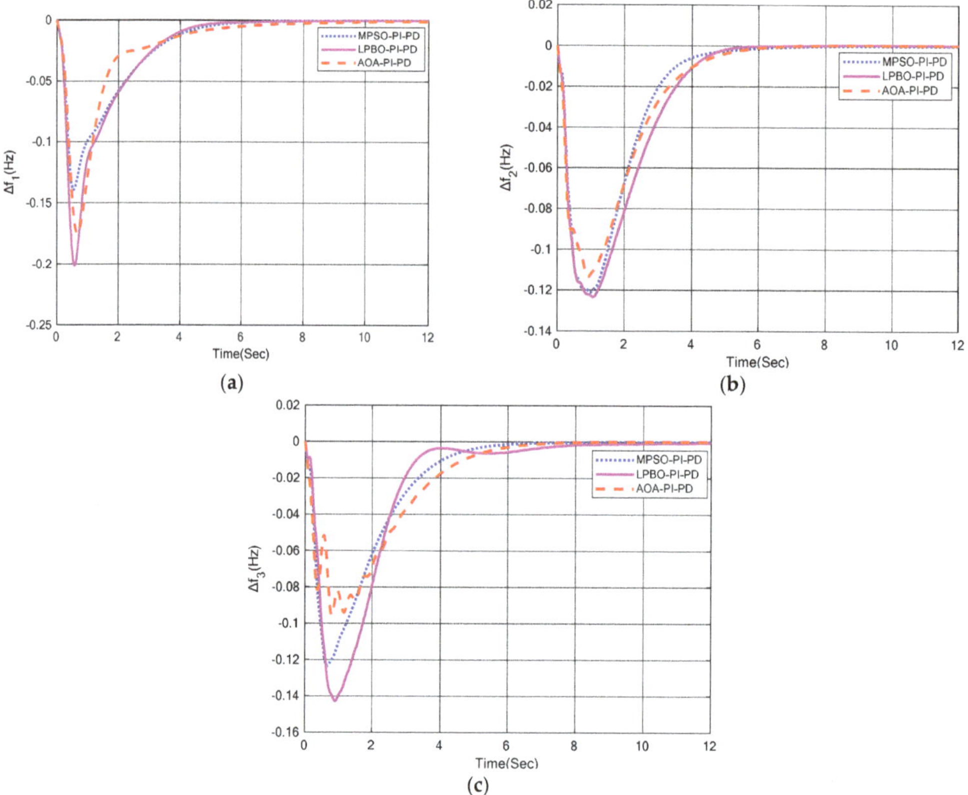

Figure 13. LFC response with PI–PD control scheme. (**a**) Δf_1; (**b**) Δf_2; (**c**) Δf_3.

Table 8 shows the summary of LFC responses of area-1, area-2, and area-3 using MPSO-PI-PD, LPBO-PI-PD, and AOA-PI-PD control techniques, respectively. For area-1, LPBO-PI-PD provided 14% and 31% quick settling times as compared to the MPSO-PI-PD and AOA-PI-PD control schemes, respectively. The overshoot percentage and steady state error were zero in each case.

Table 8. LFC responses (area-3).

Area	Control Scheme	Settling Time	% Overshoot	Undershoot	s-s Error
	MPSO-PI-PD	5.43	0	−0.14	0
Area-1	LPBO-PI-PD	4.65	0	−0.20	0
	AOA-PI-PD	6.73	0	−0.175	0
	MPSO-PI-PD	5.04	0	−0.120	0
Area-2	LPBO-PI-PD	4.87	0	−0.122	0
	AOA-PI-PD	5.46	0	−0.115	0
	PSO-PI-PD	5.40	0	−0.122	0
Area-3	LPBO-PI-PD	7.16	0	−0.143	0
	AOA-PI-PD	6.40	0	−0.095	0

Further, MPSO-PI-PD exhibited 30% and 20% better undershoot responses as compared to the LPBO-PI-PD and AOA-PI-PD control techniques, respectively. For area-2, LPBO-PI-PD yielded 3.3% and 11% quick settling times, as compared to the MPSO-PI-PD and AOA-PI-PD control schemes, respectively. The overshoot percentage and steady state error were zero in each case. Further, AOA-PI-PD exhibited 4.16% and 5.74% better undershoot responses as compared to the MPSO-PI-PD and LPBO-PI-PD control schemes, respectively. For area-3, MPSO-PI-PD provided 25% and 16% quick settling times as compared to the LPBO-PI-PD and AOA-PI-PD control techniques, respectively. The overshoot percentage and steady state error were again zero in each case. Further, AOA-PI-PD exhibited 22% and 34% better undershoot responses as compared to the MPSO-PI-PD and LPBO-PI-PD control schemes, respectively. Figure 14 shows the terminal voltage responses of area-1, area-2, and area-3 using MPSO-PI-PD, LPBO-PI-PD, and AOA-PI-PD control techniques in a three-area IPS, respectively.

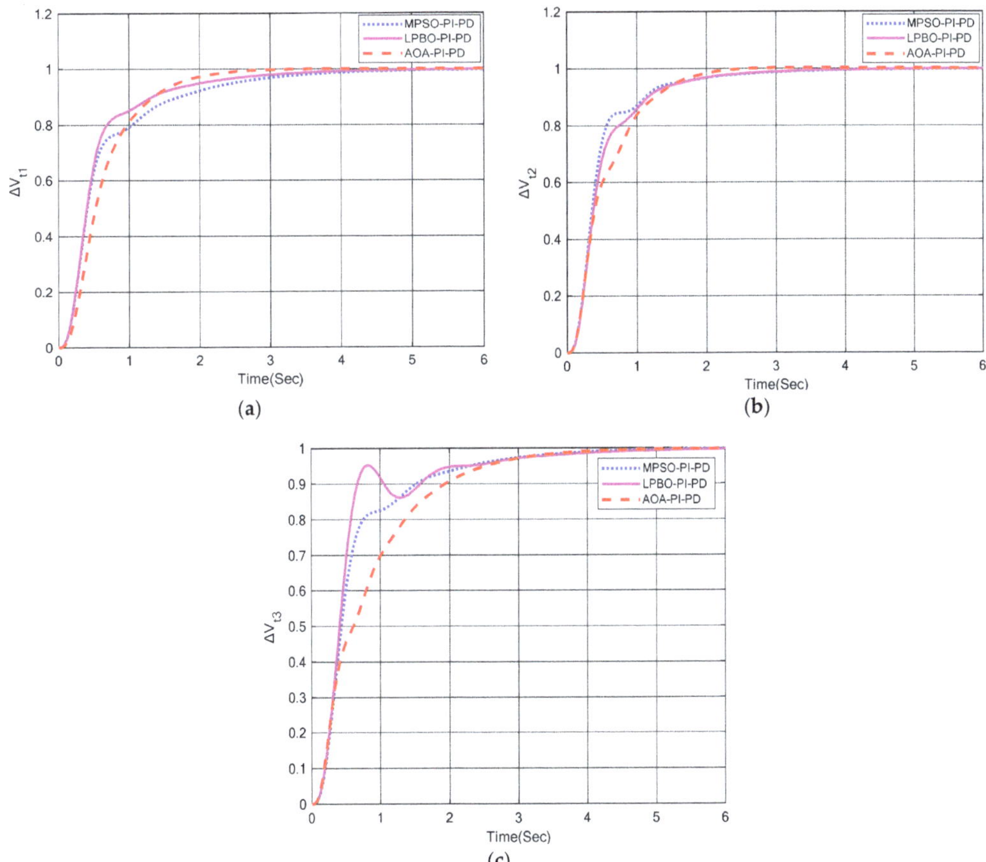

Figure 14. AVR response with PI-PD control scheme. (**a**) V_{t1}; (**b**) V_{t2}; (**c**) V_{t3}.

Table 9 shows the summary of AVR responses of area-1, area-2, and area-3 using the MPSO-PI-PD, LPBO-PI-PD, and AOA-PI-PD control schemes, respectively. For area-1, AOA-PI-PD provided 26% and 2% quick rise times as compared to the MPSO-PI-PD and LPBO-PI-PD control techniques, respectively. Moreover, AOA-PI-PD yielded 38% and 29% fast settling times as compared to the MPSO-PI-PD and LPBO-PI-PD control schemes, respectively. Further, it was observed that the percentage of overshoot and steady state error were almost zero with each tuning technique. For area-2, MPSO-PI-PD offered 3% and 13% quick rise times as compared to the LPBO-PI-PD and AOA-PI-PD control techniques, respectively. Moreover, AOA-PI-PD provided 21% and 19% fast settling times as compared to the MPSO-PI-PD and LPBO-PI-PD control schemes, respectively. Further, it can be seen that the overshoot percentage and steady state error were almost zero with each tuning technique. For area-3, LPBO-PI-PD produced 64% and 73% quick rise times as compared to the MPSO-PI-PD and AOA-PI-PD control techniques, respectively. Moreover, AOA-PI-PD provided 0.3% and 5.45% fast settling times as compared to the MPSO-PI-PD and LPBO-PI-PD control schemes, respectively. Further, it can be seen that the overshoot percentage and steady state error were negligible with each tuning technique. Figure 15 shows the graphical comparison of the performance parameters of the MPSO-PI-PD, LPBO-PI-PD, and AOA-PI-PD control techniques in a three-area interconnected system.

Table 9. AVR responses (area-3).

Area	Control Scheme	Rise Time	Settling Time	% Overshoot	s-s Error
	MPSO-PI-PD	1.53	3.48	5.8225×10^{-6}	0
Area-1	LPBO-PI-PD	1.15	3.01	4.5973×10^{-4}	0
	AOA-PI-PD	1.13	2.15	0.083	0
	MPSO-PI-PD	0.95	2.44	0	0
Area-2	LPBO-PI-PD	0.98	2.37	0	0
	AOA-PI-PD	1.09	1.92	0.37	0
	MPSO-PI-PD	1.32	3.30	0	0
Area-3	LPBO-PI-PD	0.48	3.48	0.001	0
	AOA-PI-PD	1.75	3.29	0	0

Figure 16 shows the tie–line power responses of area-1, area-2, and area-3 using the MPSO-PI-PD, LPBO-PI-PD and AOA-PI-PD control schemes in a three-area IPS, respectively. It can be inferred that PI-PD-based control schemes including MPSO-PI-PD, LPBO-PI-PD, and AOA-PI-PD yielded satisfactory tie–line powers responses with negligible undershoots and overshoot percentages in the three-area IPS.

5.3. Sensitivity Analysis

In this section, the robustness of the proposed nature-inspired computation-based PI-PD control techniques were tested with large variations in the system parameters of the three-area IPS with combined LFC-AVR. The generator time constant (T_g) and turbine time constant (T_t) were varied to Â $\pm$ 50% of their nominal values. The newer values of T_g and T_t after Â $\pm$ 50% variations are given in Appendix B. The optimum parameters of the PI-PD control scheme were the same as those used in Case 2. The AVR and LFC responses of the PI-PD control scheme with variations in T_t and T_g are depicted in Figures 17 and 18, respectively. Tables 10 and 11 show the summary of the performance parameters of LFC and AVR responses under parametric variations. From the obtained results, it is evident the overshoot percentages and steady state error were almost zero in each case. The AVR responses are almost indistinguishable to each other, despite the variation in system parameters. Figure 19 shows the graphical comparison of the performance parameters under this scenario. It is clearly observed that the system response under Â $\pm$ 50% variations was very identical to response with nominal values. This indicates that the proposed LPBO-PI-PD control technique was very realistic and robust under variations in the system parameters. These results clearly reveal that the re-tuning of the proposed controller is not necessary with large variations of at least Â $\pm$ 50%.

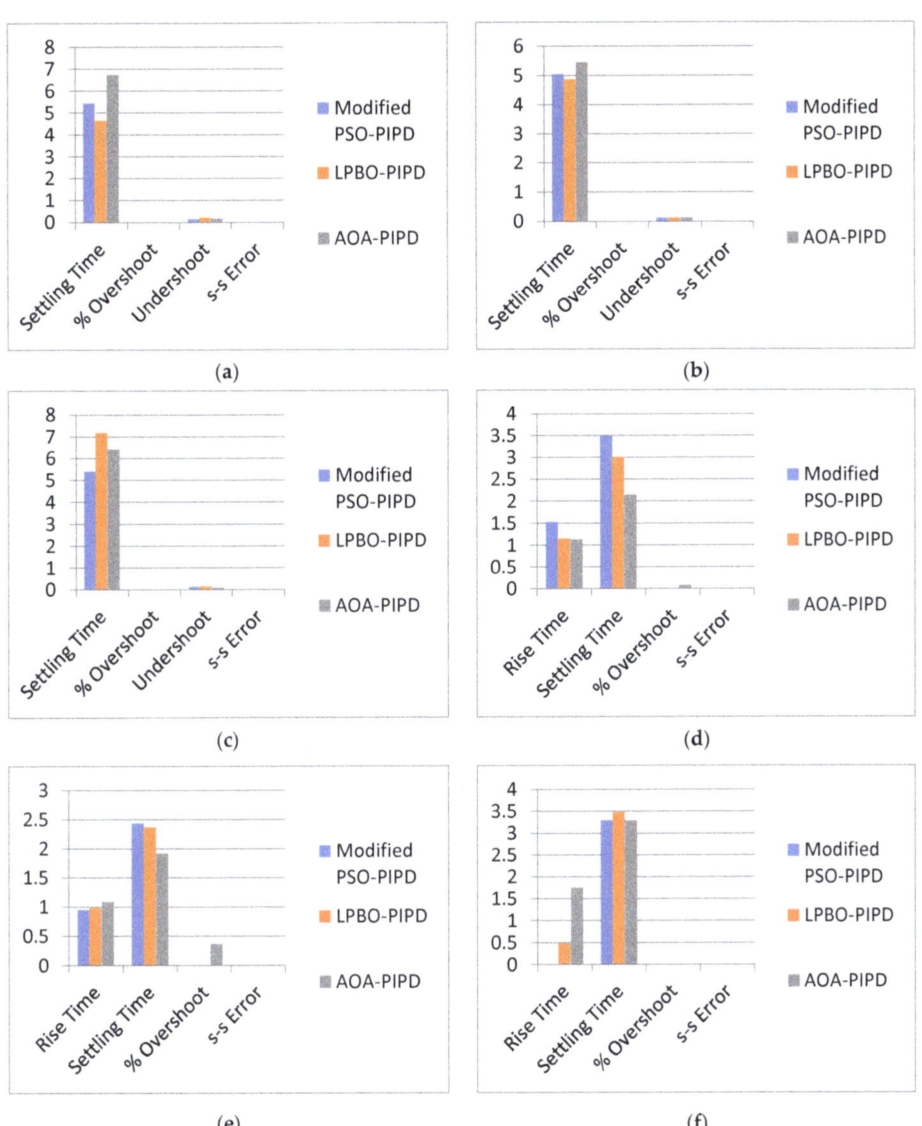

Figure 15. Graphical comparison of performance parameters. (**a**) Δf_1; (**b**) Δf_2; (**c**) Δf_3;(**d**) V_{t1}; (**e**) V_{t2}; (**f**) V_{t3}.

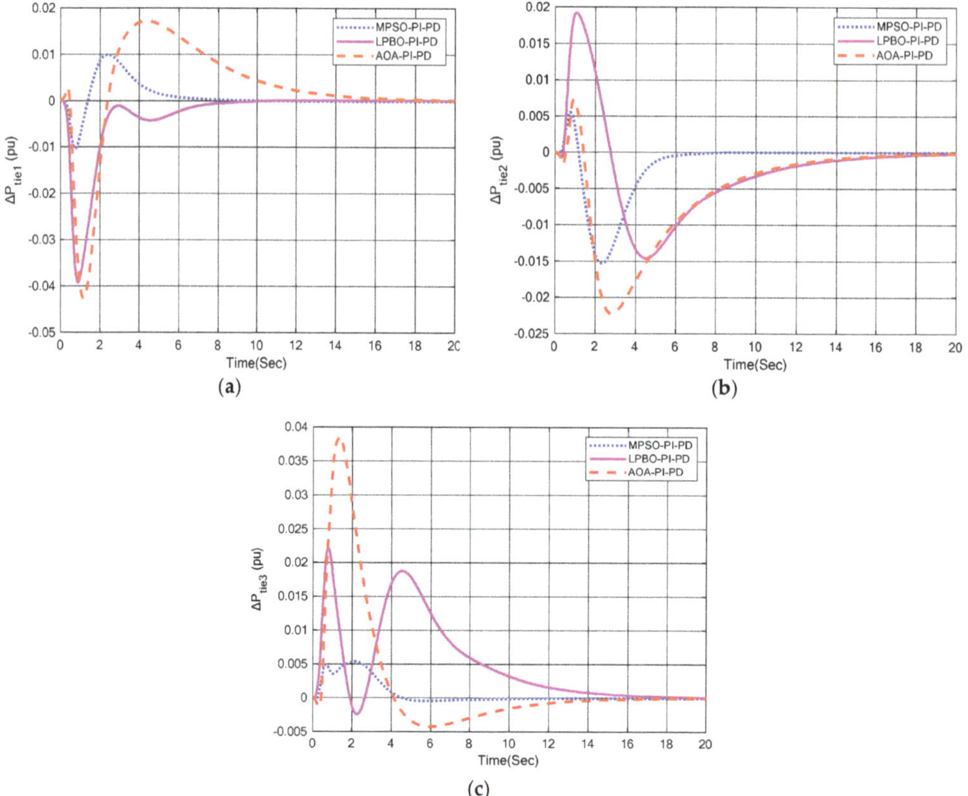

Figure 16. Tie−line power responses with PI−PD control scheme. (**a**) ΔP_{tie1}; (**b**) ΔP_{tie2}; (**c**) ΔP_{tie3}.

Table 10. Settling time responses of PI-PD control scheme with variations in system parameters.

Parameters/Variation	Settling Time (LFC and AVR)					
	Δf_1	Δf_2	Δf_3	V_{t1}	V_{t2}	V_{t3}
Nominal T_g, T_t	4.65	4.87	7.16	3.01	2.37	3.48
T_{g1}, T_{g2}, $T_{g3/}+50\%$	4.60	4.76	7.02	2.74	2.11	3.56
T_{g1}, T_{g2}, $T_{g3/}-50\%$	4.71	4.95	7.32	3.25	2.59	3.56
T_{t1}, T_{t2}, $T_{t3/}+50\%$	4.63	5.01	7.18	3.03	2.38	3.48
T_{t1}, T_{t2}, $T_{t3/}-50\%$	4.60	4.71	7.11	2.99	2.36	3.48

Table 11. Overshoot and undershoot responses of PI-PD control scheme with variations in system parameters.

Parameters/Variation	%Overshoot (LFC and AVR)						%Undershoot (LFC)		
	Δf_1	Δf_2	Δf_3	V_{t1}	V_{t2}	V_{t3}	Δf_1	Δf_2	Δf_3
Nominal T_g, T_t	0	0	0	0	0	0	−0.2	−0.122	−0.143
T_{g1}, T_{g2}, $T_{g3/}+50\%$	0	0	0	0	0	3.4	−0.185	−0.13	−0.155
T_{g1}, T_{g2}, $T_{g3/}-50\%$	0	0	0	0	0	0	−0.215	−0.125	−0.13
T_{t1}, T_{t2}, $T_{t3/}+50\%$	0	0	0	0	0	0	−0.245	−0.125	−0.15
T_{t1}, T_{t2}, $T_{t3/}-50\%$	0	0	0	0	0	0	−0.16	−0.125	−0.135

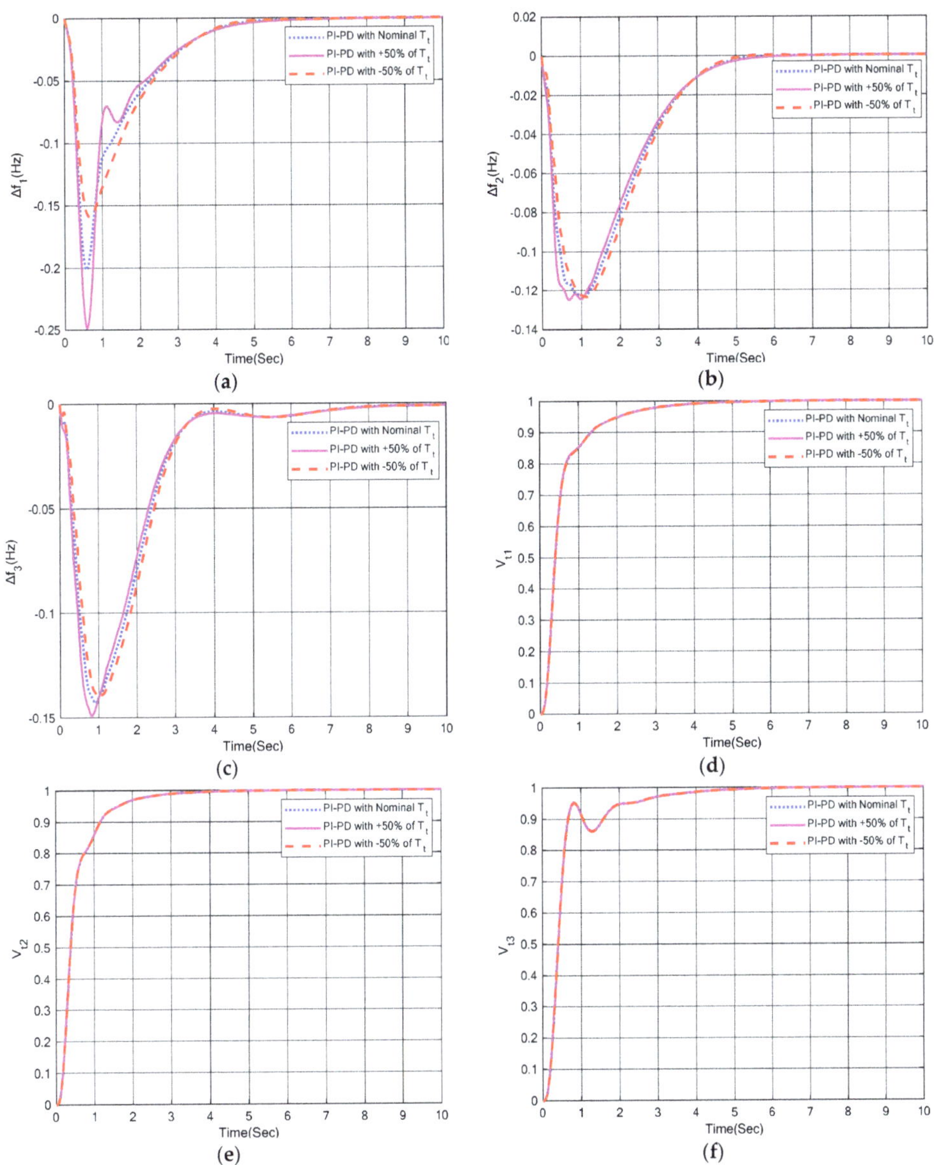

Figure 17. LFC and AVR responses with variations in T_t. (**a**) Δf_1 in area-1; (**b**) Δf_2 in area-2; (**c**) Δf_3 in area-3; (**d**) V_{t1} in area-1; (**e**) V_{t2} in area-2 and (**f**) V_{t3} in area-3.

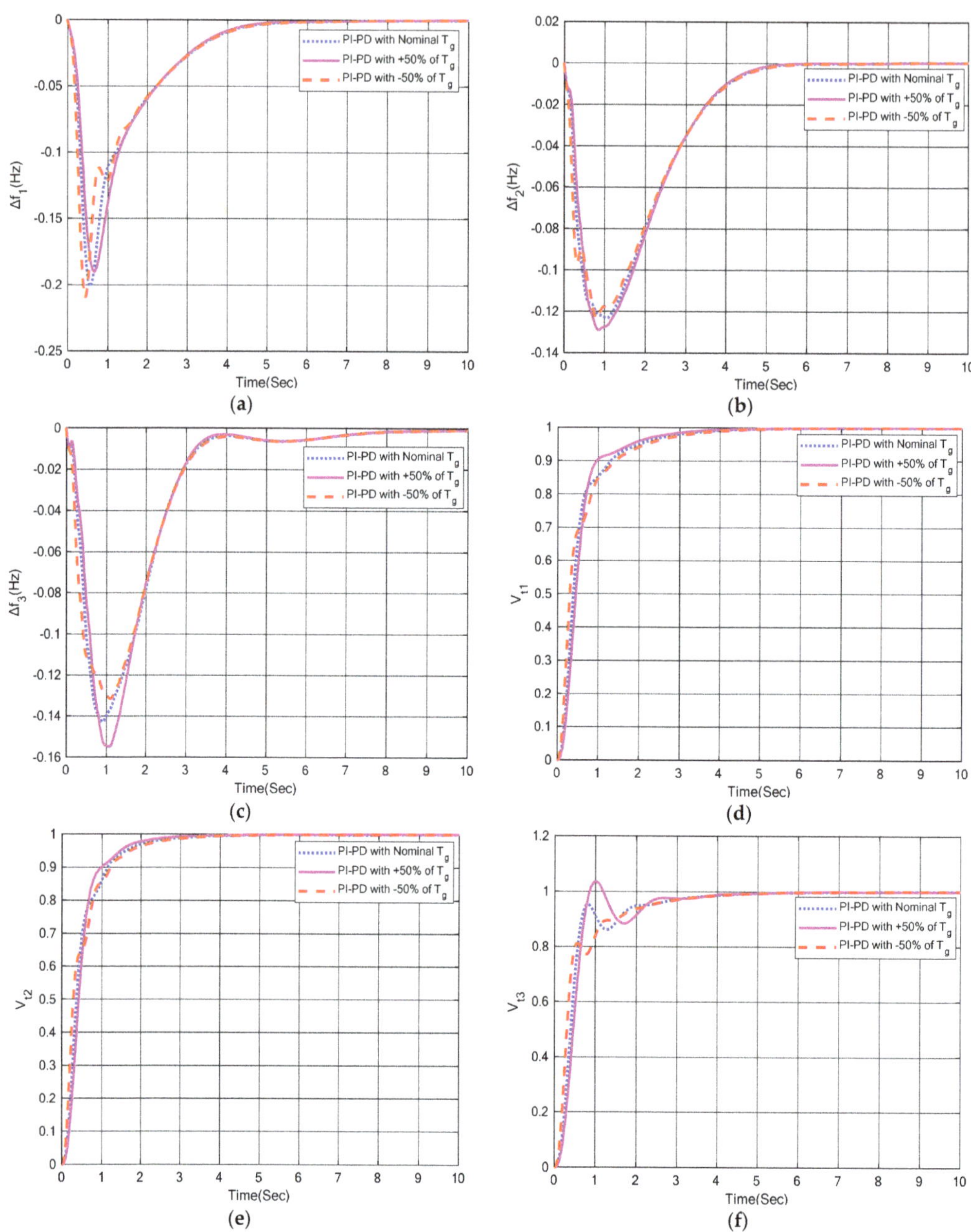

Figure 18. LFC and AVR responses with variations in T_g. (**a**) Δf_1 in area-1; (**b**) Δf_2 in area-2; (**c**) Δf_3 in area-3; (**d**) V_{t1} in area-1; (**e**) V_{t2} in area-2 and (**f**) V_{t3} in area-3.

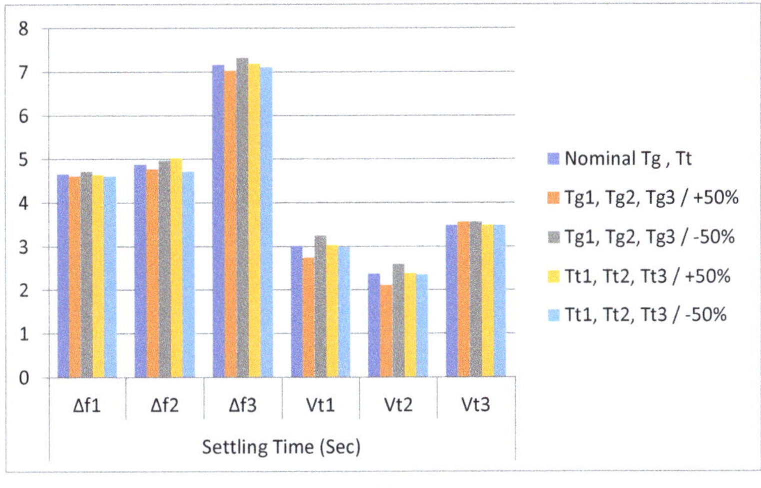

(**a**)

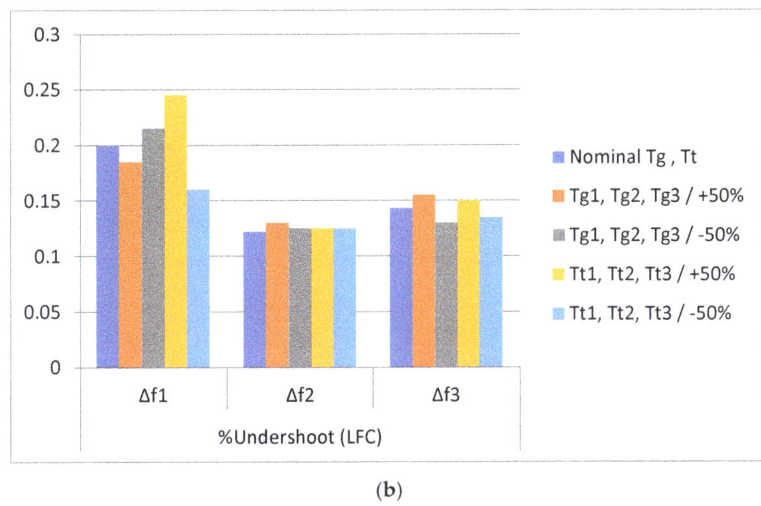

(**b**)

Figure 19. Graphical comparison of performance parameters in three-area IPS with variations in T_g and T_t. (**a**) Settling time; (**b**) undershoot.

6. Conclusions and Future Work

The multi-area IPS included numerous control areas, which are connected through the AC tie–line. The transient and steady state performance of a multi-area IPS with AVR-LFC was thoroughly investigated in this research.Nature-inspired computation including MPSO-, LPBO-, and AOA-based PI-PD control technique was proposed for the optimization of the multi-area system. From the findings of Case 1, it is concluded that all proposed schemes provided relatively better undershoot responses as compared to the NLTA-PID controller [5] for LFC. Particularly, the AOA-PI-PD control scheme exhibited 60% and 56% better undershoots in the area-1 and area-2 LFC, respectively, as compared to the NLTA-PID controller. Similarly, NLTA-PID provided 18% and 17% overshoot in the area-1 and area-2 AVR, respectively, but the proposed PI-PD control scheme completely eliminated the overshoot percentage with each tuning algorithm. The results of Case 2 reveal that LPBO-PI-PD provided 14% and 31% quick settling times in area-1, whereas 3.3% and

11% quick settling times were provided in the area-2 LFC as compared to the MPSO-PI-PD and AOA-PI-PD control techniques, respectively. In the area-3 LFC, MPSO-PI-PD provided relatively lower settling times (25% and 16%) as compared to the LPBO-PI-PD and AOA-PI-PD control schemes, respectively. MPSO-PI-PD provided relatively better undershoot (30% and 20%) in the area-1 LFC, whereas the AOA-PI-PD control technique provided better undershoot in the area-1 (4.16% and 5.74%) and area-2 (22% and 34%) LFC, respectively. Moreover, the AOA-PI-PD control scheme provided 26% and 2% quick rise times, and 38% and 29% fast settling times in the area-1 AVR as compared to the MPSO-PI-PD and LPBO-PI-PD control schemes, respectively. Further, MPSO-PI-PD provided 3% and 13% fast rise times in the area-2 AVR as compared to the LPBO-PI-PD and AOA-PI-PD control schemes, respectively. AOA-PI-PD provided 21% and 19% quick settling times in the area-2 AVR, and 0.3% and 5.45% fast settling times in the area-3 AVR as compared to the MPSO-PI-PD and AOA-PI-PD control schemes, respectively. For the area-3 AVR, LPBO-PI-PD provided 64% and 73% fast rise times as compared to the MPSO-PI-PD and AOA-PI-PD control schemes, respectively. Finally, the resilience of the PI-PD control technique was assessed by varying the system parameters ($\hat{A} \pm 50\%$), and a comprehensive sensitivity analysis was carried out to confirm its robustness. The results confirm the superiority of the proposed PI-PD control scheme when applied to multi-area IPS with combined LFC and AVR. Keeping in mind the value of the present work, IPS with a combined LFC-AVR can be analyzed by incorporating multi-source and various energy storage devices to enhance the dynamic response of the power systems. Further, neuro-fuzzy and hybrid ANN controllers can also be utilized for multi-area multi-sources IPS. It will be worth employing PI-PD, neuro-fuzzy, or hybrid ANN to multi-area IPS under nonlinearity constraints. Moreover, very recently introduced nature-inspired computing techniques such as dandelion optimization, artificial rabbits optimization, and sea-horse optimization can be explored to find the optimal parameters of controllers in such types of application.

Author Contributions: Conceptualization, T.A., S.A.M., A.T.A.; Data curation, T.A., S.A.M., A.D., H.M.; Formal analysis, T.A., S.A.M., A.D., H.M., A.T.A., I.A.H.; Investigation, A.T.A., I.A.H.; Methodology, T.A., S.A.M., A.D., H.M., A.T.A., I.A.H.; Project administration, A.T.A., I.A.H.; Resources, T.A., S.A.M., A.D., H.M.; Supervision, A.T.A., I.A.H.; Validation, T.A., S.A.M., A.D., H.M.; Visualization, A.D., H.M., A.T.A., I.A.H.; Writing—original draft, T.A., S.A.M., A.D., H.M.; Writing—review & editing, T.A., S.A.M., A.D., H.M., A.T.A., I.A.H.; Funding Acquisition, I.A.H. All authors have read and agreed to the published version of the manuscript.

Funding: This research was funded by Norwegian University of Science and Technology.

Institutional Review Board Statement: Not applicable.

Informed Consent Statement: Not applicable.

Acknowledgments: The authors would like to acknowledge the support of Norwegian University of Science and Technology for paying the Article Processing Charges (APC) of this publication. Special acknowledgment to Automated Systems & Soft Computing Lab (ASSCL), Prince Sultan University, Riyadh, Saudi Arabia. In addition, the authors wish to acknowledge the editor and anonymous.

Conflicts of Interest: The authors declare no conflict of interest.

Appendix A

Table A1. Area-2 [5].

Sr. No.	Area-1		Area-2	
	System's Parameter	Value	System's Parameter	Value
1	B_1	1	B_2	1
2	R_1	2.4	R_2	1.2
3	K_{G1}	1	K_{G2}	1

Table A1. *Cont.*

Sr. No.	Area-1		Area-2	
	System's Parameter	Value	System's Parameter	Value
4	T_{G1}	0.08	T_{G2}	0.12
5	K_{t1}	1	K_{t2}	1
6	T_{t1}	0.3	T_{t2}	0.15
7	ΔP_{D1}	0.02	ΔP_{D2}	0.02
8	K_{p1}	120	K_{p2}	100
9	T_{p1}	20	T_{p2}	10
10	K_{a1}	10	K_{a2}	10
11	T_{a1}	0.1	T_{a2}	0.1
12	K_{e1}	1	K_{e2}	1.5
13	T_{e1}	0.4	T_{e2}	0.6
14	K_{g1}	1	K_{g2}	1.5
15	T_{g1}	1	T_{g2}	1.5
16	K_{s1}	1	K_{s2}	1
17	T_{s1}	0.01	T_{s2}	0.01
18	G_1	1.5	G_6	1.5
19	G_2	0.3	G_7	0.3
20	G_3	0.1	G_8	0.1
21	G_4	1.4	G_9	1.4
22	G_5	0.5	G_{10}	0.5
23	T_{12}	0.545	T_{21}	0.545

Appendix B

Table A2. System parameters of area-1, area-2, and area-3.

Sr. No.	Area-1		Area-2		Area-3	
	System's Parameter	Value	System's Parameter	Value	System's Parameter	Value
1	B_1	1	B_2	1	B_3	1
2	R_1	2.4	R_2	1.20	R_3	1.20
3	K_{G1}	1	K_{G2}	1	K_{G3}	1
4	T_{G1}	0.08	T_{G2}	0.12	T_{G3}	0.12
5	K_{t1}	1	K_{t2}	1	K_{t3}	1
6	T_{t1}	0.3	T_{t2}	0.15	T_{t3}	0.15
7	ΔP_{D1}	0.02	ΔP_{D2}	0.02	ΔP_{D3}	0.02
8	K_{p1}	120	K_{p2}	100	K_{p3}	100
9	T_{p1}	20	T_{p2}	10	T_{p3}	10
10	K_{a1}	10	K_{a2}	10	K_{a3}	10
11	T_{a1}	0.1	T_{a2}	0.1	T_{a3}	0.1
12	K_{e1}	1	K_{e2}	1.5	K_{e3}	1.8
13	T_{e1}	0.4	T_{e2}	0.6	T_{e3}	0.8
14	K_{g1}	1	K_{g2}	1.5	K_{g3}	1.8
15	T_{g1}	1	T_{g2}	1.5	T_{g3}	1.8
16	K_{s1}	1	K_{s2}	1	K_{s3}	1
17	T_{s1}	0.01	T_{s2}	0.01	T_{s3}	0.01
18	G_1	1.5	G_6	1.5	G_{11}	1.5
19	G_2	0.3	G_7	0.3	G_{12}	0.3
20	G_3	0.1	G_8	0.1	G_{13}	0.1
21	G_4	1.4	G_9	1.4	G_{14}	1.4
22	G_5	0.5	G_{10}	0.5	G_{15}	0.5
23	T_{12}	0.545	T_{21}	0.545	T_{31}	0.545
24	T_{13}	0.545	T_{23}	0.545	T_{32}	0.545

Table A3. System parameters after $\hat{A} \pm 50\%$ variations in T_g and T_t.

Sr. No.	Area-1		Area-2		Area-3	
	System's Parameter	Value	System's Parameter	Value	System's Parameter	Value
1	T_{g1} (+50%)	1.5	T_{g2} (+50%)	2.25	T_{g3} (+50%)	2.7
	T_{g1} (Nominal)	1	T_{g2} (Nominal)	1.5	T_{g3} (Nominal)	1.8
	T_{g1} (−50%)	0.5	T_{g2} (−50%)	0.75	T_{g3} (−50%)	0.9
2	T_{t1} (+50%)	0.45	T_{t2} (+50%)	0.225	T_{t3} (+50%)	0.225
	T_{t1} (Nominal)	0.3	T_{t2} (Nominal)	0.15	T_{t3} (Nominal)	0.15
	T_{t1} (−50%)	0.15	T_{t2} (−50%)	0.075	T_{t3} (−50%)	0.075

References

1. Kalyan, C.N.S.; Goud, B.S.; Reddy, C.R.; Bajaj, M.; Sharma, N.K.; Alhelou, H.H.; Siano, P.; Kamel, S. Comparative Performance Assessment of Different Energy Storage Devices in Combined LFC and AVR Analysis of Multi-Area Power System. *Energies* **2022**, *15*, 629. [CrossRef]
2. Ghosh, A.; Ray, A.K.; Nurujjaman; Jamshidi, M. Voltage and frequency control in conventional and PV integrated power systems by a particle swarm optimized Ziegler–Nichols based PID controller. *SN Appl. Sci.* **2021**, *3*, 1–13. [CrossRef]
3. Prakash, A.; Parida, S.K. Combined Frequency and Voltage Stabilization of Thermal-Thermal System with UPFC and RFB. In Proceedings of the 2020 IEEE 9th Power India International Conference (PIICON), Sonepat, India, 28 February–1 March 2020; pp. 1–6. [CrossRef]
4. Fayek, H.H.; Rusu, E. Novel Combined Load Frequency Control and Automatic Voltage Regulation of a 100% Sustainable Energy Interconnected Microgrids. *Sustainability* **2022**, *14*, 9428. [CrossRef]
5. Nahas, N.; Abouheaf, M.; Sharaf, A.M.; Gueaieb, W. A Self-Adjusting Adaptive AVR-LFC Scheme for Synchronous Generators. *IEEE Trans. Power Syst.* **2019**, *34*, 5073–5075. [CrossRef]
6. Gupta, A.; Chauhan, A.; Khanna, R. Design of AVR and ALFC for single area power system including damping control. In Proceedings of the 2014 Recent Advances in Engineering and Computational Sciences (RAECS 2014), Chandigarh, India, 6–8 March 2014; pp. 6–8. [CrossRef]
7. Chandrashekar, M.J.; Jayapal, R. AGC and AVR implementation in a deregulated power system using optimized controller with Renewable integrated DC link. In Proceedings of the 2019 1st International Conference on Advanced Technologies in Intelligent Control, Environment, Computing & Communication Engineering (ICATIECE 2019), Bangalore, India, 19–20 March 2019; pp. 355–364. [CrossRef]
8. Rajbongshi, R.; Saikia, L.C. Coordinated performance of interline power flow controller and superconducting magnetic energy storage in combined ALFC and AVR system under deregulated environment. *J. Renew. Sustain. Energy* **2018**, *10*, 044102. [CrossRef]
9. Kalyan, C.N.S.; Rao, G.S. Combined frequency and voltage stabilisation of multi-area multisource system by DE-AEFA optimised PID controller with coordinated performance of IPFC and RFBs. *Int. J. Ambient Energy* **2020**, 1–17. [CrossRef]
10. Kalyan, C.N.S.; Rao, G.S. Frequency and voltage stabilisation in combined load frequency control and automatic voltage regulation of multiarea system with hybrid generation utilities by AC/DC links. *Int. J. Sustain. Energy* **2020**, *39*, 1009–1029. [CrossRef]
11. Sahani, A.K.; Raj, U.; Shankar, R.; Mandal, R.K. Firefly Optimization Based Control Strategies for Combined Load Frequency Control and Automatic Voltage Regulation for Two-Area Interconnected Power System. *Int. J. Electr. Eng. Inform.* **2019**, *11*, 747–758. [CrossRef]
12. Lal, D.K.; Barisal, A.K. Combined load frequency and terminal voltage control of power systems using moth flame optimization algorithm. *J. Electr. Syst. Inf. Technol.* **2019**, *6*, 1–24. [CrossRef]
13. Grover, H.; Verma, A.; Bhatti, T.S. Load frequency control & automatic voltage regulation for a single area power system. In Proceedings of the 2020 IEEE 9th Power India International Conference (PIICON 2020), Sonepat, India, 28 February–1 March 2020; pp. 1–5. [CrossRef]
14. Salman, G.; Jafar, A.S.; Ismael, A.I. Application of artificial intelligence techniques for LFC and AVR systems using PID controller. *Int. J. Power Electron. Drive Syst. (IJPEDS)* **2019**, *10*, 1694–1704. [CrossRef]
15. Kouba, N.E.Y.; Menaa, M.; Hasni, M.; Boudour, M. Optimal control of frequency and voltage variations using PID controller based on Particle Swarm Optimization. In Proceedings of the 2015 4th International Conference on Systems and Control (ICSC), Sousse, Tunisia, 28–30 April 2015; pp. 424–429. [CrossRef]
16. Safiullah, S.; Rahman, A.; Lone, S.A. State-observer based IDD controller for concurrent frequency-voltage control of a hybrid power system with electric vehicle uncertainties. *Int. Trans. Electr. Energy Syst.* **2021**, *31*, e13083. [CrossRef]
17. Mohammadikia, R.; Nikoofard, A.; Tavakoli-Kakhki, M. Application of MPC for an automatic voltage regulator and load frequency control of interconnected power system. In Proceedings of the 2020 28th Iranian Conference on Electrical Engineering (ICEE), Tabriz, Iran, 4–6 August 2020. [CrossRef]
18. Eke, I.; Saka, M.; Gozde, H.; Arya, Y.; Taplamacioglu, M.C. Heuristic optimization based dynamic weighted state feedback approach for 2DOF PI-controller in automatic voltage regulator. *Eng. Sci. Technol. Int. J.* **2021**, *24*, 899–910. [CrossRef]

19. Wagle, R.; Sharma, P.; Sharma, C.; Gjengedal, T.; Pradhan, C. Bio-inspired hybrid BFOA-PSO algorithm-based reactive power controller in a standalone wind-diesel power system. *Int. Trans. Electr. Energy Syst.* **2021**, *31*. [CrossRef]

20. Prasad, S.; Purwar, S.; Kishor, N. Load frequency regulation using observer based non-linear sliding mode control. *Int. J. Electr. Power Energy Syst.* **2018**, *104*, 178–193. [CrossRef]

21. Panigrahi, N.; Nayak, A.; Mishra, S.R. Morphology of the Alumina Nanoparticles for the Arrangement of the KF Stacked Nano–γ–Al2O3 as Catalyst for Conversion of Biomass to Fuel. In *Advances in Energy Technology*; Springer: Berlin/Heidelberg, Germany, 2021; pp. 125–132.

22. Daraqz, A.; Malik, S.A.; Waseem, A.; Azar, A.T.; Haq, I.U.; Ullah, Z.; Aslam, S. Automatic generation control of multi-source interconnected power system using FOI-TD controller. *Energies* **2021**, *14*, 5867. [CrossRef]

23. Guo, J. Application of a novel adaptive sliding mode control method to the load frequency control. *Eur. J. Control* **2020**, *57*, 172–178. [CrossRef]

24. Arya, Y.; Kumar, N. Design and analysis of BFOA-optimized fuzzy PI/PID controller for AGC of multi-area traditional/restructured electrical power systems. *Soft Comput.* **2016**, *21*, 6435–6452. [CrossRef]

25. Daraz, A.; Malik, S.A.; Mokhlis, H.; Haq, I.U.; Laghari, G.F.; Mansor, N.N. Fitness Dependent Optimizer-Based Automatic Generation Control of Multi-Source Interconnected Power System with Non-Linearities. *IEEE Access* **2020**, *8*, 100989–101003. [CrossRef]

26. Raj, U.; Shankar, R. Deregulated Automatic Generation Control using Novel Opposition-based Interactive Search Algorithm Cascade Controller Including Distributed Generation and Electric Vehicle. *Iran. J. Sci. Technol. Trans. Electr. Eng.* **2020**, *44*, 1233–1251. [CrossRef]

27. Kose, E. Optimal Control of AVR System with Tree Seed Algorithm-Based PID Controller. *IEEE Access* **2020**, *8*, 89457–89467. [CrossRef]

28. Kalyan, C.N.S.; Goud, B.S.; Reddy, C.R.; Ramadan, H.S.; Bajaj, M.; Ali, Z.M. Water Cycle Algorithm Optimized Type II Fuzzy Controller for Load Frequency Control of a Multi-Area, Multi-Fuel System with Communication Time Delays. *Energies* **2021**, *14*, 5387. [CrossRef]

29. Tripathy, S.; Debnath, M.K.; Kar, S.K. Optimal Design of Fractional Order 2DOF-PID Controller for Frequency Control in Interconnected System. In *Green Technology for Smart City and Society*; Springer: Singapore, 2020; pp. 23–33. [CrossRef]

30. Sahu, P.C.; Prusty, R.C.; Sahoo, B.K. Modified sine cosine algorithm-based fuzzy-aided PID controller for automatic generation control of multiarea power systems. *Soft Comput.* **2020**, *24*, 12919–12936. [CrossRef]

31. Ajithapriyadarsini, S.; Mary, P.M.; Iruthayarajan, M.W. Automatic generation control of a multi-area power system with renewable energy source under deregulated environment: Adaptive fuzzy logic-based differential evolution (DE) algorithm. *Soft Comput.* **2019**, *23*, 12087–12101. [CrossRef]

32. Çelik, E. Design of new fractional order PI–fractional order PD cascade controller through dragonfly search algorithm for advanced load frequency control of power systems. *Soft Comput.* **2020**, *25*, 1193–1217. [CrossRef]

33. Daraz, A.; Malik, S.A.; Mokhlis, H.; Haq, I.U.; Zafar, F.; Mansor, N.N. Improved-Fitness Dependent Optimizer Based FOI-PD Controller for Automatic Generation Control of Multi-Source Interconnected Power System in Deregulated Environment. *IEEE Access* **2020**, *8*, 197757–197775. [CrossRef]

34. Daraz, A.; Malik, S.A.; Haq, I.U.; Khan, K.B.; Laghari, G.F.; Zafar, F. Modified PID controller for automatic generation control of multi-source interconnected power system using fitness dependent optimizer algorithm. *PLoS ONE* **2020**, *15*, e0242428. [CrossRef] [PubMed]

35. Biswas, S.; Roy, P.K.; Chatterjee, K. FACTS-based 3DOF-PID Controller for LFC of Renewable Power System Under Deregulation Using GOA. *IETE J. Res.* **2021**, 1–14. [CrossRef]

36. Fathy, A.; Kassem, A.M.; Abdelaziz, A.Y. Optimal design of fuzzy PID controller for deregulated LFC of multi-area power system via mine blast algorithm. *Neural Comput. Appl.* **2018**, *32*, 4531–4551. [CrossRef]

37. Daraz, A.; Malik, S.A.; Azar, A.T.; Aslam, S.; Alkhalifah, T.; Alturise, F. Optimized Fractional Order Integral-Tilt Derivative Controller for Frequency Regulation of Interconnected Diverse Renewable Energy Resources. *IEEE Access* **2022**, *10*, 43514–43527. [CrossRef]

38. Zhao, S.; Zhang, T.; Ma, S.; Chen, M. Dandelion Optimizer: A nature-inspired metaheuristic algorithm for engineering applications. *Eng. Appl. Artif. Intell.* **2022**, *114*. [CrossRef]

39. Alsattar, H.A.; Zaidan, A.A.; Zaidan, B.B. Novel meta-heuristic bald eagle search optimisation algorithm. *Artif. Intell. Rev.* **2019**, *53*, 2237–2264. [CrossRef]

40. Qais, M.H.; Hasanien, H.M.; Alghuwainem, S. Transient search optimization: A new meta-heuristic optimization algorithm. *Appl. Intell.* **2020**, *50*, 3926–3941. [CrossRef]

41. Rahman, C.M.; Rashid, T.A. A new evolutionary algorithm: Learner performance based behavior algorithm. *Egypt. Inform. J.* **2020**, *22*, 213–223. [CrossRef]

42. Fathy, A.; Alharbi, A.G.; Alshammari, S.; Hasanien, H.M. Archimedes optimization algorithm based maximum power point tracker for wind energy generation system. *Ain Shams Eng. J.* **2021**, *13*, 101548. [CrossRef]

43. Zou, H.; Li, H. Tuning of PI–PD controller using extended non-minimal state space model predictive control for the stabilized gasoline vapor pressure in a stabilized tower. *Chemom. Intell. Lab. Syst.* **2015**, *142*, 1–8. [CrossRef]

44. Raja, G.L.; Ali, A. New PI-PD Controller Design Strategy for Industrial Unstable and Integrating Processes with Dead Time and Inverse Response. *J. Control. Autom. Electr. Syst.* **2021**, *32*, 266–280. [CrossRef]
45. Kaya, I. Optimal PI–PD Controller Design for Pure Integrating Processes with Time Delay. *J. Control. Autom. Electr. Syst.* **2021**, *32*, 563–572. [CrossRef]
46. Irshad, M.; Ali, A. Robust PI-PD controller design for integrating and unstable processes. *IFAC-PapersOnLine* **2020**, *53*, 135–140. [CrossRef]
47. Peram, M.; Mishra, S.; Vemulapaty, M.; Verma, B.; Padhy, P.K. Optimal PI-PD and I-PD Controller Design Using Cuckoo Search Algorithm. In Proceedings of the 5th International Conference on Signal Processing and Integrated Networks, Delhi, India, 22–23 February 2018; pp. 643–646. [CrossRef]
48. Deniz, F.N.; Yüce, A.; Tan, N. Tuning of PI-PD Controller Based on Standard Forms for Fractional Order Systems. *J. Appl. Nonlinear Dyn.* **2018**, *8*, 5–23. [CrossRef]
49. Zheng, M.; Huang, T.; Zhang, G. A New Design Method for PI-PD Control of Unstable Fractional-Order System with Time Delay. *Complexity* **2019**, *2019*, 1–12. [CrossRef]
50. Ali, T.; Adeel, M.; Malik, S.A.; Amir, M. Stability Control of Ball and Beam System Using Heuristic Computation Based PI-D and PI-PD Controller. *Tech. J. Univ. Eng. Technol.* **2019**, *24*, 21–29.
51. Tian, D.; Shi, Z. MPSO: Modified particle swarm optimization and its applications. *Swarm Evol. Comput.* **2018**, *41*, 49–68. [CrossRef]

 *sustainability*

MDPI

Article

Line Overload Alleviations in Wind Energy Integrated Power Systems Using Automatic Generation Control

Kaleem Ullah [1], Abdul Basit [1,*], Zahid Ullah [2], Rafiq Asghar [3], Sheraz Aslam [4,*] and Ayman Yafoz [5]

1   US-Pakistan Center for Advanced Study in Energy, University of Engineering and Technology Peshawar, Peshawar 25000, Pakistan
2   Department of Electrical Engineering, University of Management and Technology Lahore, Sialkot Campus, Sialkot 51310, Pakistan
3   Faculty of Electrical & Electronics Engineering, Universiti Malaysia Pahang (UMP), Pekan 26600, Malaysia
4   Department of Electrical Engineering, Computer Engineering and Informatics, Cyprus University of Technology, Limassol 3036, Cyprus
5   Department of Information Systems, Faculty of Computing and Information Technology, King Abdulaziz University, Jeddah 21589, Saudi Arabia
*   Correspondence: abdul.basit@uetpeshawar.edu.pk (A.B.); sheraz.aslam@cut.ac.cy (S.A.)

Citation: Ullah, K.; Basit, A.; Ullah, Z.; Asghar, R.; Aslam, S.; Yafoz, A. Line Overload Alleviations in Wind Energy Integrated Power Systems Using Automatic Generation Control. *Sustainability* **2022**, *14*, 11810. https://doi.org/10.3390/su141911810

Academic Editor: Nien-Che Yang

Received: 24 August 2022
Accepted: 15 September 2022
Published: 20 September 2022

Publisher's Note: MDPI stays neutral with regard to jurisdictional claims in published maps and institutional affiliations.

Abstract: Modern power systems are largely based on renewable energy sources, especially wind power. However, wind power, due to its intermittent nature and associated forecasting errors, requires an additional amount of balancing power provided through the automatic generation control (AGC) system. In normal operation, AGC dispatch is based on the fixed participation factor taking into account only the economic operation of generating units. However, large-scale injection of additional reserves results in large fluctuations of line power flows, which may overload the line and subsequently reduce the system security if AGC follows the fixed participation factor's criteria. Therefore, to prevent the transmission line overloading, a dynamic dispatch strategy is required for the AGC system considering the capacities of the transmission lines along with the economic operation of generating units. This paper proposes a real-time dynamic AGC dispatch strategy, which protects the transmission line from overloading during the power dispatch process in an active power balancing operation. The proposed method optimizes the control of the AGC dispatch order to prevent power overflows in the transmission lines, which is achieved by considering how the output change of each generating unit affects the power flow in the associated bus system. Simulations are performed in Dig SILENT software by developing a 5 machine 8 bus Pakistan's power system model integrating thermal power plant units, gas turbines, and wind power plant systems. Results show that the proposed AGC design efficiently avoids the transmission line congestions in highly wind-integrated power along with the economic operation of generating units.

Keywords: automatic generation control; wind energy; transmission line security; dispatch strategies; Pakistan power system

---

1. Introduction

Wind power plants are directly connected to various voltage levels of power systems and the majority provide a sustainable connection to high voltage transmission grids. However, wind power is highly reliant on intermittent wind speed that incurs stochasticity and inaccurate accuracy, resulting in forecasting errors [1]. The forecasting errors of wind power have pertinent effects on power system operations by incurring a mismatch between supply and load demand that leads to deviation of generation and power exchanges from their scheduled values. The power system planners employ various scheduling mechanisms to deploy the extra amount of reserve, keeping the power grids in a balance condition. Conventionally, the additional reserves are provided manually or through automatic generation control (AGC) considering only the economic operation in the account [2]. As a result, large fluctuations in the line power flow occur, which subsequently increases the risk of transmission lines overloading during the power balancing operation [3–5]. This may further deteriorate the power system security and causes permanent damage to the transmission line network. Contrarily, if enough margin capacity for the fluctuation running is set up in the transmission lines, the available capacity for the energy market transactions is reduced and high economic losses can occur. Therefore, the overloading of the transmission line network is becoming a serious problem threatening the stable operation of the power system in large-scale wind energy-based power systems.

The problem has been highlighted in several research articles [6–8]. In this regard, a comprehensive review of the deployment of AGC for conventional and modern power systems is proposed in [9]. Similarly, the integration of wind and electric vehicles in power systems is studied using the AGC strategies [10]. Various models are developed and implemented [11,12] in different kinds of electricity markets to alleviate the line overload in the massive renewable integrated power systems. Two broad models are normally used for the alleviations of line overloads, namely cost-free means and not-cost-free means. The cost-free means included actions like outraging of congested lines or operation of FACTS devices because the marginal costs (and not the capital costs) involved in their usage are nominal. The not-cost-free means include the re-dispatching of the generation amounts by backing down some generators while others increase their output and, as a result, the generators no longer operate at equal incremental costs. This study has explored the latter model, in which the dispatched power from the generating units is rescheduled to minimize line congestion. The authors in [13] proposed an enhanced power flow management scheme for the renewable energy-based power system to avoid transmission line overload during maximum power flow fluctuations. The study illustrated a detailed dispatching mechanism for the AGC system to reschedule the power flow in the heavily loaded line, utilizing the remained capacity of light-loaded lines, which act as a buffer of fluctuating power. However, the proposed AGC model does not operate effectively in the case when all lines are set to target, since the model is based on the utilization of the capacity of the light-loaded line. The authors in [14] used a direct method to reschedule the generation of the power plant units, alleviating the line overloads after any N-1 contingencies. In this method, the power at the receiving bus of the overloaded line is appropriately modified by an amount equal to the overloaded power during each iteration of the load flow solutions. However, the method uses a hit-and-miss procedure to measure the correct bus power adjustment and, therefore, consumed enough time.

Non-linear programming methods are employed in several research articles to find a coordinated control action to eliminate the transmission line overload in renewable integrated power systems [15,16]. The authors in [17] utilized a learning machine learning approach with classical constraints for the economic dispatch model to tackle the problem of line overload alleviations. In [18], an economic dispatch model was developed for the generating units based on the direct acyclic graph (DAG), which provides an optimal remedy for the large network having different interconnected areas. The study in [19] proposed a dispatch model for line overload alleviation, utilizing a technique based on mixed linear programming to minimize switch opening as a solution to reduce overloads. The authors in [20] reduced the overloading of the transmission lines using a fuzzy logic model that tried to recreate the network operators' actions; however, the generation cost has not been considered. In [21], the authors used the metaheuristics of fuzzy logic and genetic algorithm for the implementation of online economic dispatch that assisted the overload study by eliminating the modeling of the entire AC system and the difficulties of the non-convergence of exact solution approaches. The authors in [22] employed fractional order integral control for AGC of a multi-source interconnected power system. The research work in [23] utilized a primal-dual interior point-based technique for the dispatch process of generating units. However, before executing the algorithm for a final solution, different stages of the simplification for the dispatch scenarios are deliberated, which are required to be executed in clusters. Such a process creates a direct impact on the scenario selection.

The aforementioned literature has tackled the overloading problem in a detailed manner. However, the studies lack a real-time dispatch strategy for the AGC system, taking into account the secure operation of transmission lines, along with the economic operation in a high wind integrated power system network. This paper proposes real-time load alleviation techniques for the AGC system by incorporating the line capacity constraints in the dispatch and scheduling process. The study analyzes the impact of the overloaded power on the transmission lines and proposes a dispatch strategy for the AGC system to keep the power fluctuation of the transmission line under the prescribed limits during the power balancing operation. Transmission lines having an anticipated load factor greater than the threshold value are selected as target heavy load lines. The overloaded power is reduced in the target lines by optimizing the dispatching ratio of the AGC system. This is done by calculating the amount of overloaded power and reducing the dispatched power of the target generating units by that amount. Meanwhile, the deficient power is injected into the grid by increasing the dispatched power from the local grid station of that overloaded bus. In this way, the area control error is regulated along with the protection of the transmission lines. However, if the line loadings are not detected, the dispatching ratio is set to be constant at the generation capacity ratio. Hence, the proposed dispatch strategy mitigates overloads of transmission elements after N-1 contingencies, ensuring the minimum risk of collapse in the massively penetrated wind power systems. To verify the efficacy of the proposed method, simulations are performed on Pakistan's 5-machine 8-bus power system model in Dig silent power factory software. The impact of the overloaded power due to a random change in the dispatch order is firstly analyzed and then the proposed method is implemented to de-overload the target lines.

The pertinent contributions of this work are as follows:

- The aggregated model of Pakistan's 5-machine 8-bus power system network is developed by integrating the 500 kV transmission lines and generating units of thermal power plants, gas turbines, and wind power plant systems.
- A model of the AGC system is designed and a dispatch strategy is developed to maximize or minimize the power flow increase among target lines at which power flow increase is detected.
- The dispatch strategy is designed by considering different constraints such as maximum, minimum capacities, and reserve availability of generating units and power capacities of transmission lines. Further, the system model is simplified to find quick solutions to contingencies issues. The proposed model of the AGC system is validated using the step input and the real-time data inputs.

The remainder of the paper Is divided as follows: Section 2 presents the modeling of the power plant units, such as thermal power plants, wind power plants, and gas turbines. Section 3 describes the proposed model of the AGC system along with the objective function and the proposed dispatch strategy for the line overload in the power balancing operation. Section 4 validates the proposed AGC system by initially applying a step input and then a real data input. Finally, Section 5 includes a summary, recommendations, and possible future works in the proposed research domain.

2. Modelling of the Power Plant Generating Units

This section contains thorough information on the modeling of power plant units, such as thermal power plant systems (THPPs), gas turbine-based power plant systems (GTPPs), and wind power plant systems (WPPs). The governor, which is designed and installed on power plant systems, provides the primary frequency response from each power plant unit. In addition, a centralized AGC system is designed for a secondary response, which is discussed in the following section.

2.1. Thermal Power Plant Model

The study considers the aggregated model of the thermal power plant system to provide the required primary and secondary reserve power in the balancing process of load and generation along with the routine operation in the power system network. The entire response of the THPPs is based on the response of the boiler influencing the system stability operation. Figure 1 presents the detailed diagram of the THPPs, which is modeled based on the investigation taken from the study [24,25] and is simplified for long-term dynamic simulation studies. It is illustrated from the diagram that the steam turbine block provides the required mechanical outpower (P_{mech}) based on the input from two blocks, which include the governor block (cv) and the steam pressure (P_t) from the boiler block. L_R represents the load reference set point, which is provided from the boiler control section and varies in response to any change in the system load. This change is counter in the boiler section by recalculating the mainstream pressure value (P_t), considering the real limitation of the turbine output and the associated delays of the store steam energy in the boiler. The load reference set point signal (L_R) is a feed-forward signal of the boiler regulating the turbine valve to match the current generation value with the reference value. Furthermore, the GRC and STC effects are integrated by considering a ramp rate limitation of 30 MW/min.

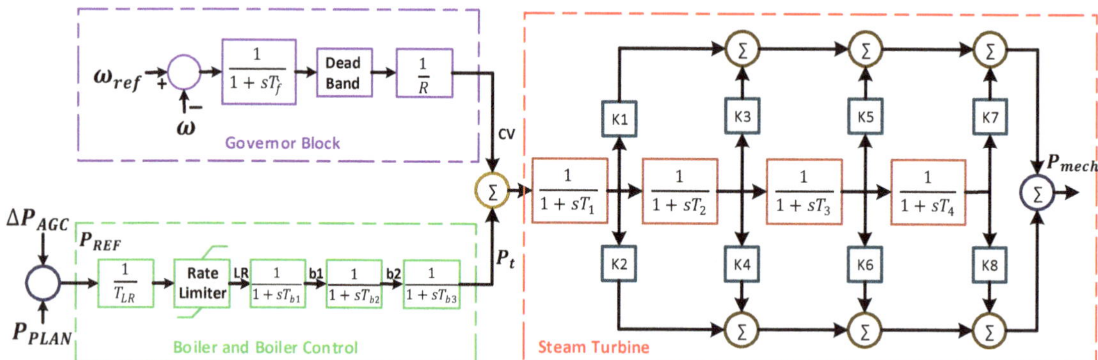

Figure 1. Thermal Power Plant Model.

The governor provides the required primary response by controlling the turbine speed valves based on speed variations of the generator and the droop setting of the governor. The droop setting in this study is fixed at 4 percent. Dead bands are added to the governor to prevent the motion of the steam valve from any small changes in speed due to the mechanical fault. b_1, b_2, and P_t are the corresponding lump storing steam series at interior pressure, while T_{b1}, T_{b2}, and T_{b3} are the associated time constants representing the time delays related to the boiler model. The response time of the boiler lies in the range of 5–6 min [26], leading to the overall response of the THPPs. This study considers a cross-compound double reheat steam turbine [24], as shown in Figure 1, which derives the output power in its mechanical form based on the inputs from the boiler model (P_t) and the governor output (cv). The four time constants (T_1, T_2, T_3, and T_4) represent the overall response of the turbine, which characterizes the charging of various volumes including high-pressure turbine bowls and the time constant for re-heater, crossover, and double reheat units. $K_1 - K_8$ are different coefficients signifying the contribution of the turbine sections to the net mechanical output power. K_1, K_2 represent very high pressure, K_3, K_4 are only for high pressure, while K_5, K_6 and K_7, K_8 represent intermediate- and low-pressure contribution values.

2.2. Gas Turbine Power Plant Model

The gas turbine power plant aggregated model is being considered in this study to provide only the primary response during the power balancing process along with the daily routine operation. The developed model, as shown in Figure 2, is based on the study taken from [25–27], which comprises different blocks including the power distribution block (PDB), power limitation block (PLB), and gas turbine dynamics block (GTDB). The governor model is designed for GTPPs integrating the low-pass filter and dead band effects to protect the model from responding to high frequency and low frequency variations, respectively. When there is any change in system load, the frequency of the system deviates from its original value, which is converted into a power demand signal (ΔP_c) by the droop characteristic signal. The resultant ΔP_c of the signal is provided as an input to the power limitation block, which utilizes the combustion technology physical constraints to impose physical barriers on the turbine response. The PLB block restricted the reference power signal to P_{max} and P_{min}, which are upper and lower power level limits. L_{max} and L_{min} are the upper and lower load reference set points in the PLB block, which ensure that no combustion constraints are violated during normal operation. A rate limiter block regulates the ramping of the power demand signal at a specific rate, preventing excessive fire during ramping up and the extinguishment of a narrow combustion flame during severe ramping down.

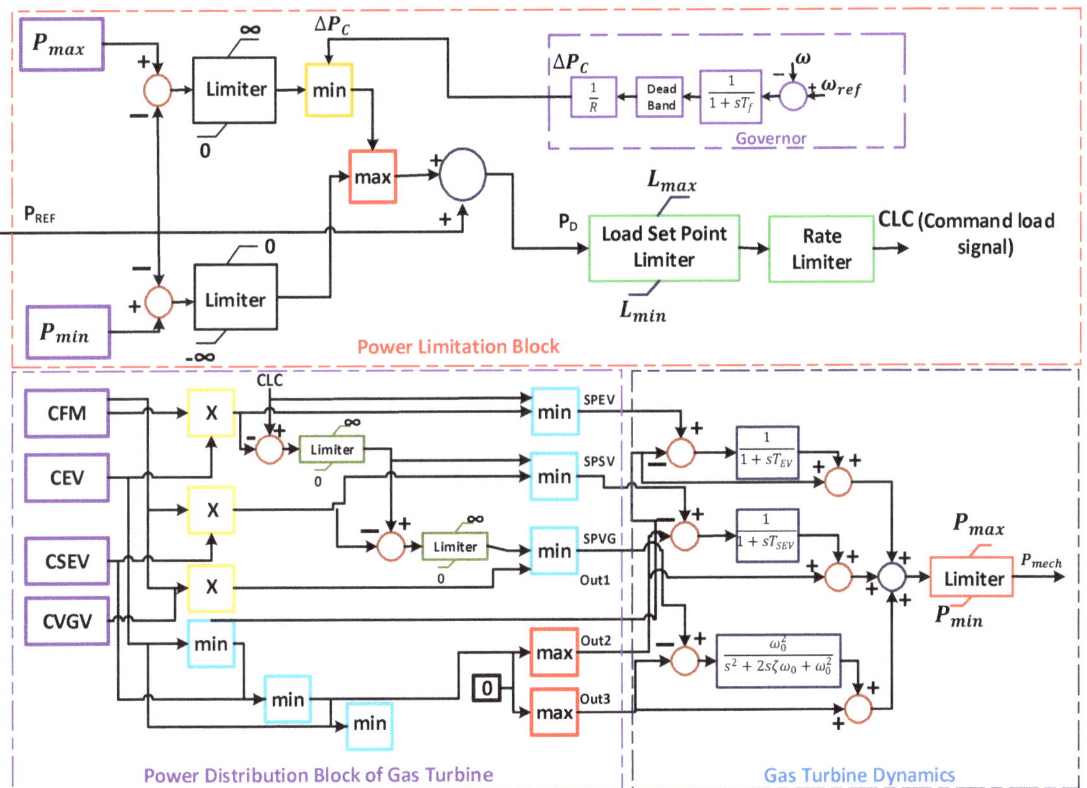

Figure 2. Gas Turbine Power Plant Model.

Command load signal (CLC) is the resultant signal generated by the PLB block, which is fed to the PDB block. The PDB block consists of two chambers fired in series. The EV chamber takes the compressed air as an input and mixes it with 50 percent of the total fuel after heating. Here, it is important to mention that the fuel flow is controlled through the CLC signal from the PDB block, while the airflow is controlled through the shaft speed and variable inlet guide vane (VIGV). The resultant mixture is entered into the high-pressure turbine, forcing it to spin, which causes the pressure to drop. The remaining 50 percent of the fuel and some additional air is added to the resultant mixture in the SEV chamber. The new mixture is then expended through the low-pressure turbine, forcing it to spin. This kind of procedure inculcates high operational flexibility, low emission, and high efficiency. The power contribution factors (CEV, CSEV, CVGV, and CFM) in the PDB control blocks represent the physical characteristics of fuel flow, airflow, and allowable temperature. The output of the power contribution factors depends on the CLC signal, air compressor, and the two combustors' capacities. The dynamics of the gas turbine dynamic block are represented by the compressor and combustor unit. The combustor EV and SEV are represented by the 1st order transfer function, while the variable inlet guide vane (VIGV) is represented by the 2nd order transfer function. T_{EV} and T_{SEV} are time constants for the burner of the EV and SEV combustors, respectively. The overall mechanical output of the turbine is a function of CFM, CEV, CSEV, CVGV, and CLC and is restricted between P_{max} and P_{min}.

2.3. Wind Power Plant Model

The study has modeled the wind power plant system to investigate its behavior during the active power balancing operation by providing the primary and secondary regulating reserves along with the normal system operation. The proposed model is developed to be operated at the power system level rather than considering the impact of a specific wind farm. This is because the aggregated performance of many wind turbines has a greater influence on the power system than the operation of a single turbine. The study considered the model proposed by the IEC61400-27-1 committee as the starting point for the model, drafted for the simulation models of wind power generation. Here, it is important to mention that the proposed model is developed for the control operation of the active power in the system, which is the focus of this study. However, the model includes all the relevant dynamics for long-term simulation studies. A detailed figure of the wind power plant system (WPPs) is shown in Figure 3, mainly consisting of three blocks, which are the wind power plant active power controller (WPPAPC), the wind turbine active power controller (WTAPC), and a generator reference current block.

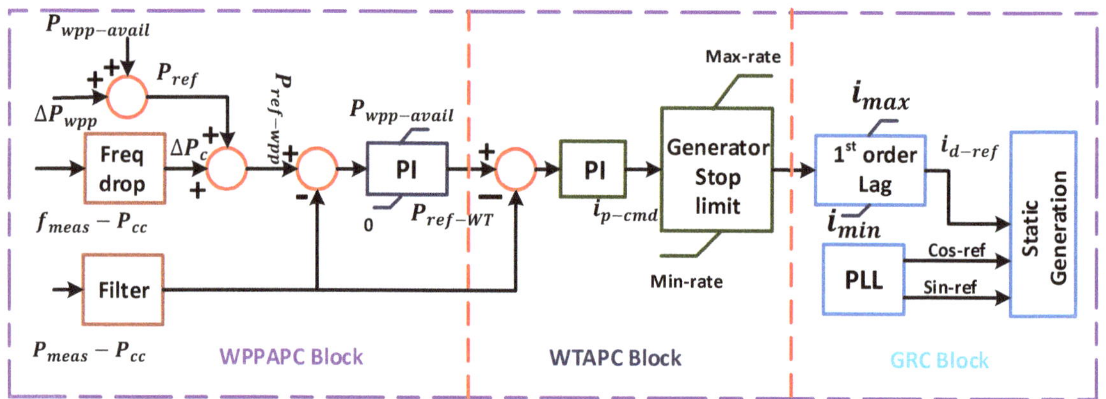

Figure 3. Wind Power Plant Model.

The WPPAPC block generates a reference signal at the wind turbine level (P_{ref_WT}) in response to any change in the reference signal at the wind power plant level (P_{ref_WPP}s). P_{ref_WPP} is a function of the reference signal (P_{ref}), governor output (ΔP_c), and measured power at PCC (P_{meas_PCC}). The governor model considers the effect of the dead bands and provides the required power change signal (ΔP_c) based on the droop characteristics signal and the available power strength. The PI controller of the WPPAPC block regulates its P_{ref} signal based on the difference between the P_{ref_WPP} and P_{meas_PCC} signals. The signal of the $P_{WPP_{avail}}$ limits the PI controller output power. Following the reference signal from the WPPAPC block, the WTAPC block generates the currently active component (I_{Pcmd}) as calculated by the PI controller and it is calculated based on the difference between the P_{ref_WT} and P_{meas_PCC} signals. This study considers the type IV wind turbine technology for the generator as being able to provide the required operational flexibility. Machine-side convertors (MSC) of such turbines are decoupled from the grid-side converter (GSC) to perform their tasks. MSC rotates the generator at a specific speed, while the GSC controls the active and reactive power flow to the grid side. The generator used in the wind turbine is modeled as a static generator considering the current source because grid-side wind turbine behavior is determined by the full-scale converter. The dynamic response of the static generator is a function of the inputs from the phase-locked loop (PLL) and reference current. In addition, a ramp-rate limitation is also added to limit the reference value based on available wind power. WPP has the fastest response time among the power plant units to any change in system loads, which is typically 2–4 s.

3. Power System AGC Model and the Proposed Control Strategy

This section presents detailed information about the proposed power system AGC model, followed by the objective functions and the proposed dispatch strategy implemented in the AGC system.

3.1. Power System AGC Model

The secure and reliable operation of the power system is determined by the automatic generation control (AGC), which continuously monitors the frequency of an interconnected power system and accordingly performs the remedy measures if required. In the AGC control, two variables determine the whole operation, which includes frequency deviations and the interchanged power. These two variables accumulatively make one equation called area control error (ACE), which is a prime step in the execution of the AGC control. The equation of ACE for an ith area of an interconnected power system is given as:

$$P_{ACE,i}[MW] = \sum_{j \,\in\, \mathcal{A}_n} \beta_i \Delta f + \Delta P_{ij} \tag{1}$$

In the above equation, $P_{ACE,i}[MW]$ is the imbalance of power in the ith area of the system following any change in the system load. Δf represents the deviations in the frequency, while β_i is the frequency bias constant of the ith area. β_i is represented by the equation $\beta_i\left[\frac{MW}{Hz}\right] = D_i + \frac{1}{R_i}$, in which $D_i\left[\frac{MW}{Hz}\right]$ characterizes the damping of the power system, while $R_i\left[\frac{MW}{Hz}\right]$ is the droop characteristics. In this study, R_i is fixed at a 4 percent value. $\Delta P_{ij}[MW]$ is the total change in the interchange power, which is the difference between the actual and scheduled interchange power and is called the tie-line error. The system frequency instantly varies following any load change in the network due to which the primary reserves are activated through a governor installed on each generator unit. Meanwhile, AGC calculates the required area control error $P_{ACE,i}$ and releases the secondary reserves to regulate it and release the primary reserve. This process brings the frequency to its nominal level. In such a way, the AGC regulator maintains the frequency of the system at its nominal level and keeps the interchange power at its scheduled value by performing the minute-to-minute balancing in the ith controlled area. The AGC operation time lies in the range of 1–10 min.

Conventionally, the AGC system regulates the area control error and tie-line interchanges, which is the only power flow control function. However, the power flow on the rest of the lines in the power system network is not regulated, which often results in the overloading of the transmission lines during the power balancing operation. The operation of AGC regulates the active power flow, which is absorbed or injected by different components of the power system. Therefore, overloading of the power lines can be avoided during the power balancing operation by considering the line limits in the AGC dispatch process. The focus of this study is to integrate the line capacities of the transmission line system in AGC dispatch, which considers the line loading limits of all the lines before distributing the dispatching power to the generating units. Hence, the power fluctuation in the transmission lines is controlled by re-scheduling the participation factor of the generating units.

Figure 4 shows the network layout of the proposed power system AGC model. An 8-bus 5-machine model of Pakistan's power system is considered in this study, consisting of different generating units of thermal power plants, gas turbines, and wind power plant systems. The generating units are connected to different buses of the 500 kV transmission system. Furthermore, to study the dynamics of the AC interconnection, an external grid is connected, emulating the characteristics of a general grid having an inertia of 16 s. It is important to mention that the unit of inertia utilized in this study is in seconds. This is because, in a per unit system, inertia is equivalent to energy per power, which means that the rate at which the power changes in the power system network is restricted. Moreover, the primary frequency response of the connected grid is 6000 MW/Hz, which means that the rate at which the external grid responds per Hz changes with system frequency. Detailed information on the connected buses and the associated lines is given in Table 1.

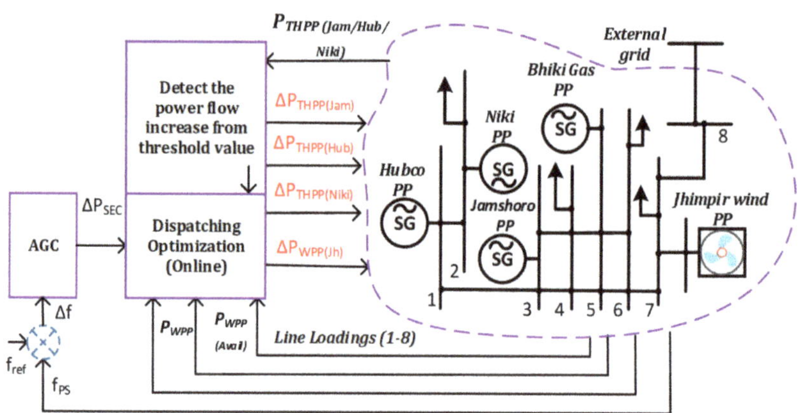

Figure 4. Proposed Power System AGC Model.

Table 1. Grid stations and associated lines of the network.

S. No.	Grid Stations (500 kV)	Location (Pakistan)	Connected Lines and Dist. (500 kV)
1	Niki (NKI)	North Karachi	HBC-NKI (63 km)
2	Shikarpur (SKP)	Shikarpur, Sindh	SKP-DU (2) (153 km), DUN-SKP (2) (159 km)
3	Dadu-New (DUN)	Sindh	JMS-DUN (2) (199 km), DUN-SKP (2)
4	Jamshoro (JMS)	Jasmshoro, Sindh	HBC-JMS (198 km), JMS-DUN (2)
5	Hubco (HBC)	Lasbela, Balochistan	HBC-NKI (63 km), HBC-JMS
6	Dadu (DU)	Dadu, Sindh	SKP-DU (2), DU-DGK (544 km)
7	Guddu New (GDN)	Kashmore, Sindh	DU-GDN (319 km)
8	DG Khan (DGK)	DG Khan	DU-DGK, DGK-EXT Grid

The objective of this study is to alleviate the overloading of the transmission lines during the power balancing operation by considering the power capacities of the lines. Therefore, to simplify the complex model of transmission line systems to achieve the aforementioned objectives, this study has ignored the other parameters of the transmission line while performing this task. The other parameters include the line lengths and the line voltages. Furthermore, the AGC implemented in this study considers a PI controller, which is used to minimize the P_{ACE}, as provided in (2).

$$\Delta P_{Sec} = K.\Delta P_{ACE} + KT \int \Delta P_{ACE} dt \ K \tag{2}$$

The values of the parameters K and T are necessary to restore the network's frequency and the interchange power to their original values. The values of these parameters are set following the conventional criteria for a coordinated secondary control system [28]. The activation rate of the reserves from the generating units of the power plants is characterized by the tracking speed of the controller known as the time constant. ΔP_{Sec} [MW] is the amount of reserve power calculated by the AGC controller to be distributed among the generating units utilizing the proposed dispatch technique. In this study, the extra amount of reserve power is utilized from different units of thermal power plants $(\Delta P_{THPP(JMS)}, \Delta P_{THPP(HBC)}, \Delta P_{THPP(NIKi)}$ [MW]) and the wind power plant (ΔP_{WPP}) [MW], considering all the constraints of the generating units and the transmission lines.

3.2. Objective Function and the Proposed Dispatch Strategy

The uncontrolled power flow in the transmission line of large-scale wind-integrated power systems replicate serious repercussions on the health of the transmission system. Power flow congestion management aims to alleviate the transmission line overload while lowering the generation cost. This is expressed mathematically as follows:
Objective No. 1:

$$Min \sum_{J=1}^{N_L} \left(\wp_J - \wp_{CP} \right) \tag{3}$$

Subjected to:
Equality Constraints:

$$P_{Gi} - P_{LDi} = \sum_{J=1}^{N_B} |\mathbb{V}_i||\mathbb{V}_j||Y_{ij}| \cos(\delta_i - \delta_j - \theta_{ij}) \tag{4}$$

$$Q_{Gi} - Q_{LDi} = \sum_{J=1}^{N_B} |\mathbb{V}_i||\mathbb{V}_j||Y_{ij}| \sin(\delta_i - \delta_j - \theta_{ij}) \tag{5}$$

Inequality Constraints:

$$P_{min} \leq P_{Gi} \leq P_{max} \tag{6}$$

$$Q_{min} \leq Q_{Gi} \leq Q_{max} \tag{7}$$

$$\Delta P_{min\ limit} \leq \Delta P_{Gi} \leq \Delta P_{max\ limit} \tag{8}$$

Figure 5 presents the proposed dispatch strategy for the developed AGC model as mentioned in Figure 4. Initially, when there is any change in the system frequency, area control error is calculated and based on that, the AGC regulator calculates the required change in reserve power (ΔP_{Sec}), which is then distributed among the different generating units to regulate the area control error (P_{ACE}). For positive regulation, the required AGC response is divided among the different generating units of the thermal power plant. However, the negative regulation wind power plant is also integrated into the AGC response along with thermal power plant units. In the event of a negative dispatch, wind power output is reduced only if all thermal power plants are running at their lower limitations, which are set at 20% of their full capacity, or if the dispatched power is approaching their lower limits. This reduces the use of reserve power from thermal generating units while allowing the wind power plant to run at full capacity.

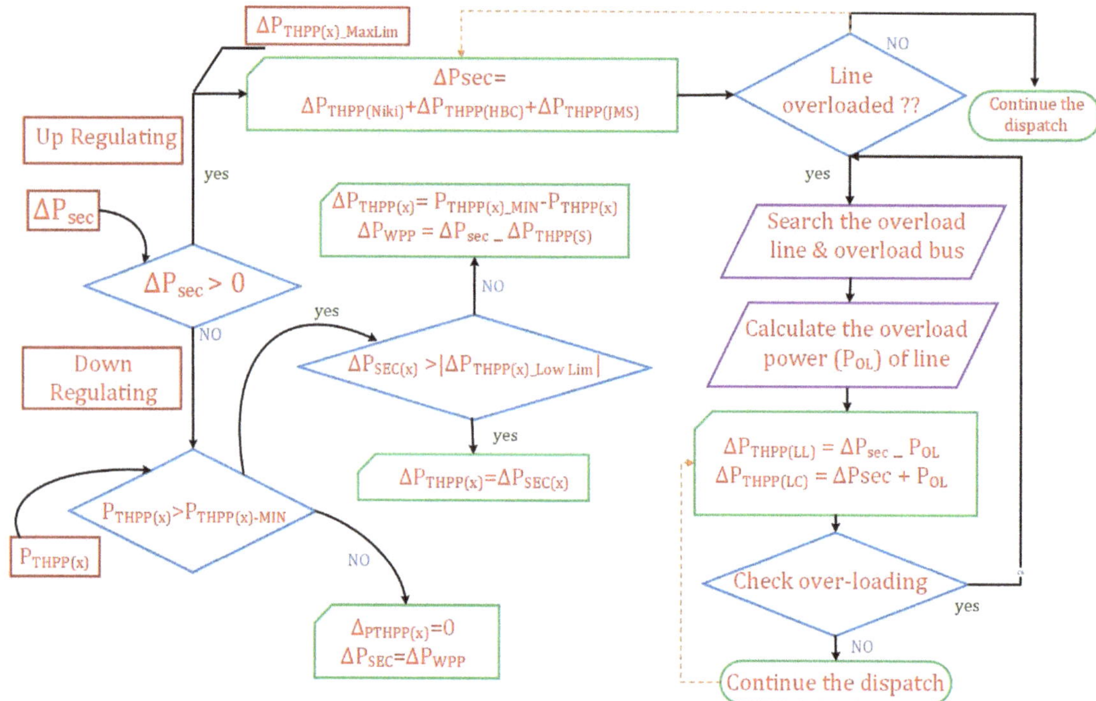

Figure 5. Proposed Dispatch Strategy for the AGC System.

Meanwhile, the line data of power flows in all the transmission lines are continuously measured during the dispatch process, and in case any of the lines are overloaded during the dispatch process, the proposed dispatch strategy performs the optimal dispatching, in which the impact of AGC dispatch on overloaded power flow lines are calculated. Transmission lines with an estimated load factor larger than the threshold value are chosen as target heavy load lines. Overloaded power is reduced in the target lines by adjusting the AGC system's dispatching ratio. This is accomplished by determining the amount of overloaded power and reducing the dispatched power of the target generating units by that amount. Meanwhile, inadequate power is injected into the grid by raising the dispatched power from the local grid station of that overloaded bus. In this way, the area control error is regulated along with the protection of the transmission lines. However, if the line loadings are not detected, the dispatching ratio is set to be constant at the generation capacity ratio. Here, line loading means an excess of power flow over a certain threshold value fixed by the operator. The estimated value or the threshold value of the transmission line loading is fixed at 90 % in this study, which means that if a line exceeds its limit of 90%, it is considered a loaded line. The line loading of each transmission line is measured using the equation given as:

$$\text{Line Loading} = \frac{\text{current flowing in the line}}{\text{Current rating of the line}} \times 100 \tag{9}$$

4. Simulation and Results

The aforementioned section presented the developed power system AGC model and the associated dispatch strategy, which contemplate the transmission line overloading during the dispatch process to perform the daily generation and load balancing operation. The section implements the dispatch strategy and analyzes the results. To carry out this objective, the study has modeled a hypothetical network of the Pakistan power system in which a specific part of the network is considered to implement and analyze the proposed control strategy. The selected part of the network is a 5-machine 8-bus model with 500 kV transmission lines connected through different buses and with different types of power plant units. The network consists of three THPPs units, GTPPs, and a WPP system. Generating units of THPPs, GTPPs, and WPPs are responsible for the primary reserves, while secondary reserves are delivered from the THPPs and WPPs following the proposed dispatch strategy. Furthermore, to investigate power changes on the external grid, the proposed power system network is coupled to an external grid that mimics the specific characteristics of a grid with a main frequency response of 6000 MW/Hz and an inertia of 16 s.

To understand the workings of the proposed power system AGC model, this study is divided into two phases. In the first phase, the effectiveness of the proposed dispatch strategy is analyzed for a step response of 150 MW when randomly applied on any bus of the power system network. In the second phase, the study implemented the same AGC model, utilizing a real-time input series for the generating units and connected loads. In both phases, the overloading of the transmission lines is analyzed during the dispatch process of the AGC model in the power balancing operation. Table 2 presents the current operating scenario, in which the power plant units are operating at different points to meet the daily load demand. Initially, before applying the step response, the load and the generation in the current scenario are balanced and, therefore, no imbalances in the power system are present, resulting in the frequency of the system being at the nominal level. The line loadings of the associated transmission lines are obtained in this scenario and are shown in Table 3. It is illustrated from the table that no line exceeded the limit of 90%, which is the maximum limit of line loading fixed in this study.

Table 2. Capacities, regulating reserves, and the initial operating points of generating units in a selected part of Pakistan's power system network.

Power Plant Models	THPP (Jamshoro)	THPP (Hubco)	THPP (Niki)	GTPP (Bhiki)	WPP (Jhimpir)	Load (Jamshoro)	Load (Niki)	Load (Dadu)	Load (Guddo-New)
Capacities	1320	1202	800	220	2800	–	–	–	–
Reserves	±100	±100	±120	0	−400	–	–	–	–
I.O.P	1188	1081.8	500	218	2520	2920.3	887.5	700	1000

Table 3. Ratings and current loading of the transmission lines in the balanced condition.

Power System Trans-Lines	DU-GDN	DU-DGK	SKP-DU	DUN-SKP	JMS-DUN	HBC-JMS	HBC-NIKI
Current ratings (KA)	1.8	0.3	1.5	1.5	1.5	1.2	0.6
Loadings (%)	51.7	17.8	31.6	40.0	40.0	66.8	74.6

4.1. Step Response Analysis

To initiate the case study of the first phase, a load step of 150 MW was applied at the Niki bus station of the proposed power system network, deviating the system frequency from its nominal value, thus creating an imbalance between the load and generation. The system frequency is regulated by activating the secondary reserves following the primary reserves through the AGC system from the thermal power plants units (($\Delta P_{\text{THPP(JMS)}}$, $\Delta P_{\text{THPP(HBC)}}$, $\Delta P_{\text{THPP(NIKi)}}$) installed at different grid stations. The response of the AGC system is based on the calculated ACE signal (P_{ACE}) in the power system network, which is fed to the PI controller to determine the indispensable secondary response (ΔP_{Sec}) from the concerned power plant units. P_{ACE} and ΔP_{Sec} signals are resultantly drawn in Figure 6a to show the response of the area control error and the required power dispatch to regulate the ACE signal. It is illustrated from the figure that the ΔP_{Sec} signal leg behind the P_{ACE} was due to the delays accompanying the AGC and the power plant units. The subsequent individual response ($\Delta P_{THPP(x)}$) of THPP units against the ACE signal is drawn in Figure 6b. For positive regulation, the required AGC response is divided among the different generating units of thermal power plants. However, for negative regulation, the wind power plant is also integrated into the AGC response along with thermal power plant units.

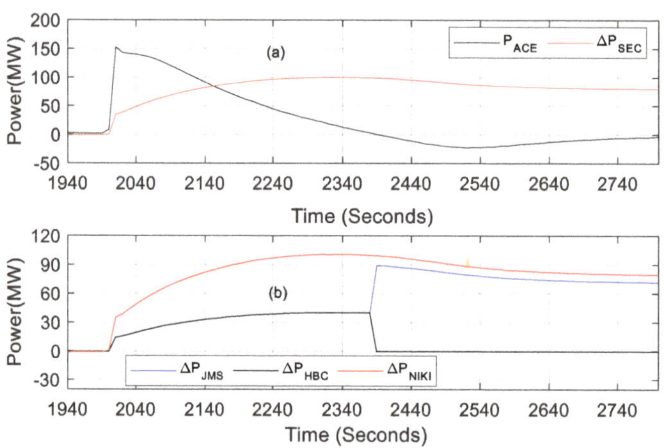

Figure 6. (**a**) ACE and the subsequent AGC response (**b**) Dispatch power from THPP units (Jamshoro, Hubco, and Niki).

From Figure 6, it can be seen clearly that when a load of 150 MW was applied at the Niki grid station, the AGC system automatically activated the reserve power from each generating unit taking part in the AGC dispatch process by keeping in view the line limit of each transmission line. In this process, all the lines approaching the Niki grid station would have a possibility to cross the threshold loading values to meet the increasing load demand at the Niki grid station. Hence, to alleviate the overloading of these lines, the dispatched power from all the buses connected to these lines should be reduced by the amount of overladed power calculated with the help of Equation (9). In this case, the transmission line approaching the Niki grid station from the Hubco grid station became overloaded. This can be seen in Figure 6b between the time duration of 2340 and 2440 s, when the loading of a line connecting the Hubco and Niki grid station is overloaded as a result of the AGC operation and the response from the thermal power plant units is rescheduled to de-overload that specific line. The resulting dispatch power of the generating units shows that the generation at the Hubco grid station is reduced to zero, while the power from the generating unit of the Jamshoro grid station is reduced to a lower level. This is because the associated transmission lines of these grid stations are overloaded and, hence, a further dispatch from these generating units can overload these transmission lines. Conversely, the dispatched power by the generating unit at the Niki grid station (Local) is increased by the amount of power that the other two generating units have decreased in power. This does not affect the power balancing operation of the AGC system and, thus, the frequency of the system remains at the nominal level. Hence, the proposed dispatch strategy effectively regulated the area control error (P_{ACE}) by not violating the maximum line limit of the transmission lines, which is fixed at 90 % in this study. This can be witnessed from the resultant response of the system frequency (Figure 7a) following the AGC response and subsequent response from the external grid, as shown in Figure 7b.

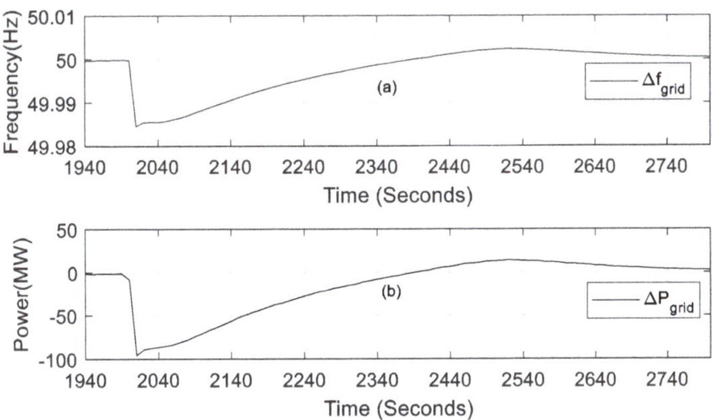

Figure 7. (**a**) Grid frequency and (**b**) Grid deviations following AGC response.

To further quantify the results, the steady-state values of the line loadings of all the transmission lines following the AGC response are presented in Table 4, and their responses are drawn in Figure 8, from which it can be illustrated that no line has crossed the limit of 90 %; hence, all the lines are operating within their nominal limits. Hence, the proposed AGC model effectively mitigated the imbalance engendered due to the load changes in the power system.

Table 4. Rating and current loading of the transmission lines following the AGC response.

Power System Trans-Lines	DU-GDN	DU-DGK	SKP-DU	DUN-SKP	JMS-DUN	HBC-JMS	HBC-NIKI
Current ratings (KA)	1.8	0.3	1.5	1.5	1.5	1.2	0.6
Loadings (%)	51.7	17.8	31.6	39.9	39.9	60.0	88.2

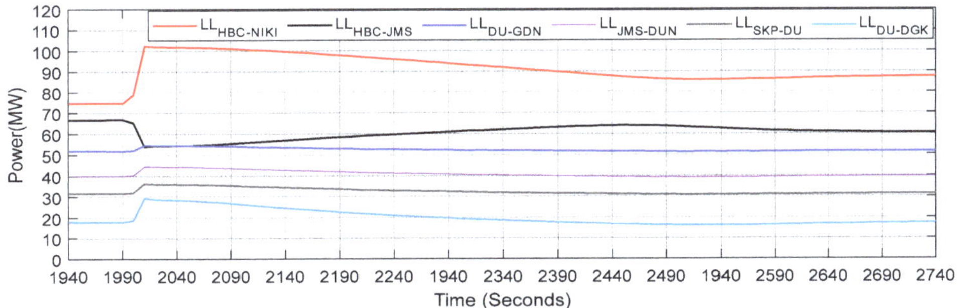

Figure 8. Line loading following AGC response.

Nevertheless, the line loadings remain within the limit, which ensures a secure and reliable operation of the power system. Furthermore, from Figure 8, it is illustrated clearly that the loading of the line connecting the HBC-NIKI grid station is reduced from the maximum limits between the time duration of 2340 and 2440 s. The resultant loading of the other lines following the AGC response is also shown in the figure, which shows that none of the lines crossed the maximum loading limits during the operation.

4.2. Real-Time Input-Based Analysis

The aforementioned section validated the proposed AGC dispatch strategy for avoiding the overloading of the transmission lines in the power balancing operation when a load step was applied at the Niki bus station. The developed dispatch strategy is further analyzed in this section utilizing a real-time input series for the generating units and connected loads. The capacities and the regulating reserves of the generating units and the connected loads are kept the same, as mentioned in Table 2. Furthermore, the initial generating power of all the power plant units is shown in Figure 9, in which three generating units are based on thermal power—one is a gas turbine and one is the wind power plant. Furthermore, the generating units based on the thermal power plant and the wind power plant contribute their power to the secondary regulation process. Here, it is important to mention that the forecast error in wind power engendered a power imbalance between the load and generation.

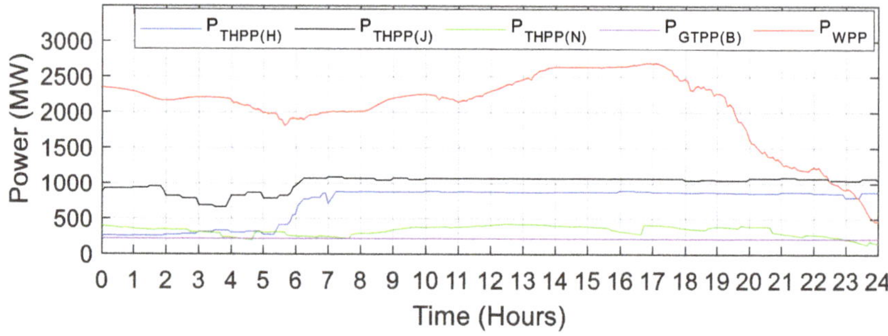

Figure 9. Initial generation from power plant units.

Figure 10 presents the initial power imbalance between the demand and generation resulting from the forecasting error of the wind power plant. The resulting imbalances are required to be mitigated by utilizing the secondary reserves from wind and thermal power plants installed at different buses. In this case, all the thermal power plants contribute to both positive and negative regulation, while the wind power plant only regulates negative power imbalances. Meanwhile, at the same time, the power flow of the transmission lines is measured by performing the load flow analysis according to the developed dispatch strategy. If any of the transmission lines cross the threshold value of the line loading as fixed by the operator, the dispatch ratio is optimized by the AGC regulator in a way that alleviates the overloaded lines by reducing the power generation from the concerned power plant units. The reduced power is equal to the amount required to de-alleviate the target-loaded lines. In addition, the deficient power is injected into the grid from the local grid station of the overloaded line. Conversely, if none of the lines cross the threshold load value, the dispatch ratio of the generating units would remain the same as it was before the loading impact. Here, the threshold value is set at 90% of the full line capacity.

Figure 11 illustrates the responses of all the generating units during the 24 h simulation period to provide the required reserve power during the AGC operation. Here, it can be seen from the figure that all the thermal power plants are actively participating in the positive and negative regulation process, while the wind power plant is only providing the regulation power during the negative regulation process. This is because the wind power plant is cheaper and always preferred to be operating at its maximum level. However, in this study, the generation capacity of wind power is reduced only if all the thermal power plants approach their lower operating level ($P_{chp,\,min(x)}$), which is set to 20% of their full capacity, or the dispatched power of all the thermal power plants touches their lower limits. In such a situation, thermal power plants does not have sufficient reserves to be inculcated in the grid. Therefore, the wind power plant starts reducing its output power to regulate the P_{ACE} as per the requirement of the network.

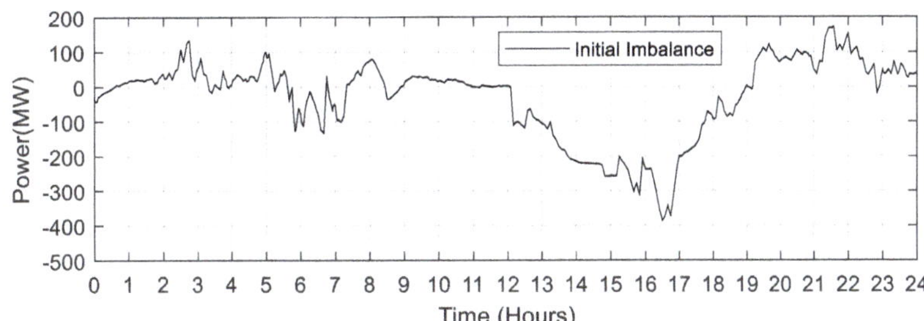

Figure 10. The initial power imbalance in the power system network.

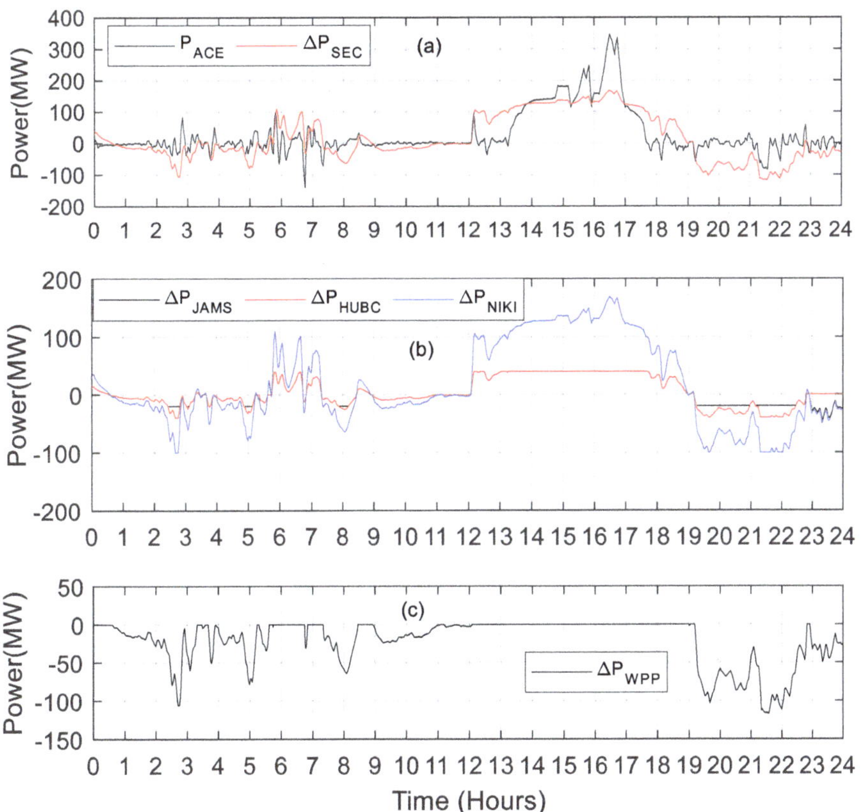

Figure 11. (**a**) ACE and the subsequent AGC response (**b**) Dispatch power from THPPs (Jamsh., Hubco & Niki), and (**c**) WPP (Jhimpir).

Along with such a regulation process, the dispatch mechanism takes into account the limitation of the transmission line capacities to avoid any overloading issues during the power balancing operation. To achieve this objective, regulating reserves are activated from those generating units, which do not overload the associated transmission lines. In this case, it can be seen from Figure 11 that the dispatched power is at a maximum from the Niki power plant and a minimum from the Jamshoro power plant units. Moreover, the reserve activation from the Hubco power plant system remained at the lowest level. This is because the line connecting the Hubco and Niki grid station is already operating at its maximum limit and further dispatches from the Hubco grid or any other grid station may overload this line. However, the power from the local generating station at the Niki bus compensated for the required regulation power. The resultant response of all the lines in the network following the AGC operation is shown in Figure 12, where it be seen that all the lines are operating below the maximum loading limit of 90%. Furthermore, it can be realized from the figure that the loading of the line connecting the Hubco to the Niki grid station remains at the maximum level and, therefore, the dispatch from the generating units installed at the associated buses remains at the lower level.

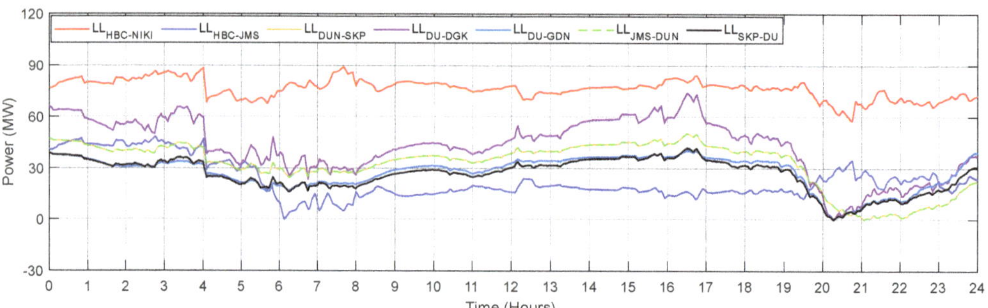

Figure 12. Line loadings following the AGC response.

Figure 13 presents the response of the grid frequency and the deviations of the AC interconnection. It is illustrated from the figure that the frequency of the system is regulated to its nominal level at different points during the simulation period. This is due to the activation of the reserve power from different generating units. The power deviations that appeared on the external grid are reduced by a substantial amount after the AGC response. Here, it is important to mention that the external grid deviations are the final imbalances, which will remain in the system following the AGC response.

The comparison of the initial and final power imbalance following the AGC response in the network is shown in Figure 14, from which it is illustrated that the initial power imbalance is reduced by a substantial amount after the activation of the reserve power from generation.

In addition, Figure 15 demonstrates the comparison of the loading power of the overloaded line during the AGC operation. From the figure, it can be concluded that the overloaded power has been significantly alleviated, bringing it to the level below the threshold value, which is fixed at 90% in this study. Hence, it can be concluded that the proposed dispatch strategy effectively performs the AGC operation and protects the transmission lines in the network.

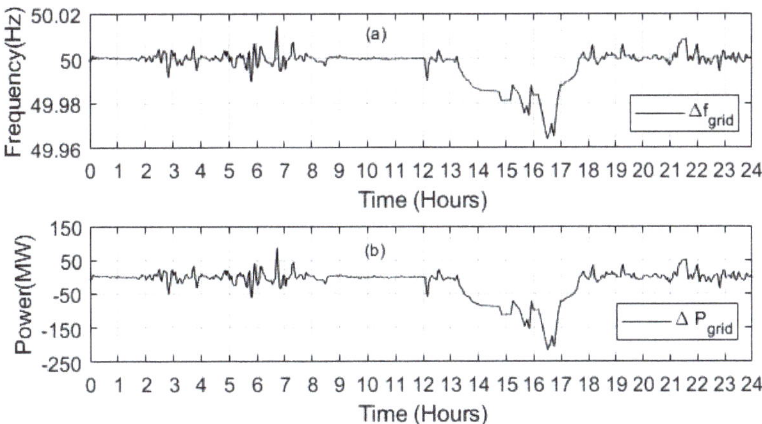

Figure 13. (**a**) Grid frequency and (**b**) Grid deviations following the AGC response.

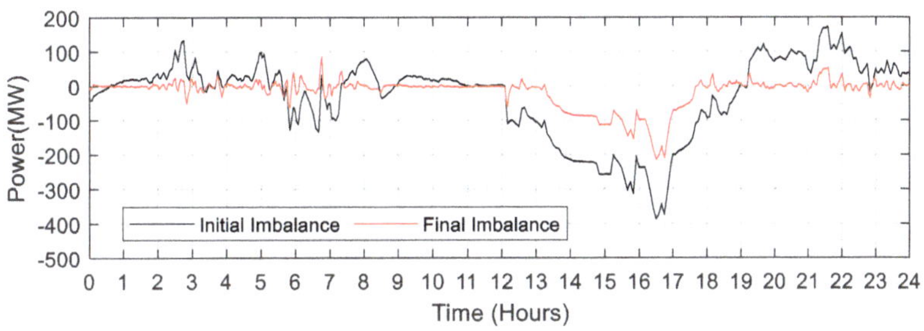

Figure 14. Comparison of initial and final power imbalances following the AGC operation.

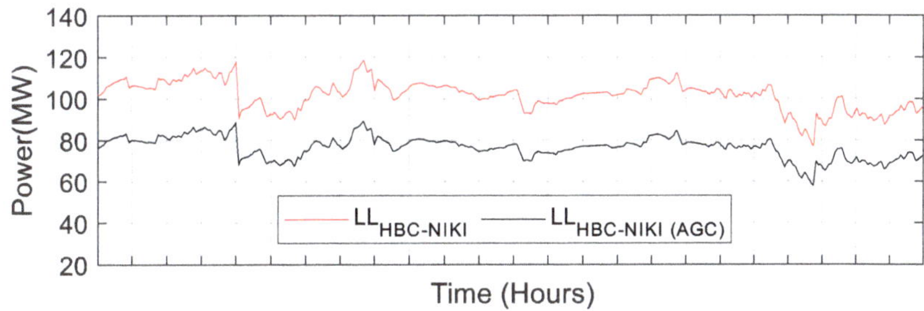

Figure 15. Comparison of the loading power of the overloaded line during AGC operation.

5. Conclusions and Future Directions

Concerns about the security and economy of the grids have increased due to developments in power system networks including the integration of large-scale wind energy sources, which engendered the power imbalance between the load and generation due to associated forecasting errors. Additional reserves are normally provided from the generating units to keep the balance in the network. However, the bulky injection of reserve power results in large fluctuations of the power flows in the associated transmission lines by not considering the transmission system constraints. This study incorporated the transmission line capacities in the real-time dispatch of the AGC system, resulting not only in an economic operation, but also avoiding transmission line overloading in real-time, thereby making the real-time operations secure and reliable. A simulation model consisting of a 5-machine 8-bus model was designed for the study in the Dig silent power factory software.

A real-time dispatch strategy was developed and implemented by performing the step response and the real-time input analysis. The results demonstrate that the suggested AGC design efficiently eliminates transmission line congestions during the daily load generation balancing operation while allowing generating units to operate economically.

In the present work, the proposed control strategy for avoiding line congestion was implemented in a single area network with minimum buses to minimize the computational burden. In the future, this work can be extended to the power system network, which has a multi-bus bar structure with interconnected power systems.

Author Contributions: Conceptualization, K.U., R.A. and S.A.; Data curation, A.B. and R.A.; Formal analysis, Z.U. and R.A.; Investigation, K.U., A.B., Z.U. and S.A.; Methodology, K.U., A.B. and Z.U.; Project administration, A.B., S.A. and A.Y.; Resources, A.Y.; Software, K.U., Z.U. and A.Y.; Validation, A.B., R.A. and S.A.; Visualization, R.A. and A.Y.; Writing – original draft, K.U., A.B. and Z.U.; Writing—review & editing, R.A., S.A. and A.Y. All authors have read and agreed to the published version of the manuscript.

Funding: This research received no external funding.

Institutional Review Board Statement: Not applicable.

Informed Consent Statement: Not applicable.

Conflicts of Interest: The authors declare no conflict of interest.

Abbreviations

Acronym/Symbol	Definition
GTPP	Gas turbine power plant
THPP	Thermal power plant
CLC	Command load signal
CIGRE	International Council on Large Electric Systems
FACTS	Flexible AC transmission system
PLB	Power limitation block
STC	Steam temperature control
$\wp_J[\mathrm{MW}]$	Active power flow in jth line
N_L, N_G	Number of loaded lines and generators
$P_{LDi}[\mathrm{MW}]$	Load demand at ith bus
Y_{ij}	Mutual admittance
θ_{ij}	Line impedance angle
CVGV	Variable inlet guide vane position compressor capacity
GTDB	Gas turbine dynamics block
CSEV	Sequential environmental burner capacity
PDB	Power distribution block
CFM	Baseload function
DAG	Direct acyclic graph
GRC	Generation rate constraints
$\wp_{CP}[\mathrm{MW}]$	Maximum capacity of the line
$P_{Gi}[\mathrm{MW}]$	Active power of the ith generator
N_B	Number of buses
δ_i, δ_j	Bus i and j voltage angles
$\Delta P_{Gi}[\mathrm{MW}]$	Dispatch power from an ith generator

References

1. Sheraz, A.; Herodotou, H.; Mohsin, S.M.; Javaid, N.; Ashraf, N.; Aslam, S. A survey on deep learning methods for power load and renewable energy forecasting in smart microgrids. *Renew. Sustain. Energy Rev.* **2021**, *144*, 110992.
2. Abdul, B.; Ahmad, T.; Ali, A.Y.; Ullah, K.; Mufti, G.; Hansen, A.D. Flexible modern power system: Real-time power balancing through load and wind power. *Energies* **2019**, *12*, 1710.
3. Morison, K.; Wang, L.; Kundur, P. Power system security assessment. *IEEE Power Energy Mag.* **2004**, *2*, 30–39. [CrossRef]
4. Kirschen, D.S. Power system security. *Power Eng. J.* **2002**, *16*, 241–248. [CrossRef]

5. Ni, M.; McCalley, J.D.; Vittal, V.; Tayyib, T. Online risk-based security assessment. *IEEE Trans. Power Syst.* **2003**, *18*, 258–265. [CrossRef]

6. Ronellenfitsch, H.; Timme, M.; Witthaut, D. A dual method for computing power transfer distribution factors. *IEEE Trans. Power Syst.* **2016**, *32*, 2. [CrossRef]

7. Farooq, S.M.; Hussain, S.; Kiran, S.; Ustun, T.S. Certificate based security mechanisms in vehicular ad-hoc networks based on IEC 61850 and IEEE WAVE standards. *J. Electron.* **2019**, *8*, 96. [CrossRef]

8. Arenas-Crespo, O.; Candelo, J.E. A power constraint index to rank and group critical contingencies based on sensitivity factors. *J. Arch. Electr. Eng.* **2018**, *67*, 2.

9. Ullah, K.; Basit, A.; Ullah, Z.; Aslam, S.; Herodotou, H. Automatic generation control strategies in conventional and modern power systems: A comprehensive overview. *Energies* **2021**, *14*, 2376. [CrossRef]

10. Ullah, K.; Basit, A.; Ullah, Z.; Albogamy, F.R.; Hafeez, G. Automatic Generation Control in Modern Power Systems with Wind Power and Electric Vehicles. *Energies* **2022**, *15*, 1771. [CrossRef]

11. Gupta, M.; Kumar, V.; Banerjee, G.K.; Sharma, N.K. Mitigating Congestion in a Power System and Role of FACTS Devices. *Adv. Electr. Eng.* **2017**, *2017*, 4862428. [CrossRef]

12. Lo, K.; Yuen, Y.; Snider, L. Congestion management in deregulated electricity markets. In Proceedings of the International Conference on Electric Utility Deregulation and Restructuring and Power Technologies, London, UK, 4–7 April 2000.

13. Ustun, T.S.; Hussain, S.M.S.; Orihara, D.; Iioka, D. IEC 61850 modeling of an AGC dispatching scheme for mitigation of short-term power flow variations. *Energy Rep.* **2022**, *8*, 381–391. [CrossRef]

14. Mohamed, A.; Jasmon, G. Realistic power system security algorithm. *IEE Proc. C Gener. Transm. Distrib.* **1988**, *135*, 2. [CrossRef]

15. Shandilya, A.; Gupta, H.; Sharma, J. Method for generation rescheduling and load shedding to alleviate line overloads using local optimization. *IEE Proc. C Gener. Transm. Distrib.* **1993**, *140*, 5. [CrossRef]

16. Abrantes, H.D.; Castro, C. A new efficient nonlinear programming-based method for branch overload elimination. *J. Electr. Power Comp.* **2022**, *30*, 6.

17. Verma, S.; Saha, S.; Mukherjee, V. Optimal rescheduling of real power generation for congestion management using teaching-learning-based optimization algorithm. *J. Electr. Syst. Inf. Tech.* **2018**, *5*, 889–907. [CrossRef]

18. Maharana, M.K.; Swarup, K.S. Graph theory based corrective control strategy during single line contingency. In Proceedings of the 2009 International Conference on Power Systems, Kharagpur, India, 27–29 December 2009; pp. 1–6.

19. Gupta, K.; Kiran, D.; Abhyankar, A.R. Flexibility in transmission switching for congestion management. In Proceedings of the 2016 National Power Systems Conference (NPSC), Bhubaneswar, India, 19–21 December 2016.

20. Pandiarajan, K.; Babulal, C. Overload alleviation in electric power system using fuzzy logic. In Proceedings of the International Conference on Computer, Communication and Electrical Technology (ICCCET), Tirunelveli, India, 18–19 March 2011.

21. Chahar, R.K.; Chuahan, A. Devising a New Method for Economic Dispatch Solution and Making Use of Soft Computing Techniques to Calculate Loss Function. In *Software Engineering*; Springer: Berlin, Germany, 2019; pp. 413–419.

22. Daraz, A.; Malik, S.A.; Waseem, A.; Azar, A.T.; Haq, I.U.; Ullah, Z.; Aslam, S. Automatic generation control of multi-source interconnected power system using FOI-TD controller. *Energies* **2021**, *14*, 5867. [CrossRef]

23. Bie, P.; Chiang, H.-D.; Zhang, B.; Zhou, N. Online multiperiod power dispatch with renewable uncertainty and storage: A two-parameter homotopy-enhanced methodology. *IEEE Trans. Power Syst.* **2018**, *33*, 6. [CrossRef]

24. Gruoup, W. Dynamic Models For Fossil Fueled Steam Units In Power System Studies. *IEEE Trans. Power Syst.* **1991**, *6*, 2.

25. Suwannarat, A. *Integration and Control of Wind Farms in the Danish Electricity System*; Institut for Energiteknik, Aalborg Universitet: Aalborg, Denmark, 2008.

26. Force, T. *Modeling of Gas Turbines and Steam Turbines in Combined Cycle Power Plants*; CIGRE Technical Brochure 238; CIGRE: Paris, France, 2003.

27. Basit, A.; Hansen, A.D.; Sørensen, P. Dynamic model of frequency control in Danish power system with large scale integration of wind power. In Proceedings of the China Wind Power Conference (CPW'13), Beijing, China, 16–18 October 2013; pp. 16–18.

28. Rebours, Y.G.; Kirschen, D.S.; Trotignon, M.; Rossignol, S. A survey of frequency and voltage control ancillary services—Part I: Technical features. *IEEE Trans. Power Syst.* **2007**, *22*, 350–357. [CrossRef]

Article

A Dragonfly Optimization Algorithm for Extracting Maximum Power of Grid-Interfaced PV Systems

Ehtisham Lodhi [1,2], Fei-Yue Wang [3], Gang Xiong [1,4,*], Ghulam Ali Mallah [5], Muhammad Yaqoob Javed [6], Tariku Sinshaw Tamir [1,2] and David Wenzhong Gao [7]

[1] The SKL for Management and Control of Complex Systems, Institute of Automation, Chinese Academy of Sciences, Beijing 100190, China; lodhi2018@ia.ac.cn (E.L.); tamir@ia.ac.cn (T.S.T.)
[2] School of Artificial Intelligence, University of Chinese Academy of Sciences, Beijing 100049, China
[3] Beijing Engineering Research Center of Intelligent Systems and Technology, Chinese Academy of Sciences, Beijing 100190, China; feiyue.wang@ia.ac.cn
[4] The Cloud Computing Center, Chinese Academy of Sciences, Dongguan 523808, China
[5] Department of Computer Science, Shah Abdul Latif University, Khairpur 66111, Pakistan; ghulam.ali@salu.edu.pk
[6] Department of Electrical & Computer Engineering, Lahore Campus, COMSATS University Islamabad, Lahore 54000, Pakistan; yaqoob.javed@cuilahore.edu.pk
[7] Department of Electrical & Computer Engineering, University of Denver, Denver, CO 80208, USA; wenzhong.gao@du.edu
* Correspondence: gang.xiong@ia.ac.cn; Tel.: +86-10-62544787

Citation: Lodhi, E.; Wang, F.-Y.; Xiong, G.; Mallah, G.A.; Javed, M.Y.; Tamir, T.S.; Gao, D.W. A Dragonfly Optimization Algorithm for Extracting Maximum Power of Grid-Interfaced PV Systems. *Sustainability* 2021, *13*, 778. https://doi.org/10.3390/su131910778

Academic Editors: Sheraz Aslam, Herodotos Herodotou and Nouman Ashraf

Received: 8 August 2021
Accepted: 22 September 2021
Published: 28 September 2021

Publisher's Note: MDPI stays neutral with regard to jurisdictional claims in published maps and institutional affiliations.

Abstract: Currently, grid-connected Photovoltaic (PV) systems are widely encouraged to meet increasing energy demands. However, there are many urgent issues to tackle that are associated with PV systems. Among them, partial shading is the most severe issue as it reduces efficiency. To achieve maximum power, PV system utilizes the maximum power point-tracking (MPPT) algorithms. This paper proposed a two-level converter system for optimizing the PV power and injecting that power into the grid network. The boost converter is used to regulate the MPPT algorithm. To make the grid-tied PV system operate under non-uniform weather conditions, dragonfly optimization algorithm (DOA)-based MPPT was put forward and applied due to its ability to trace the global peak and its higher efficiency and shorter response time. Furthermore, in order to validate the overall performance of the proposed technique, comparative analysis of DOA with adaptive cuckoo search optimization (ACSO) algorithm, fruit fly optimization algorithm combined with general regression neural network (FFO-GRNN), improved particle swarm optimization (IPSO), and PSO and Perturb and Observe (P&O) algorithm were presented by using Matlab/Simulink. Subsequently, a voltage source inverter (VSI) was utilized to regulate the active and reactive power injected into the grid with high efficiency and minimum total harmonic distortion (THD). The instantaneous reactive power was adjusted to zero for maintaining the unity power factor. The results obtained through Matlab/Simulink demonstrated that power injected into the grid is approximately constant when using the DOA MPPT algorithm. Hence, the grid-tied PV system's overall performance under partial shading was found to be highly satisfactory and acceptable.

Keywords: photovoltaic (PV); partial shading; maximum power point tracking (MPPT); dragonfly optimization algorithm (DOA); adaptive cuckoo search optimization (ACSO); fruit fly optimization algorithm combined with general regression neural network (FFO-GRNN); improved particle swarm optimization (IPSO); voltage source inverter (VSI); total harmonic distortion (THD)

1. Introduction

Renewable energy resources have emerged as an important source of energy over the last few decades. The wind, solar, fuel cells, biomass, geothermal, and hydrothermal are leading energy resources. The wind, hydrothermal, and geothermal energy resources are highly localized compared with other energy resources [1]. The sunlight is a supreme, abundant, and viable means of renewable energy that can accommodate growing public energy demands [2,3]. The PV system is also attractive owing to its scalability, simple architecture, lack of fuel cost, low carbon footprint, being free of noise, and friendliness to the environment [4,5]. PV systems can be categorized into two major groups, i.e., utility-interfaced PV systems and standalone PV systems [6]. There are many urgent issues to tackle that are associated with PV systems. Partial shading is the most severe issue of the PV system, as it distinctly diminishes the efficiency and output power of the PV array. During partial shading, the P-V trajectory will become more distinct and complex due to the availability of numerous peaks. Moreover, nonlinear behavior is also observed in the I-V characteristic curve because of variation in temperature and irradiance [7,8]. The characteristic curves have a special, single MPP at which the system works with supreme efficiency. The block diagram of grid-interfaced PV systems is presented in Figure 1, which usually comprises PV arrays and boost converters to improve DC voltage. The system is then connected with a three-phase inverter to convert DC voltage to AC voltage, and then sent into the power grid. The boost converter is exceptionally helpful to increase the DC voltage generated from the PV array [9]. The switching of the boost converter is regulated by the MPPT method based on the duty cycle. The MPP can be calculated by using various MPPT methodologies. A brief survey of these techniques is presented in [10]. The proficiency of these techniques can be evaluated by tracking speed, maximum power achieved, complexity, the number of sensors required, and time to reach the MPP.

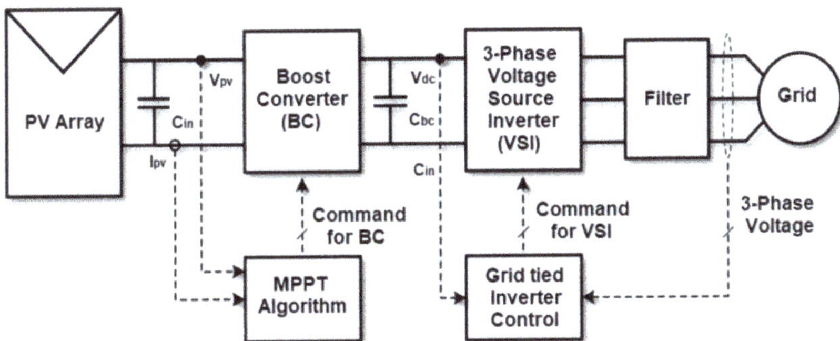

Figure 1. The block diagram of the proposed Grid-Connected PV System.

The MPPT techniques can be categorized into conventional and intelligent techniques. The progress in this field of MPPT algorithms continues due to multiple optimization solutions [11]. To certify the better operation under the condition of uniform irradiance, some conventional MPPT techniques are proposed, such as Perturb and Observe (P&O) [12], Hill climbing (HC) [13], Incremental conductance (INC) [14]. Fractional short circuit current is also proposed and performed well under normal weather conditions [15], and modified P&O-based MPPT is implemented under single-solution optimization [16]. These approaches experience the oscillations in the MPP that consequences overall sustained loss in power and efficiency of the PV system. Furthermore, under partial shading, the above-mentioned conventional techniques are stuck at local peaks and fail to reach the global peak, surely reducing the PV system's proficiency and effectiveness.

To resolve the precedent drawbacks in conventional MPPT techniques, researchers proposed numerous algorithms by employing different methodologies in the literature. The intelligent MPPT control uses artificial intelligence-based techniques, like artificial neural networks (ANN) techniques [17], genetic algorithm [18], fuzzy logic controller (FLC), and neuro-fuzzy algorithms [19], which are proposed to track the global MPP with short response times and handle the nonlinear relationship between variables. Training neural networks requires thousands of datasets to be modeled for non-uniform weather conditions [20]. However, applying these MPPT methodologies to acquire the maximum power is a costly and time-consuming task. It will also implement a control system much harder for grid-interfaced PV systems [21].

On the other hand, evolutionary techniques are usually employed, and among them, the particle swarm optimization (PSO) technique is usually preferred [22,23]. It is a simple and efficient technique that predominantly deals with multimodal and irregular issues. However, this cannot reach the global maxima owing to random numbers rooted in particles velocity. Consequently, the efficiency of this method is affected, which leads to premature convergence. Recently proposed soft-computing MPPT algorithms include flower optimization [24], cat swarm optimization [25], artificial bee colony [26], grey wolf optimization [27], and moth flame optimization [28]. The MPPT controllers have shown reasonable performance and better response to track the MPP under partial shading. However, these techniques exhibit some drawbacks, including comprehensive mathematical modeling, tuning of multiple parameters, and requiring a large population to perform optimization tasks. If the size of the sample population decreases, the success rate to determine the global MPP also drops.

The hybrid MPPT algorithms are also used in the literature to locate the MPP for improving the search speed with fewer steady-state oscillations [29–32]. In Ant colony optimization (ACO)-integrated P&O optimization-based MPPT algorithm, the P&O was employed for local search and ACO for global search [29]. In [30], diversification was added by combining differential evolution (DE) with PSO algorithm-based MPPT. The DE is used even while PSO is employed for odd iterations to reach MPP. The GWO-based MPPT is used in combination with the P&O algorithm for improving the convergence time. The GWO was employed in the early stage, while P&O was used at the final stage [31]. The deterministic PSO combined with the INC-based MPPT algorithm was presented in [32]. The INC was used to search local mode, while the PSO deterministically updated the velocity without using random numbers. However, the important benefit of the evolutionary algorithm was lost due to the removal of randomization. Hence, it may not trace the global MPP. Furthermore, suppose the irradiance level is less than 500 W/m^2. In that case, the interaction between PV array and grid network will be more complicated, and undesirable reactive power will occur, which diminishes the overall PV system [33]. The overall efficiency of grid-interfaced PV systems is consequent of a combination of the following components: PV array (approx. 18–44%), inverter (approx. 95%), and MPPT (approx. 98%), as described in [34]. Hence, modern control systems work on the MPPT side to improve grid-interfaced PV systems' efficiency.

After a comprehensive analysis, it was noticed that mostly global MPPT methodologies deal with partial shading for a standalone PV system and have not often discussed partial shading for grid-interfaced PV systems. Therefore, this paper suggested a dragonfly optimization algorithm (DOA)-based MPPT methodology to overcome these issues. The novelty of this research work is to employ the DOA-based MPPT technique working under partial shading conditions for grid-interfaced PV systems. The comprehensive analysis was presented based on a novel DOA MPPT technique to trace the global MPP under partial shading conditions. Furthermore, the results of DOA, compared with those of P&O, PSO, ACSO, IPSO, and FFO-GRNN algorithms, prove the advancement in proficiency, reliability, and robustness of DOA to reach global MPP. A dual-level interfacing scheme based on a boost converter and three-phase VSI was presented to interface the PV system with the grid. The switching of the boost converter was controlled by duty cycle through the DOA-based global MPPT technique. The VSI comprises two regulating loops, i.e., an external DC voltage and an internal current normalization loop. The voltage loop normalizes the output power from the PV module to the grid and stabilizes the grid's power flow. However current control loop was used to regulate the injected current to the grid and keep it in phase with the grid voltage to achieve a unified power factor. Finally, the accuracy of the projected scheme was verified successfully by simulation in MATLAB/Simulink. The important contributions of the research are listed below:

- The proficient and enhanced dragonfly optimization algorithm (DOA) was implemented.
- The suggested MPPT method can track the global MPP with fewer iterations under partial shading.
- The proposed DOA's applicability was supported by the performance comparison with existing PSO, improved PSO, and P&O algorithm.
- The proposed DOA effectively applied to the PV-interfaced grid with the help of VSI that can efficiently transfer energy between the PV array and grid side.

The rest of this paper is organized as follows. In Section 2, PV modeling under partial shading is built. Then, the DOA-based global MPPT technique is presented in Section 3. The inverter control methodology is illustrated in Section 4. Subsequently, the simulation results are shown and analyzed in Section 5. Eventually, Section 6 concludes the paper.

2. Modeling of PV Array and Partial Shading

2.1. PV Array Modelling

The solar cell is mostly composed of silicon (Si) crystalline material that conducts electricity when sunlight falls on solar cell and it converts the sunlight into electrical energy. In literature, the common models used for PV cells are based on one-diode and two-diode electrical circuit equivalent models. However, the one-diode model is preferred over the other because of its simple structure and easy implementation [35]. The equivalent model for solar cells is presented in Figure 2. By utilizing Kirchhoff's current rule (KCL), we can acquire the load current as presented in Equation (1):

$$I = I_{sc} - I_R\left(e^{\frac{(I.R_S + V)q}{K.A.T}} - 1\right) - \left(\frac{V + I.R_s}{R_p}\right) \tag{1}$$

where I is the output current of PV cell; V represents the output voltage, I_{sc} is used to show the short-circuit cell's current, q represents the electronic charge, I_R is the reverse saturation current, K is Boltzmann's constant, T is temperature of the PV Module and A is the ideality factor of diode, R_s represents the resistance in series, and R_p is the resistance in parallel. However, the value of I bypass current is very low and approaches zero. The resistance in parallel is also of huge quantity. Hence, Equation (1) can easily be transformed into Equation (2):

$$I = I_{sc} - I_R\left(e^{\frac{(I.R_S + V)q}{K.A.T}} - 1\right) \tag{2}$$

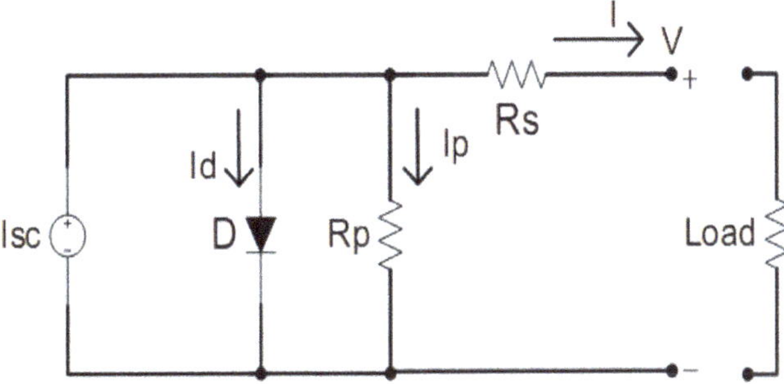

Figure 2. Single photovoltaic cell-equivalent circuit diagram.

Many PV cells are connected in series to form a PV module that is capable of delivering higher power. Usually, a single module is comprised of (36, 72, 96, and 128) cells connected in series and are currently available in markets [9]. These PV modules, in turn, associated in parallel and series combinations to increase the current and voltage, respectively, to form the PV string. The combination of PV strings is referred to as a PV array. These

PV arrays have the capacity to produce power according to the desired demand. The BP MSX 120 panel was used here and the specifications of the module under the standard conditions (25 °C and 100 W/m^2 irradiance) are shown in Table 1, and those of the 12-KW PV system are presented in Table 2. Assume that N_s is the number of cells arranged in series scheme and N_p is the number of cells settled in parallel. Hence, Equation (2) can be converted into Equation (3).

$$I = I_{sc}.N_p - I_R.N_s \left(e^{\dfrac{\left(\frac{N_p}{N_s} I.R_S + V\right)q}{K.A.T.N_s}} - 1 \right) \tag{3}$$

Table 1. The BP-MSX 120 Module specification (25 °C and 1000 W/m^2).

Parameter	Values
Number of cells in series	72
Short circuit current	3.87 A
Maximum current	3.56 A
Open circuit voltage	42.1 V
Maximum voltage	33.7 V
Maximum power	120 W

Table 2. Characteristics of 12 KW PV system.

Parameter	Values
No. of series modules in a string	10
No. of parallel modules in a string	10
Voltage at output	337 V
Current at output	35.6 A
Max power at output	12K W

2.2. Behavior of PV Array under Partial Shading

When environmental circumstances like irradiance and temperature are not varying with the passage of time, then PV output power is assumed to be stable. If there is a cloudy sky or other obstacle in the way of PV modules, radiation to the PV array declines. As a result, the power diminishes and a variation in the nature of typical PV characteristic graph is observed. The output characteristics of the PV system are dissimilar during uniform irradiation and under partial shading. The extent of sun radiation on PV array declines during partial shading. Therefore, the power output of array decreases, and a variation in the performance can be observed in the PV's characteristic curve [6]. When the intensity of radiation is similar, there will be only one power peak for all the modules. However, when the intensity of irradiance is dissimilar, the PV's characteristic curve has various local peaks but just one global peak. Herein, the PV system was composed of three PV arrays connected in series. The different levels of irradiance were applied on the PV array for establishing partial shading, and are shown in Figure 3 and organized in Table 3. The output characteristics (P-V and I-V) graphs of the PV system for partial-shading case-1 and partial-shading case-2 corresponding to divers irradiance levels are presented in Figure 4a,b respectively. Both cases had multiple local peaks and had only one global peak. Moreover, local peaks are indicated with the help of red dots while the global peaks are indicated with the help of green dots.

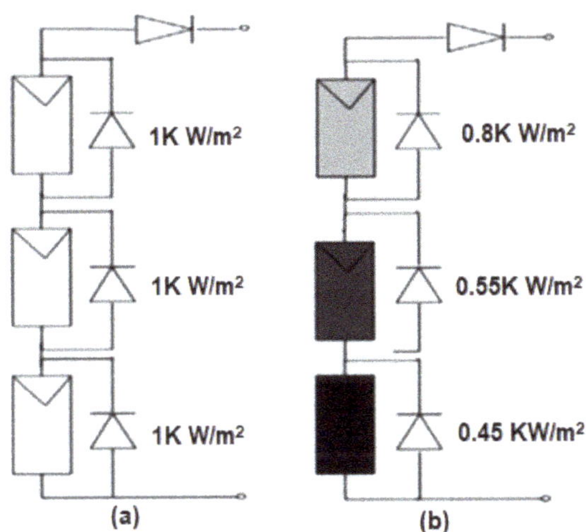

Figure 3. Irradiance configuration for: (**a**) normal condition, (**b**) partial-shading condition.

Table 3. PV array irradiance level for different partial-shading cases.

PV Arrays Cases	Irradiance (W/m²)			Maximum Output Power (W)
	1st Array	**2nd Array**	**3rd Array**	
Case-1	600	800	1000	9250
Case-2	800	550	450	4240

2.3. Boost Converter Modelling

The DC-DC boost converter is usually employed as an interfacing bridge between the PV array and inverter. The typical circuit diagram of the boost converter is shown in Figure 5. This can be used to raise the PV array voltage to the appropriate level for grid synchronization and trace global MPP by resorting to DOA. The sizing of the boost converter (value of inductor, input capacitor, and output capacitor) parameters is presented in Appendix A. From [36] and [37], the value of the inductor was measured by utilizing Equation (4), where D_m D_m is the value of duty cycle, switching frequency is shown by $f_s f_s$, while output maximum voltage is presented by V_{om} , and ΔI_r is the inductance ripple current.

$$L \geq \frac{V_{om} \cdot D_m (1 - D_m)}{f_s \cdot \Delta I_r} \tag{4}$$

The input capacitor and the group of PV arrays were arranged in parallel combinations. It is the capacitor at the input of the boost converter. This capacitance was calculated in [37] and represented by Equation (5), where the value of the current at max power is I_{om}, D_m represents duty cycle, and V_{pv_mmpp} is the output voltage of the system at MPP.

$$C_{in} \geq \left[\frac{I_{om} \cdot (D_m)^2}{0.02(1 - D_m) f_s \cdot V_{pv\_mmpp}} \right] \tag{5}$$

The capacitor in parallel with the load is called an output capacitor or DC-link capacitor. It is the capacitor at the output of the boost converter. The most vital function of this capacitor is to confine voltage to the predetermined level and reduce the ripple content from the PV source [38]. To measure the size of the output capacitor, Equation (6) was employed, where D_m represents duty cycle, V_{out} is the output voltage of the boost converter, R_{out} is output load of boost converter, and ΔV_{out} is ripple output voltage.

$$C_{out} \geq \left[\frac{V_{out} \cdot D_m}{f_s \cdot \Delta V_{out} \cdot R_{out}} \right] \tag{6}$$

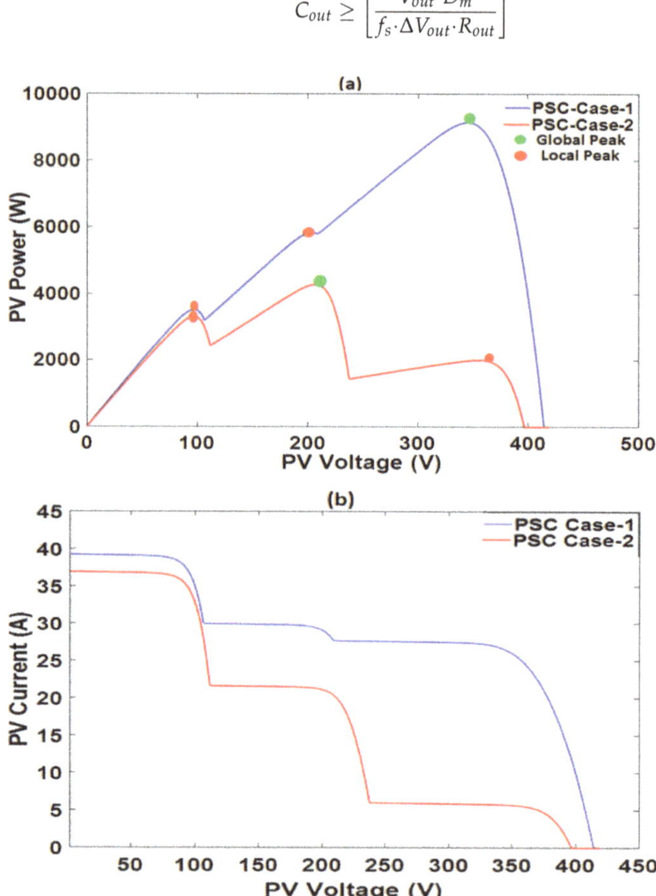

Figure 4. (**a**) P-V characteristics of the array, (**b**) I-V characteristics of the array under partial shading.

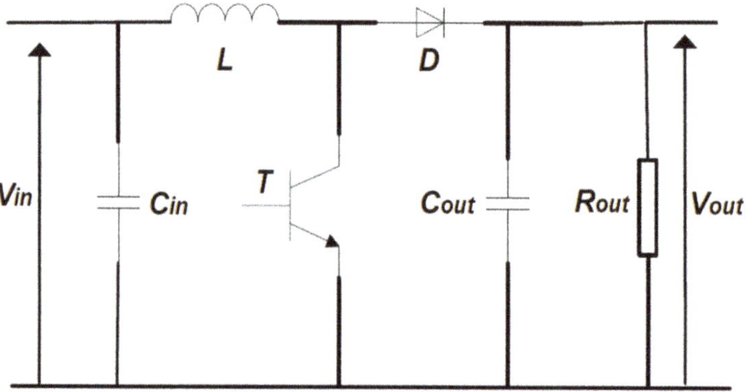

Figure 5. The typical circuit diagram of the boost converter.

3. MPPT Algorithms

3.1. Dragonfly Optimization Algorithm (DOA)

The dragonfly optimization algorithm (DOA) is a kind of evolutionary algorithm and an intelligent search-optimization technique. The idea is originated in the static and dynamic behavior of dragonflies. The small cluster of dragonflies that hunt for prey in a small locality are categorized as a static swarm. The movement of flies is characterized by abrupt and rapid changes in their respective flight paths. On the other hand, a large cluster of flies that maintains a constant direction of motion over a lengthy distance with the aim to migrate from one point to another is categorized as a dynamic swarm. Since the aim of these dragonflies in a swarm is similar to the optimization problem, the static behavior of swarm dragonflies (DFs) are used to characterize the exploitation phase, while the dynamic behavior of swarm DFs is used to characterize the exploration phase. This establishes the foundation of DOA. To derive the mathematical model for signifying the flies' motion in a cluster, five features of dragonflies, i.e., separation, alignment, cohesion, food, and enemy are described. The Sep_i, Alg_i, Coh_i, Af_i, and Ee_i are used to represent separation, alignment, cohesion, food, and enemy features of an i individual dragonfly in a cluster. The acronym list of DOA variables is shown in Table 4. These feature of flies can be represented by mathematical equations as follows:

Table 4. Acronym list for DOA variables.

Symbol	Acronym
Sep_i	Separation of the ith individual dragonfly
Alg_i,	Alignment of the ith individual dragonfly
Coh_i,	Cohesion of the ith individual dragonfly
Af_i	Food attraction
Ee_i	Enemy position
Δx_i	Step size of DF movement
w	Inertial weight
a	Separation weight
b	Alignment weight
b	Cohesion weight
d	Food factor
e	Enemy factor

To maximize the search space and avoid collision, the distance between adjacent DFs is necessary within the given locality. Let i is represent the number of individuals in a cluster with 'n' neighbors. The Separation Sep_i of the i individual in a cluster is shown with the help of Equation (7), as shown below., where x is the current position of DF and x_k is the position of the k_{th} neighboring DF.

$$Sep_i = \sum_{k=1}^{n}(x - x_k) \tag{7}$$

Matching the velocities of the individual with other DFs with in the same locality is based on alignment term Alg_i as represented in Equation (8). Here, V_k is the velocity of the kth neighboring DF.

$$Alg_i = \frac{\sum_{k=1}^{n} V_k}{n} \tag{8}$$

All the individuals of DF in a cluster are inclined to move in the direction of the mass center of neighboring DFs. The cohesion feature Coh_i of DF is determined by Equation (9):

$$Coh_i = \frac{\sum_{k=1}^{n}(V_k)}{n} \tag{9}$$

All the individuals in a cluster tend to move in the direction of food, as it is essential for living. The attraction for food Af_i feature at location x_{food} is acquired by following Equation (10).

$$Af_i = x_{food} - x \tag{10}$$

All the individuals in a cluster tend to move away from the enemy. The enemy feature Ee_i at location of enemy x_e can be calculated by Equation (11).

$$Ee_i = x_e + x \tag{11}$$

The behavior of DFs in a cluster is influenced by the combining all the five attributes. The updated location of the individual DFs is calculated by step Δx_i and denoted in Equation (12).

$$x_i = x_i + \Delta x_i \tag{12}$$

$$\Delta x_i = w\Delta x_i + (a.Sep_i + b.Alg_i + c.Coh_i + d.Af_i + e.Ee_i) \tag{13}$$

where w is the inertial weight, and a, b, and c are the separation, alignment, and cohesion weights, respectively, whereas d is used to represent the food factor and the enemy factor is represented by e. The Sep_i, Alg_i, Coh_i, Af_i, and Ee_i are used to represent separation, alignment, cohesion, food, and enemy features of an i individual dragonfly in a cluster. The variation in the explorative and exploitative behaviors of the DFs can be realized by using different values of parameters. The flow chart of DOA is also displayed in Figure 6.

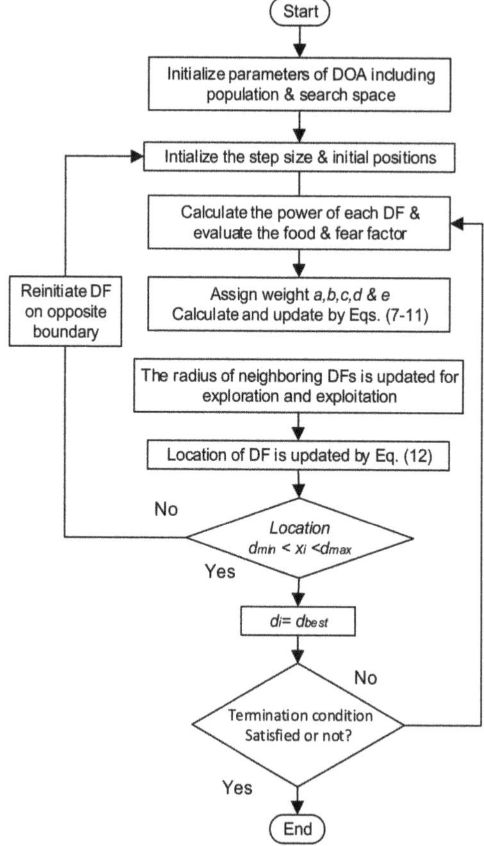

Figure 6. Flow diagram of the proposed DOA algorithm.

3.2. Application of DOA for MPPT Problem

The DOA can also be applied to track the global MPP of the PV system under partial shading. The particles are initialized around the search space. The existing location of DF is considered as the duty cycle. The constraints used for DOA is tabulated in Table 5. The pseudo code for the application of DOA for the MPPT problem is also presented in Figure 7. The important phases involved in the execution of DOA are described as below:

1. Firstly, there is a need to initialize the particles around the search space between d_{min} and d_{max}, and the step value (Δx_i) for particles is initialized properly. The duty cycle is considered as the particle position and its value is randomly chosen between 0.2 and 0.9.
2. During the second step, the boost converter is triggered by utilizing the control algorithm against each particle position and the best output power that is assumed to be the fitness (cost) function is calculated. Then, the food source and enemy location are updated. The cost function is monitored for changes and if there is any variation in power due to partial shading.
3. Subsequently, the a, b, c, d, and e values are updated. The separation, alignment, cohesion, food, and enemy features for individual DFs are calculated by using Equations (7)–(11). For exploration and exploitation, the radius of neighboring dragonflies is updated.
4. At this moment, the step and position of particle is calculated by using Equations (12) and (13) respectively. If the position of dragonflies lies outside the search space, then DOA is initiated at opposite boundary.
5. Finally, if the termination condition (the best optimal position of dragonflies to operate on global MPP) is met or satisfied, then this algorithm will stop. It also restarts the search process if a sudden change occurs in the input power.

Table 5. Constraints for DOA.

Parameter	Symbol	Value
Quantity of particles	k	4
Separation weight	a	0.2
Alignment weight	b	0.1
Cohesion constant	c	0.9
Food factor	d	0.5
Enemy constant	e	1

3.3. Comparison of DOA with Other MPPT Techniques

The suggested DOA technique was compared with other widely used conventional and intelligent MPPT algorithms, including the Perturb and Observe (P&O) algorithm, the Particle Swarm Optimization (PSO) algorithm, an improved version of the PSO algorithm (IPSO), the adaptive cuckoo search optimization (ACSO) algorithm, and fruit fly optimization combined with a general-regression neural network (FFO-GRNN). The P&O approach experienced the oscillations and failed to reach global MPP. Subsequently, it caused an overall sustained loss in power and efficiency of the PV system [12]. The PSO algorithm was inspired by the swarm behavior of particles and is a kind of evolutionary technique. Moreover, it is a simple and efficient technique, predominantly dealing with irregular issues [22]. However, during local power peaks, when particles were caught by undesirable states in the course of searching and evolution processes, the exploration capacity was quickly lost. As a result, the efficiency of PSO was affected and this led towards premature convergence. Therefore, in order to overcome these shortcomings, an Improved PSO (IPSO) algorithm is introduced [6]. The IPSO has shown better performance as compared with the PSO algorithm. The ACSO-based MPPT algorithm was employed to determine the MPP during non-uniform weather conditions, and it showed better performance than Cuckoo search optimization. Furthermore, the FFO algorithm with GRNN was also utilized to trace the global MPP under partial shading conditions. It showed better searching ability and efficiency as compared with other MPPT algorithms [17]. All these aforementioned techniques were applied on the PV system to study their performance behavior under non-uniform irradiance and partial-shading conditions. Detailed comparisons among these MPPT techniques are presented in the simulation results and discussion section of this paper.

Initialization of dragonflies' population x_i (i=1,2, 3,....,n)

Initialization of step value x_i (i=1,2, 3,....,n)

While end condition is not satisfied

Determine the objective values of all dragonflies

Upgrade the food source and enemy

Upgrade a, b, c, d and e

Compute Sep, Alg, Coh, Af, Ee by using Eqs. (7-11)

Upgrade the neighboring radius

If a dragonfly has at least one neighboring dragonfly

Upgrade the position of vector by Eq. (12)

Upgrade the vector velocity by Eq. (13)

Else

Upgrade the position vector using Levy flight

End if

Examine and correct the new positions based on

Variable boundaries

End while

Figure 7. Pseudocode for the DOA algorithm.

4. Inverter Control Methodology

The fundamental purpose of the inverter is to associate the PV array with the power grid. In a similar time, the inverter is employed to keep up the voltage at output of the boost converter, i.e., the DC link of the inverter and controlling the power (active and reactive), which are sent to the grid under partial shading. The various parts relating to the work of this system are presented in Figure 8. The performance of the overall system can be expressed by Equation (14), where U_{abc} and I_{abc} are the grid voltages and currents, and e_{abc} are the voltages of converters.

$$U_{abc} = e_{abc} + R.I_{abc} + L\frac{d}{dt}(U_{abc})$$ (14)

The stationary *abc* and the synchronously rotating *dq* reference frames were used for the implementation of this methodology. The vector control, grid voltage, and current are portrayed as vectors in the α-β reference frame. The process of changing the stationary three-phase *abc* coordinates system to the rotating *dq* coordinate frame system is called the *d-q* transformation. This change can be performed in two steps. The clark and inverse clark transformations are utilized to change over the factors into α-β reference stationary edge and vice versa. Essentially, for transformation of the value from the stationary α-β reference frame into the rotating *d-q* reference, the park and inverse-park transformations are required. The alteration of axes for vector-control frames is displayed in Figure 9. By employing *abc* to *dq* transformation, voltages and currents can be expressed by Equation

(15) for the *dq* frame of reference revolving at ω. The Equations (16) and (17) present voltage in the *dq* reference frame, as shown below.

$$\begin{pmatrix} U_d \\ U_q \end{pmatrix} = R \begin{pmatrix} I_d \\ I_q \end{pmatrix} + L \begin{pmatrix} I_d \\ I_q \end{pmatrix} + L \begin{pmatrix} 0 & -\omega \\ \omega & 0 \end{pmatrix} \begin{pmatrix} I_d \\ I_q \end{pmatrix} + \begin{bmatrix} e_d \\ e_q \end{bmatrix} \tag{15}$$

$$U_d - e_d = L\frac{d}{dt}(I_d) + R.I_d - \omega L I_q \tag{16}$$

$$U_q - e_q = L\frac{d}{dt}(I_q) + R.I_q - \omega L I_d \tag{17}$$

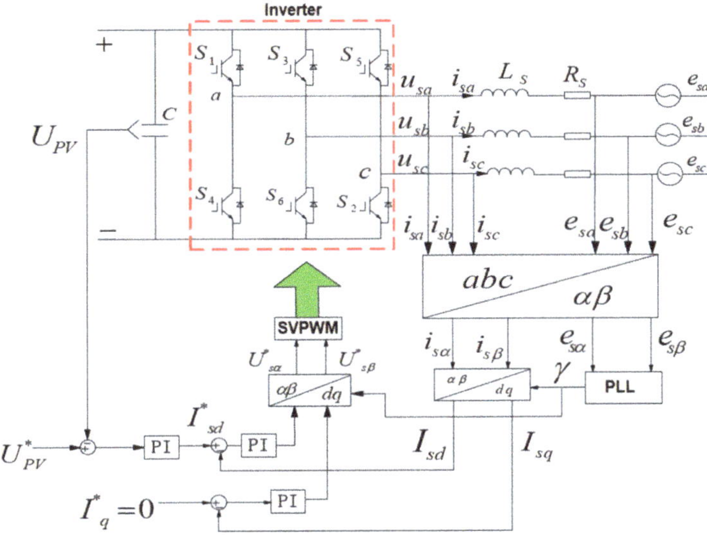

Figure 8. The block diagram of the controller for the grid-connected PV system.

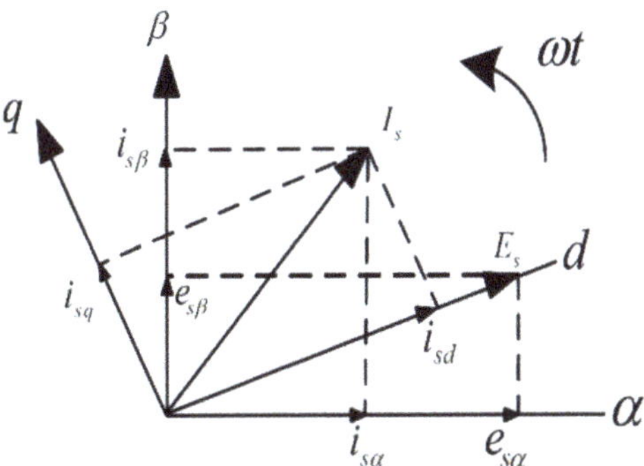

Figure 9. The alteration of axes for vector-control frames.

The synchronization of the grid plays a significant part in grid-interfaced PV systems. The PLL control strategy is employed to synchronize the output signal with the reference input signal according to phase and frequency [39]. The block of three-phase PLL block can be accessed by employing MATLAB/Simulink.

4.1. Voltage and Current Control Strategy

The approach used for regulating the inverter contains two controlling loops, as presented in Figure 8. The regulatory approach has an internal current and an external voltage-control loop. The function of the internal current-regulating loop is to regulate the injected current to the grid and keep it in phase with the grid voltage to provide power factor unity and reduce the harmonics. The AC grid current references are delivered by an external voltage-regulating loop. The external voltage-regulatory loop is used to normalize DC voltage and stabilizes the grid's power flow. The synchronous frame-of-reference dq control adopts (abc to dq) reference frame conversion to transfer the grid voltage and current to the dq reference frame. The voltage, after transformation, detects the phase and frequency of the grid.

In Figure 8, the value of U^*_{PV} is correlated with a given value of U_{PV}, and the PI regulator achieves the difference for the static control of DC voltage. Since Isd can achieve DC voltage regulation, the reference value of the active current inner loop I^*sd is given by the output of the external DC voltage loop, thus realistically controlling the active power of the grid-connected inverter output. The reference value of the inner-loop reactive current I^*qs is determined by the amount of reactive power required by the grid network. When $I^*qs = 0$, the reactive power output of the grid inverter is zero, and only the active power is transmitted to the grid, which is then operated in the unit power-factor state. The internal current-loop controller and external voltage-controller parameters can be measured by the transfer-function model of the PID tuning algorithm by employing Matlab/Simulink [40]. The values of Kp and Ks for the voltage PI controller were 9.5 and 20. The values of Kp and Ks for the current PI controller were 0.5 and 50. The period of sampling time was taken as $1\,\mu$ second.

4.2. SVPWM Technology

The three-phase inverter comprised six power switches from S1 to S6, as shown in Figure 8, all of which were regulated by the standards of space-vector pulse-width modulation (SVPWM). The SVPWM methods are characterized by consistent amplitude, but the duty cycle of every period is different. The voltage in the abc frame ought to be presented in the dq frame for the SVPWM [40]. Voltages might be characterized as vectors set in search space. These vectors are used for the switching of the inverter compared with switch combinations.

4.3. Grid Connected Filter

Since providing a sinusoidal line current to the grid without harmonic distortion was the utmost significant objective of this research, an inverter capable of filtering must be connected with the power grid. The filter can also minimize the switching losses. Therefore, to design the filter that meets these requirements as much as possible, 10% rated current output was used as the ripple output current. The value of the inductor and capacitor can be calculated by Equations (18) and (19), respectively [39].

$$L \geq \frac{V_{dc-side}}{16.f_s.\Delta I_L} \tag{18}$$

$$C \geq \frac{10}{3*2\pi*f*(V_{rated})^2} \tag{19}$$

5. Simulation Results and Discussion

The credibility of the proposed DOA-based MPPT algorithm was verified through two different cases, i.e., partial-shading case-1 and case-2. For the implementation of partial shading, diverse irradiance was applied to three PV arrays associated with the series connection. Those diverse levels of irradiance are given in Table 3. The MATLAB/Simulink was used for analyzing these MPPT methodologies. Furthermore, the suggested DOA technique was compared with the conventional P&O algorithm, PSO algorithm, IPSO algorithm, ACSO algorithm, and FFO-GRNN algorithm.

5.1. Partial Shading Case-1

In this case, partial shading occurred when PV arrays did not experience identical irradiance. The diverse irradiance (600 W/m^2, 800 W/m^2, and 1000 W/m^2) was applied to three PV arrays associated with the series connection. The power voltage characteristic curve for this case is visualized in Figure 4a with two local and one global peak at the power levels of 3578 W, 5813 W, and 9250 W respectively. In this section, the research was carried out according to the maximum power attained using the abovementioned MPPT techniques under partial-shading case-1.

Overall, the performance of the DOA-based MPPT algorithm was outclassed in every aspect. Both DOA and FFO-GRNN procedures had comparable prospects in the attainment of global MPP. The suggested DOA entirely nullified the oscillations while achieving global MPP. The steady power achieved by DOA, FFO-GRNN, IPSO, ACSO, PSO, and P&O was 9189 W, 9003 W, 8889 W, 8982 W, 8767 W, and 5196 W, respectively. Under these circumstances, the power, voltage, and current plots of the concerned techniques are displayed in Figure 10a–c, respectively. The oscillations were higher in the P&O technique, and local peaks could easily trap them. Hence, a significant decrease in efficiency was observed. The PSO method could reach the local peak shortly, but consecutive oscillation caused huge energy and proficiency loss, reaching the global peak after 0.47 s with oscillations. The FFO-GRNN technique also easily captured the global maxima and exhibited a tracking time of around 0.33 s. The IPSO technique was not too disturbed by the local peaks and stabilized at global maxima peak within 0.38 s, and it displayed fewer fluctuations compared with PSO. The tracking time to achieve the maximum peak for the ACSO technique was around 0.46 s. The proposed DOA was not affected by the local peak and tracked the global peak without a loss in power. The DOA was the fastest one to stabilize at a global peak, within 0.28 s, as depicted in Figure 10a.

Among these MPPT techniques, DOA achieved the highest efficiency of about 99.34%; subsequently, FFO-GRNN realized 97.32%, while the IPSO method achieved 97.10%. After that, ACSO attained an efficiency of 96.09%. Moreover, PSO attained 94.77% efficiency. The least efficiency of about 56.17% was attained by P&O. In terms of convergence speed, DOA effectively traced the global peak within 0.28 s while FFO-GRNN tracked after 0.33 s. The proposed DOA technique appeared to stabilize at a global peak in fewer iterations and showed a very quick response. The detailed comparative analysis of these concerned techniques under these circumstances is also presented in Table 6.

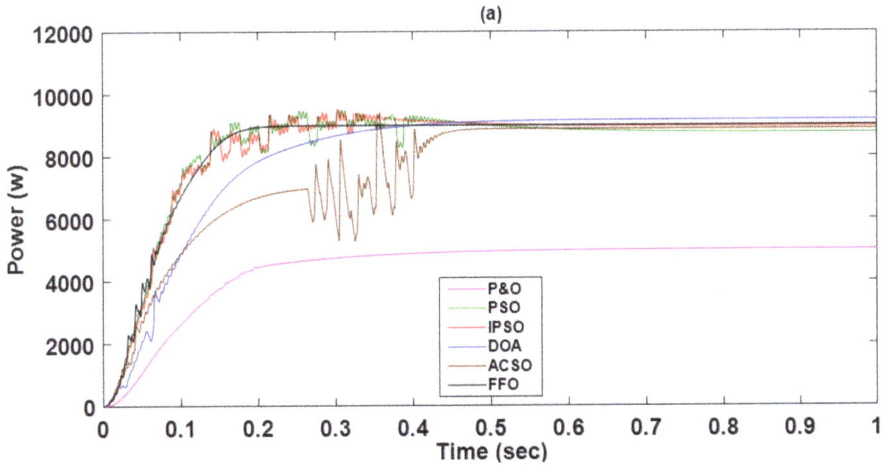

Figure 10. *Cont.*

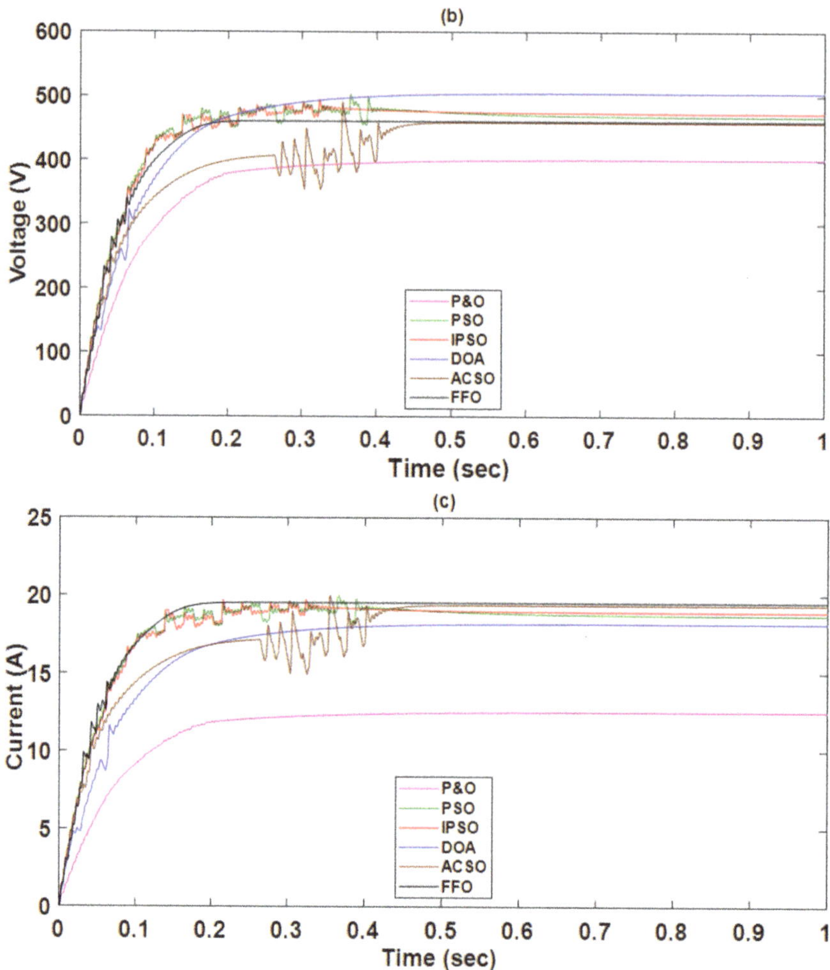

Figure 10. (**a**) Power comparison, (**b**) Voltage comparison, (**c**) Current comparison of P&O, PSO, IPSO, and DOA algorithms under partial-shading case-1.

Table 6. Performance analysis of different MPPT algorithms.

MPPT Techniques	Sensed Variables	Steady State Error	Tracking Speed	GMPP Tracking	Tracking Accuracy	Efficiency	Complexity	Cost
P&O	V, I	High	Fast	No	Low	Less	Low	Cheap
PSO	V, I	Moderate	Fast	Yes	Medium	High	Medium	Moderate
ACSO	V, I	Less	Fast	Yes	High	High	High	Expensive
IPSO	V, I	Less	Fast	Yes	High	High	High	Expensive
FFO-GRNN	V, I	Less	Fast	Yes	High	High	High	Expensive
DOA	V, I	Less	Fast	Yes	High	High	High	Expensive

5.2. Partial Shading Case-2

To make this investigation and relative study further inclusive, we considered an additional partial-shading case, in which diverse irradiance (800 W/m^2, 500 W/m^2, and 450 W/m^2) was applied to three modules associated in series connection. The power-voltage curve for this case is visualized in Figure 4a with two local peaks and only one global peak at the power levels of 2022 W, 3357 W, and 4240 W, respectively. In this section, the research was carried out according to maximum power attained using the abovementioned MPPT techniques under partial-shading case-2.

In general, the suggested DOA technique showed better and more efficient results compared with other concerned methodologies. The output power achieved by DOA, FFO-GRNN, IPSO, ACSO, PSO, and P&O techniques was 4211 W, 4094 W, 4032 W, 3979 W, 3948 W, and 2216 W, respectively. Under these circumstances, the power, voltage, and current plots of the concerned techniques are displayed in Figure 11a–c, respectively. The proposed DOA technique was not disturbed by the local peaks and stabilized at global peak power within 0.32 s. It displays fewer fluctuations compared with other MPPT methodologies. The FFO-based GRNN technique also showed better efficiency and touched the global peak after 0.30 s. The IPSO technique showed better efficiency and output power results than PSO, experienced oscillation, and stabilized at the global peak after 0.35 s. The ACSO algorithm achieved a global peak within 0.33 s, just after the IPSO technique. Figure 11a also demonstrates that PSO was affected by local peaks. Its scatter plot exhibited large oscillations in output power, caused loss in energy and efficiency, and reached the global peak after 0.44 s. Moreover, under these circumstances, the conventional P&O technique was stuck at the local peak. It failed to reach the global peak, which will surely reduce the proficiency and effectiveness of the PV system.

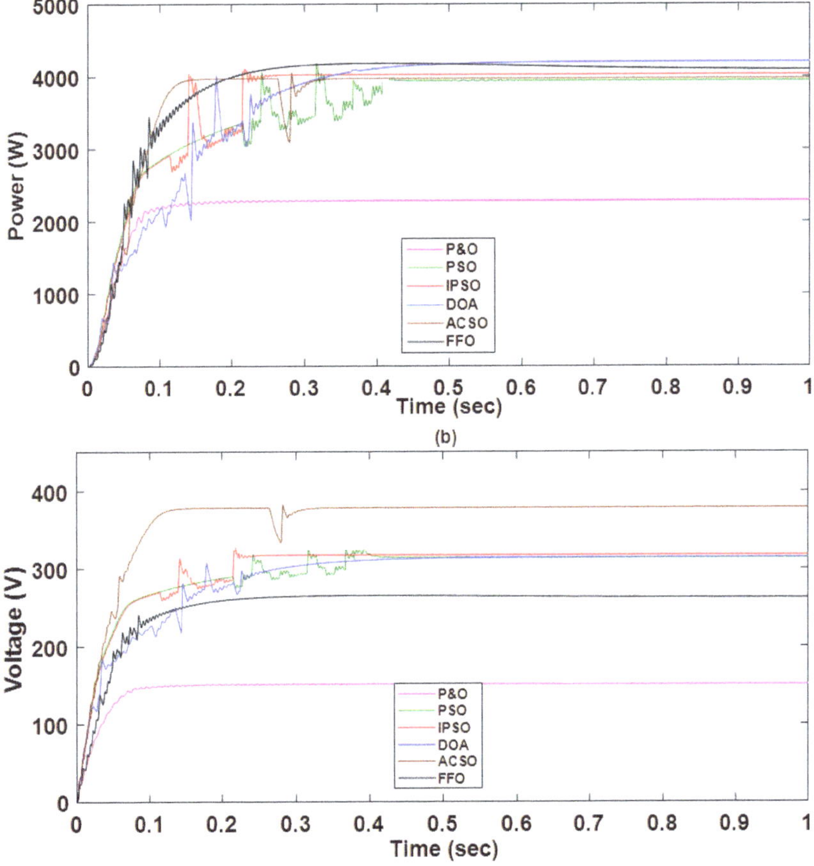

Figure 11. *Cont.*

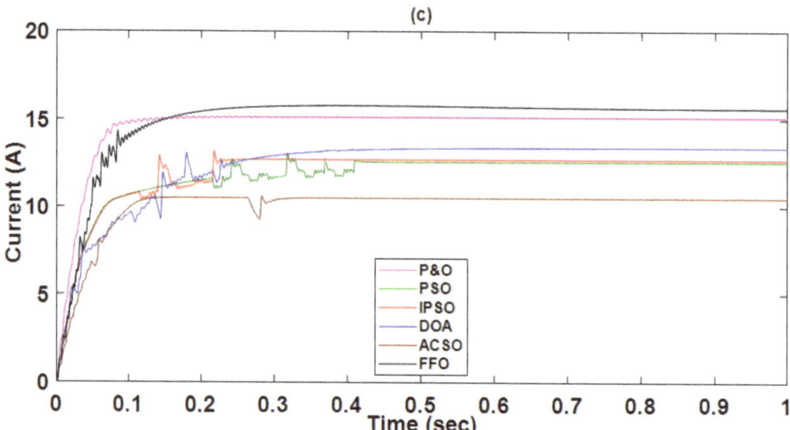

Figure 11. (a) Power comparison, (b) Voltage comparison, (c) Current comparison of P&O, PSO, IPSO, and DOA algorithms under partial-shading case-2.

Among these MPPT techniques, DOA achieved the highest efficiency, of about 99.31%. Next, FFO-GRNN maintained 97.32% efficiency. The IPSO method attained a 95.09% efficiency rate, while the ACSO technique upheld an efficacy of around 93.84%, whereas PSO attained 93.11% efficiency. Moreover, the P&O technique showed very little efficiency of around 52.26%. The DOA appeared to stabilize at a global peak in fewer iterations and showed a very quick response. The proposed DOA-based MPPT technique has improved tracking ability, faster convergence rate, and reduced power loss in a steady state. Thereby, these characteristics make DOA an excellent option to be utilized under different shading conditions. The comparison of these concerned techniques regarding convergence time, maximum power tracked, and efficiency are also tabulated in Table 6. The overall performance analysis of these MPPT algorithms regarding the steady-state error, tracking speed, complexity, accuracy, and cost are recorded in Table 7.

Table 7. Comparison of different parameters concerned MPPT techniques under partial shading.

MPPT Techniques	Irradiance Cases	Converge Time (s)	Max Traced Power (W)	Global Max Power (W)	Global MPP Located	MPPT Accuracy	Percent Error
P&O	Case-1	0.18	5196	9250	No	56.17%	43.82%
	Case-2	0.12	2216	4240	No	52.26%	47.73%
PSO	Case-1	0.48	8767	9250	Yes	94.77%	5.22%
	Case-2	0.44	3948	4240	Yes	93.11%	6.88%
ACSO	Case-1	0.46	8889	9250	Yes	96.09%	3.91%
	Case-2	0.33	3979	4240	Yes	93.84%	6.16%
IPSO	Case-1	0.38	8982	9250	Yes	97.10%	2.89%
	Case-2	0.35	4032	4240	Yes	95.09%	4.90%
FFO-GRNN	Case-1	0.33	9003	9250	Yes	97.32%	2.68%
	Case-2	0.30	4094	4240	Yes	96.62%	3.38%
DOA	Case-1	0.29	9189	9250	Yes	99.34%	0.65%
	Case-2	0.32	4211	4240	Yes	99.31%	0.68%

5.3. PV Array Interfaced with Grid Network

After investigating the simulation results of the DOA algorithm compared with those of other concerned MPPT algorithms, it demonstrated advancement in reliability, efficiency, and robustness of DOA to achieve the global MPP. Therefore, this section simulates the grid interfaced PV system by utilizing the DOA algorithm and analyzed it under partial-shading case-1. The PV arrays received different irradiances (600 W/m^2, 800 W/m^2, and 1000 W/m^2), while the temperature was constant at 25 °C. The simulation setup for the grid-interfaced PV system under partial shading by the DOA MPPT technique can be seen in Figure 12. It can be observed from Figure 10a,b that the values of the output power and a voltage obtained from the boost converter were 9189 W and 490 V, respectively. The response time of DOA was also much shorter relative to conventional P&O, PSO, IPSO ACSO, and FFO with the GRNN algorithms, although the existence of partial shading makes it hard to stay at the best optimal point.

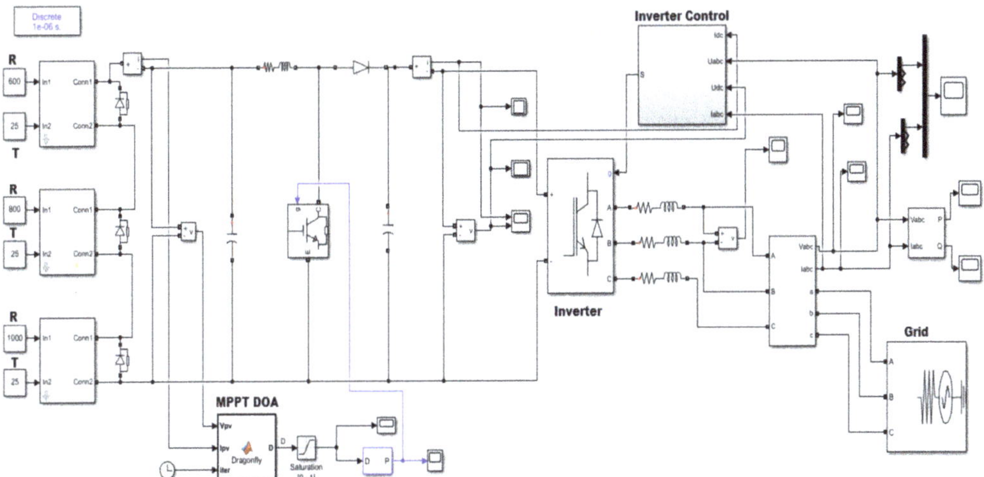

Figure 12. Simulation setup for the grid-interfaced PV system under partial shading by employing DOA MPPT Technique.

The frequency of the grid was assumed to be 50 Hz, and the sinusoidal current output from grid connection can be viewed in Figure 13a, between −18 A and +18 A. Figure 13b presents the zoom view of three-phase sinusoidal current. The peak-to-peak voltage of the analog grid was 311 V, which can also be seen in Figure 13c, while Figure 13d presents the zoom view of three-phase voltage. The amplitude of peak voltage was switched between 311 V and +311 V. Figure 13e presents the voltage and current of the power grid, and Figure 13f presents the zoom view of the grid voltage and current. Both the voltage and current were in phase, which is very important in maintaining the unity of power factor. The inverters have to diagnose voltage anomalies in the grid-interfaced PV system. In this research, the requirement for grid voltage was 311 V (peak-to-peak) or 220 V (RMS voltage). For instance, it can be observed in Figure 13c that, initially, the voltage disturbance was small, but after a short span of time (about 0.08 s), it quickly retained the three-phase voltage of the grid network at 311 V (peak-to-peak). Hence, it also fulfilled the relative requirement of the grid.

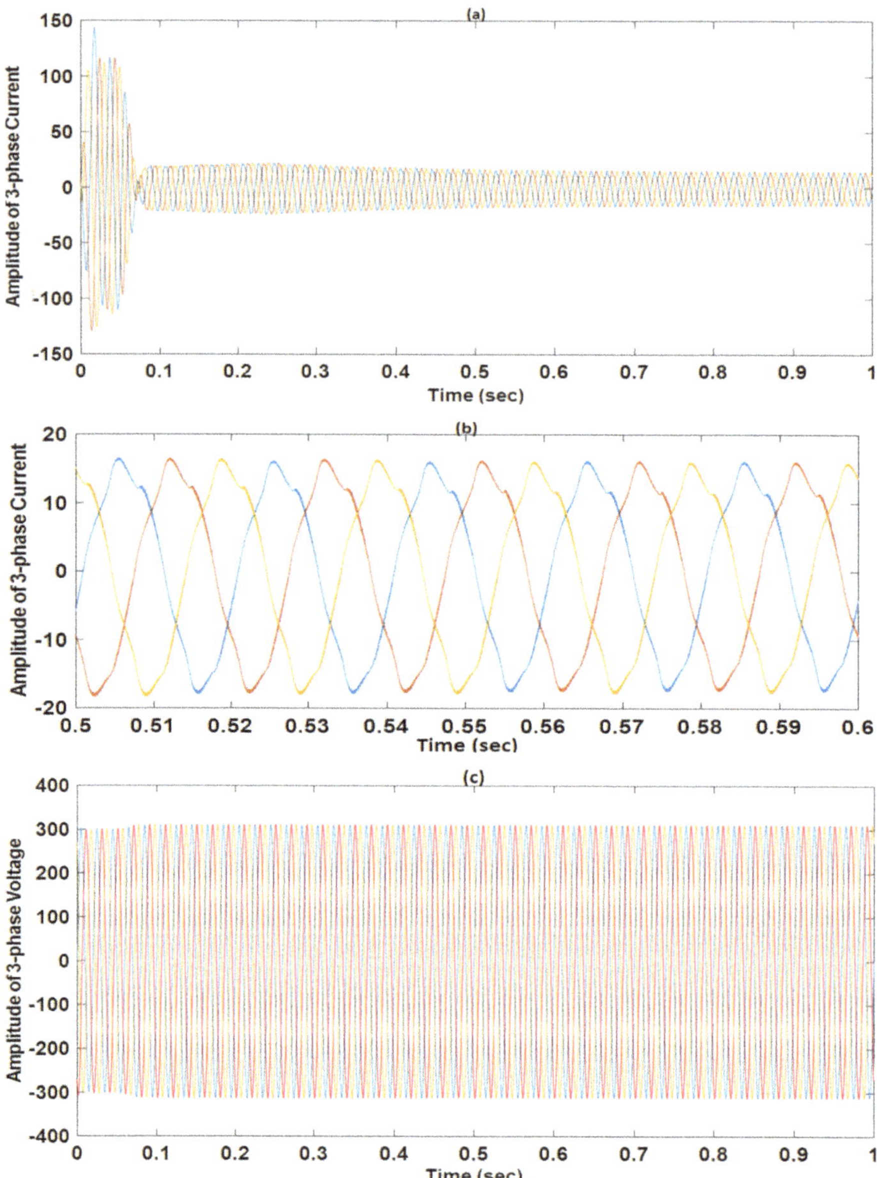

Figure 13. *Cont.*

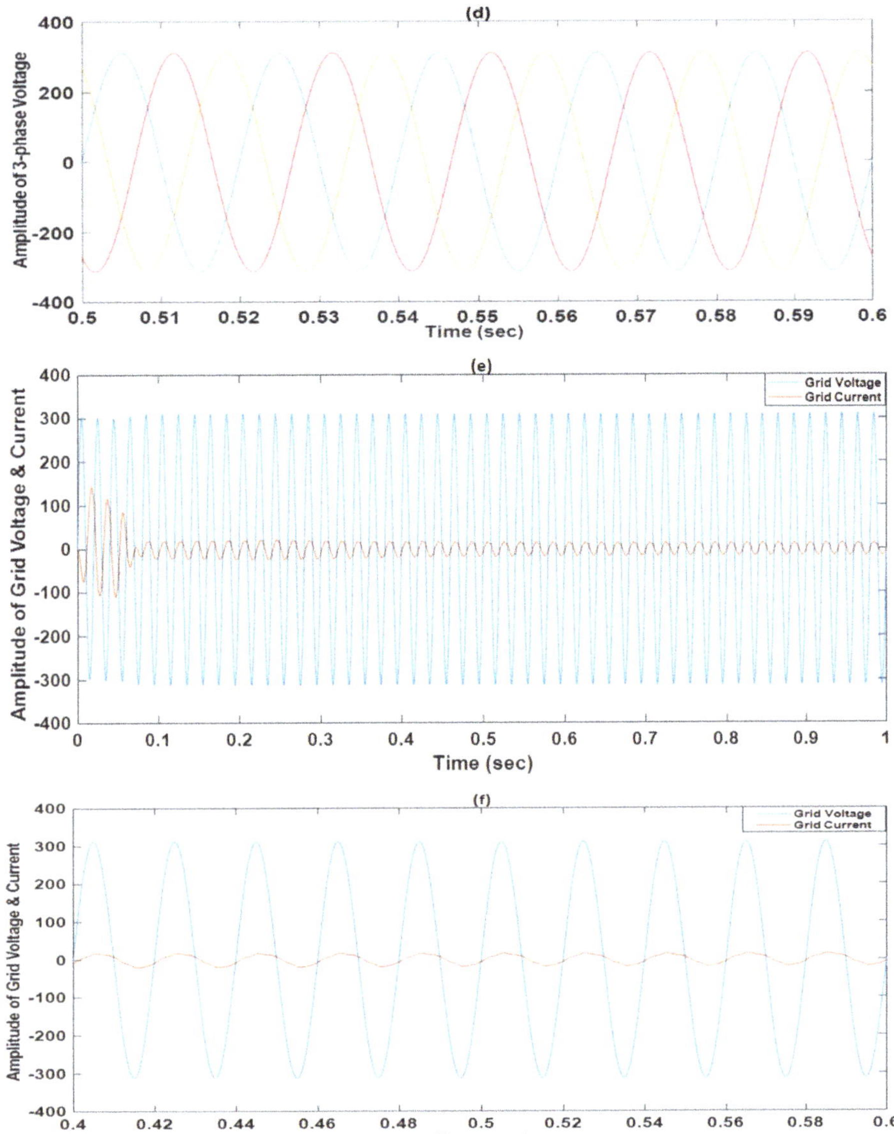

Figure 13. (a) Waveforms of the 3-phase grid current, (b) Zoomed view of the 3-phase grid current, (c) Waveforms for the 3-phase grid voltage, (d) Zoomed view of the 3-phase grid voltage, (e) Amplitude of the grid voltage and current, (f) Zoomed-view amplitude of the grid voltage and current.

The parameters in the dq transformation are visualized in Figure 14a. We can see in this figure that the value of Id was around 30 Amperes, while the value of Iq was around zero Amperes. The value of voltage Ud was around 385 V, while the value of Uq was around zero Volts. The values of DC voltage and current are shown in Figure 14b. The figure demonstrates that the value of the current was around 20 A, while the value of the voltage was around 530V. This is the same voltage generated by the boost converter and given to the input side of the inverter as DC voltage. The input current to the inverter was around 20 A. The value of the active power generated by the grid can be viewed in Figure 14c, whose value was around 9.2 KW. The reactive power also reached zero, as shown in Figure 14d. As the received power approached zero, the power factor was close to unity.

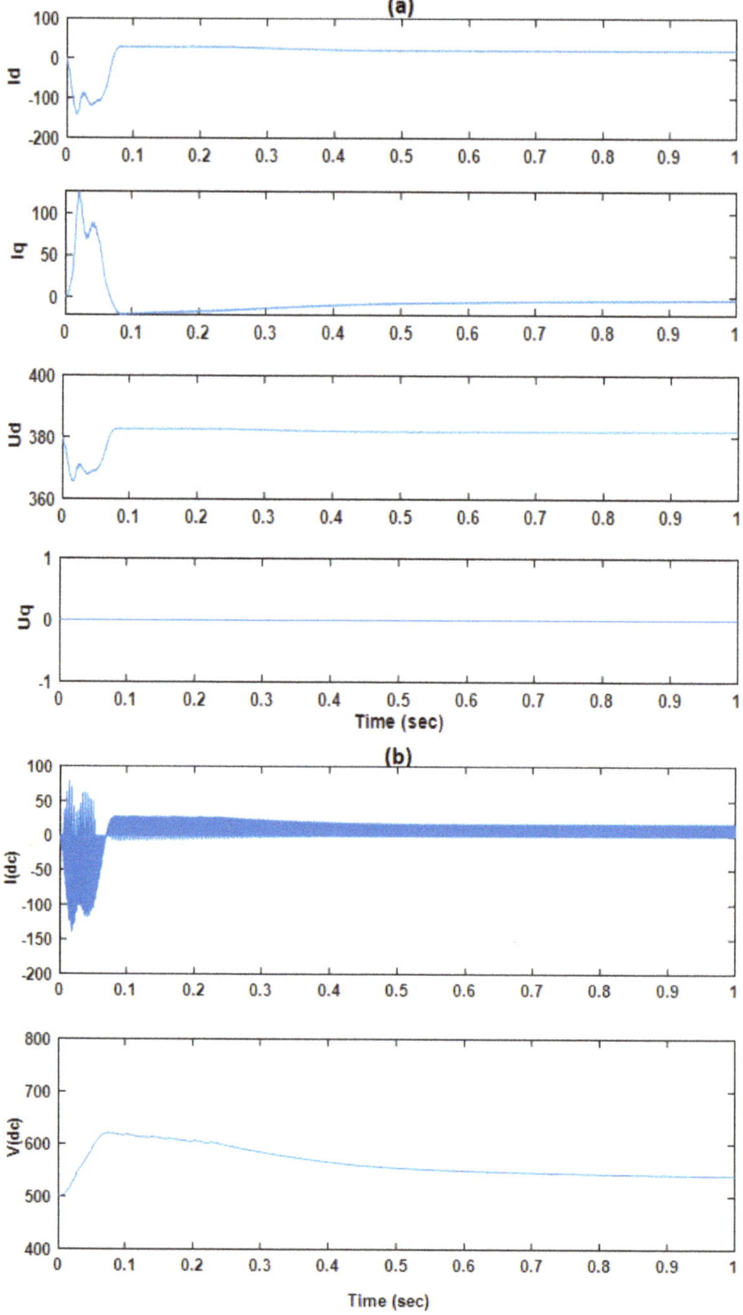

Figure 14. *Cont.*

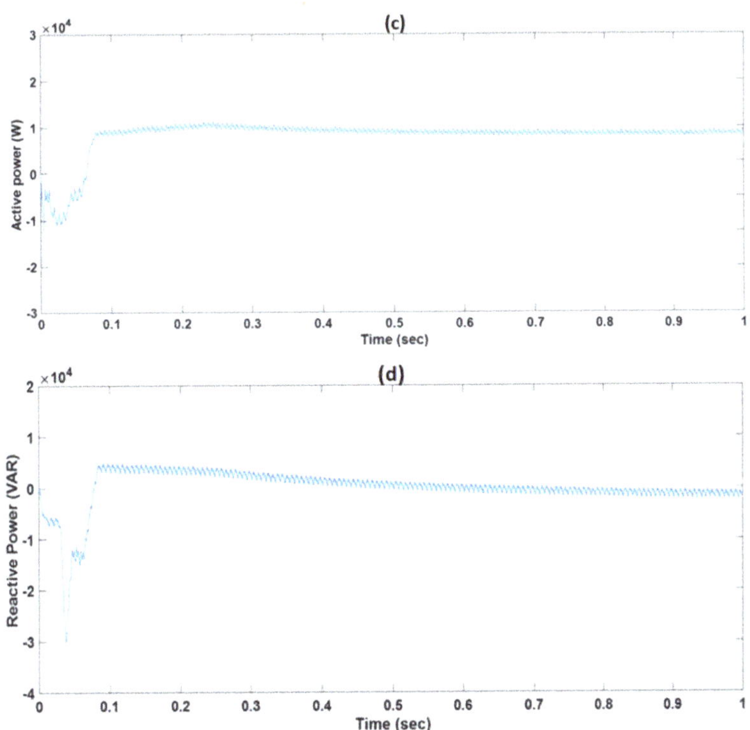

Figure 14. (**a**) Current and voltages from *dq* transformation, (**b**) DC link voltage and current, (**c**) Active power received by the grid. (**d**) Reactive power received by the grid.

The THD of the grid-interfaced PV system is the most significant power quality component. The notable concern of the grid is to comply with the THD requirement of grid interconnection standards. The Fast Fourier Transform (FFT) analysis was utilized to observe the fundamental and harmonic components in the output injected-current waveform. The THD of the grid-associated current was analyzed and compared with the standard of IEEE 519. In line with the IEEE standard, THD of the grid-connected current must be below 5% of the fundamental current frequency at the rated inverter output, since the high-order harmonics of current will cause adverse effects on a variety of equipment related to the power grid. The THD of the system is shown in Figure 15a. The magnitude of fundamental harmonics was more than 3.25%. In comparison, the magnitude of the 3rd and 5th harmonic components were 0.38% and 2.78%, respectively. The magnitude of the 7th harmonics was 2.07%, and the 9th was around 0.17%. It is obvious from Figure 15a that after the 7th harmonics, the current decreased significantly. In these circumstances, the THD was 3.76%, which is less than 5%, and meets the criteria of IEEE standard 519 for distribution into the grid. Therefore, the system performance of THD is reasonable.

According to the 929 standards of IEEE, the power factor must be greater than 0.85 (leading or lagging). The grid-linked PV inverter is intended to normalize the grid current with a unity power factor. In this way, an inverter is responsible for regulating the power factor. Herein, we can learn from Figure 15b that power factor reached unity. The PV system ought to be synchronized with the power grid. Meanwhile, the frequency range must not exceed the limit (49.2–50.6) Hz for the slight PV system, as defined by the 929 standards of IEEE. In this study, frequency was also within these limits, as displayed in Figure 15c. Consequently, frequency follows the standard requirement.

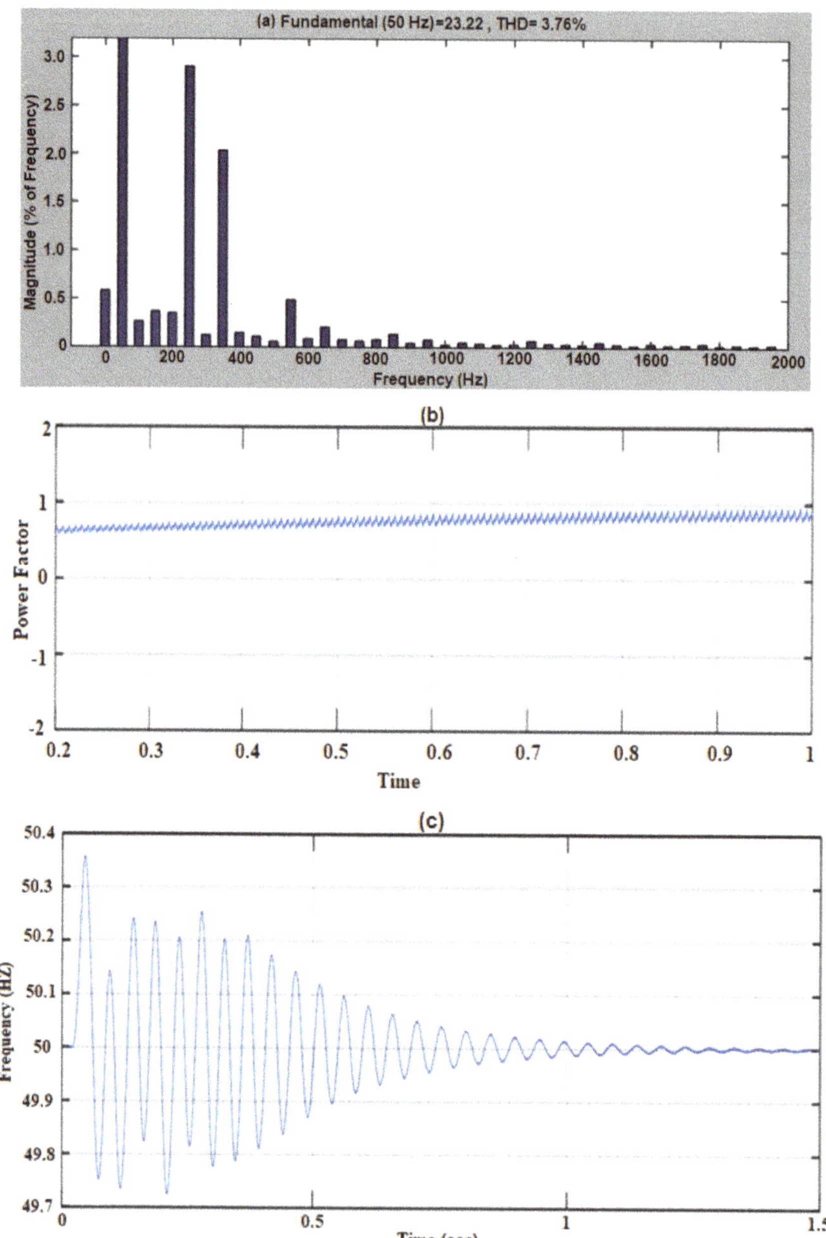

Figure 15. (**a**) Plot showing the percentage of the THD, (**b**) Power factor of a grid, (**c**) Frequency of a grid network under partial shading.

6. Conclusions

This section outlines the key areas introduced in this study and a recommended plan for applying the DOA-based global MPPT technique for grid-interfaced PV systems.

The dual-stage arrangement for grid-interfaced PV systems was presented in this research work. This arrangement consisted of a DC-DC boost converter and inverter to

Sustainability **2021**, *13*, 778

link the PV array with the grid network. The boost converter switching was regulated by utilizing the dragonfly optimization algorithm (DOA) based on the duty cycle. The proposed DOA-based MPPT technique was capable of finding and following the global peak, irrespective of the variation in shading pattern, with less oscillations around the steady state. The DOA-based MPPT technique was evaluated and compared with widely used MPPT techniques, such as P&O, PSO, IPSO, ACSO, and FFO-GRNN algorithms, under different shading patterns. The simulation results showed that the suggested DOA technique significantly out-performed the competing techniques in terms of response time, oscillations reduction, robustness, accuracy convergence speed, and power efficiency. The VSI was utilized to regulate the active and reactive power injected into grid simultaneously. To maintain the unity power factor operation, instantaneous reactive power flow was adjusted to zero. The three-phase PLL was employed to lock the inverter phase and frequency with the grid. Finally, the simulation of the overall grid-associated PV system was carried out by means of MATLAB/Simulink. The simulation results showed that THD the grid current, power factor, and frequency were within the IEEE's standard limits. Hence, the proposed DOA-based MPPT is an effective technique to practically enhance the output and overall performance of the grid-interfaced PV systems under all shading conditions.

Author Contributions: Conceptualization, E.L., F.-Y.W., M.Y.J.; methodology, E.L., M.Y.J.; software, E.L., T.S.T.; validation, E.L., M.Y.J.; formal analysis, G.A.M., D.W.G.; investigation, T.S.T., D.W.G.; resources, G.X., F.-Y.W.; data curation, E.L., M.Y.J.; writing—original draft preparation, E.L.; writing—review and editing, F.-Y.W., G.X., T.S.T., D.W.G.; visualization, G.A.M., D.W.G.; supervision, F.-Y.W., G.X.; project administration, F.-Y.W., G.X.; funding acquisition, F.-Y.W., G.X. All authors have read and agreed to the published version of the manuscript.

Funding: The work is funded by the National Key R&D Program of China (2018AAA0101502) and the Science and technology project of SGCC (State Grid Corporation of China): fundamental theory of human-in-the-loop hybrid-augmented intelligence for power grid dispatch and control.

Institutional Review Board Statement: Not applicable.

Informed Consent Statement: Not applicable.

Data Availability Statement: Not applicable.

Conflicts of Interest: The authors declare no conflict of interest.

Appendix A

In this section, the design of boost converter (value of inductor, input capacitor and output capacitor) is discussed in detail. The parameters for the sizing of boost converter is shown in Table A1.

Table A1. Design parameters for the sizing of boost converter.

Parameters	Values
Input Voltage	337 V
Output Voltage	540 V
PV maximum Power	12,000 W
Frequency	10 K Hz
Inductor ripple current	10%

The value of inductor is measured by Equation (A1), where D_m is the value of duty cycle at MPP, switching frequency is shown by f_s, while output maximum voltage is presented by V_{om} and ΔI_r ΔIlrippleis the inductance ripple current. The inductor L is calculated with the following Equation (A1) [39]:

$$L \geq \frac{V_{om} \cdot D_m (1 - D_m)}{f_s \cdot \Delta I_r} \tag{A1}$$

where V_{om} is 540 V, f_s is 10,000 and the value of duty cycle D_m is obtained from:

$$\frac{V_{out}}{V_{im}} = \frac{T}{t_{off}} = \frac{1}{1 - D_m}$$

$$D_m = 1 - \frac{V_{in}}{V_{out}} = 1 - \frac{337}{540}$$

$$D_m = 0.3759 = 0.38$$

The output maximum current is given by:

$$I_{om} = \frac{P_{out}}{V_{out}}$$

The converter is supposed no loss. The PV input power is the output power.

$$I_{om} = \frac{12,000}{540} = 22.22 \; A$$

The resistive load is given by:

$$R_{load} = \frac{V_{out}}{I_{om}}$$

$$R_{load} = \frac{540}{22.22} = 24.3 \; \Omega$$

Hence, inductor L is calculated by putting values in Equation (4).

$$L \geq \frac{540 * 0.38 * (1 - 0.38)}{10,000 * 2 * 35.6 * 0.1}$$

$$L \geq 1.78 \; mH$$

The input Capacitor is calculated from Equation (A2). Where the value of current at max power is I_{om}, D_m represents duty cycle and V_{pv_mmpp} is the output voltage of the system at MPP.

$$C_{in} \geq \left[\frac{I_{om}.(D_m)^2}{0.02(1 - D_m) f_s. V_{pv\_mmpp}} \right] \tag{A2}$$

$$C_{in} \geq \left[\frac{22.22 * (0.38)^2}{0.02 * (1 - 0.38) * 10,000 * 337} \right]$$

$$C_{in} \geq 76.7 \; \mu F$$

To measure the size of output capacitor, Equation (A3) is employed. Where V_{out} is the output voltage of the boost converter, D_m represents duty cycle, R_{out} is output load of boost converter, ΔV_{out} is ripple output voltage.

$$C_{out} \geq \left[\frac{V_{out}.D_m}{f_s.\Delta V_{out}.R_{out}} \right] \tag{A3}$$

$$C_{out} \geq \left[\frac{540 * 0.38}{10000 * 5.4 * 24.3} \right]$$

$$C \geq 156 \; \mu F$$

References

1. Li, K.; Liu, C.; Jiang, S.; Chen, Y. Review on hybrid geothermal and solar power systems. *J. Clean. Prod.* **2020**, *250*, 119481. [CrossRef]
2. Agathokleous, R.A.; Kalogirou, S.A. Status, barriers and perspectives of building integrated photovoltaic systems. *Energy* **2020**, *191*, 116471. [CrossRef]
3. Fathabadi, H. Novel stand-alone, completely autonomous and renewable energy based charging station for charging plug-in hybrid electric vehicles (PHEVs). *Appl. Energy* **2020**, *260*, 114194. [CrossRef]
4. Vezin, T.; Meunier, S.; Quéval, L.; Cherni, J.; Vido, L.; Darga, A.; Dessante, P.; Kitanidis, P.; Marchand, C. Borehole water level model for photovoltaic water pumping systems. *Appl. Energy* **2020**, *258*, 114080. [CrossRef]
5. Aziz, A.S.; Tajuddin, M.F.N.; Adzman, M.R.; Mohammed, M.F.; Ramli, M.A. Feasibility analysis of grid-connected and islanded operation of a solar PV microgrid system: A case study of Iraq. *Energy* **2020**, *191*, 116591. [CrossRef]
6. Lodhi, E.; Jing, S.; Lodhi, Z.; Shafqat, R.N.; Ali, M. Rapid and Efficient MPPT Technique with Competency of High Accurate Power Tracking for PV System. In Proceedings of the 2017 4th International Conference on Information Science and Control Engineering (ICISCE), Changsha, China, 21–23 July 2017; pp. 1099–1103. [CrossRef]
7. Ishaque, K.; Salam, Z. A review of maximum power point tracking techniques of PV system for uniform insolation and partial shading condition. *Renew. Sustain. Energy Rev.* **2013**, *19*, 475–488. [CrossRef]
8. Grgić, I.; Bašić, M.; Vukadinović, D. Optimization of electricity production in a grid-tied solar power system with a three-phase quasi-Z-source inverter. *J. Clean. Prod.* **2019**, *221*, 656–666. [CrossRef]
9. Lodhi, E.; Lina, W.; Pu, Y.; Javed, M.Y.; Lodhi, Z.; Zhijie, J.; Javed, U. Performance Evaluation of Faults in a Photovoltaic Array Based on V-I and V-P Characteristic Curve. In Proceedings of the 2020 12th International Conference on Measuring Technology and Mechatronics Automation (ICMTMA), Phuket, Thailand, 28–29 February 2020; pp. 85–90. [CrossRef]
10. Belhachat, F.; Larbes, C. A review of global maximum power point tracking techniques of photovoltaic system under partial shading conditions. *Renew. Sustain. Energy Rev.* **2018**, *92*, 513–553. [CrossRef]
11. Khan, I. Impacts of energy decentralization viewed through the lens of the energy cultures framework: Solar home systems in the developing economies. *Renew. Sustain. Energy Rev.* **2020**, *119*, 109576. [CrossRef]
12. Lodhi, E.; Lodhi, Z.; Shafqat, R.N.; Chen, F. Performance analysis of two widely used Maximum Power Point Tracking Algorithms for PV Applications. *IOP Conf. Ser. Mater. Sci. Eng.* **2017**, *220*, 012029. [CrossRef]
13. Liu, H.-D.; Lin, C.-H.; Pai, K.-J.; Lin, Y.-L. A novel photovoltaic system control strategies for improving hill climbing algorithm efficiencies in consideration of radian and load effect. *Energy Convers. Manag.* **2018**, *165*, 815–826. [CrossRef]
14. Bounechba, H.; Bouzid, A.; Snani, H.; Lashab, A. Real time simulation of MPPT algorithms for PV energy system. *Int. J. Electr. Power Energy Syst.* **2016**, *83*, 67–78. [CrossRef]
15. Motahhir, S.; Chalh, A.; El Ghzizal, A.; Derouich, A. Development of a low-cost PV system using an improved INC algorithm and a PV panel Proteus model. *J. Clean. Prod.* **2018**, *204*, 355–365. [CrossRef]
16. Peng, L.; Zheng, S.; Chai, X.; Li, L. A novel tangent error maximum power point tracking algorithm for photovoltaic system under fast multi-changing solar irradiances. *Appl. Energy* **2018**, *210*, 303–316. [CrossRef]
17. Mirza, A.F.; Ling, Q.; Javed, M.Y.; Mansoor, M. Novel MPPT techniques for photovoltaic systems under uniform irradiance and Partial shading. *Sol. Energy* **2019**, *184*, 628–648. [CrossRef]
18. Yang, B.; Yu, T.; Zhang, X.; Li, H.; Shu, H.; Sang, Y.; Jiang, L. Dynamic leader based collective intelligence for maximum power point tracking of PV systems affected by partial shading condition. *Energy Convers. Manag.* **2019**, *179*, 286–303. [CrossRef]
19. Mohamed, M.A.; Diab, A.A.Z.; Rezk, H. Partial shading mitigation of PV systems via different meta-heuristic techniques. *Renew. Energy* **2019**, *130*, 1159–1175. [CrossRef]
20. Camilo, J.C.; Guedes, T.; Fernandes, D.; Melo, J.; Costa, F.; Filho, A.J.S. A maximum power point tracking for photovoltaic systems based on Monod equation. *Renew. Energy* **2019**, *130*, 428–438. [CrossRef]
21. Venkateswari, R.; Sreejith, S. Factors influencing the efficiency of photovoltaic system. *Renew. Sustain. Energy Rev.* **2019**, *101*, 376–394. [CrossRef]
22. Lodhi, E.; Shafqat, R.N.; Kerrouche, K.D.; Lodhi, Z. Application of Particle Swarm Optimization for Extracting Global Maximum Power Point in PV System under Partial Shadow Conditions. *Int. J. Electron. Electr. Eng.* **2017**, *5*, 223–229. [CrossRef]
23. Li, H.; Yang, D.; Su, W.; Lu, J.; Yu, X. An Overall Distribution Particle Swarm Optimization MPPT Algorithm for Photovoltaic System under Partial Shading. *IEEE Trans. Ind. Electron.* **2019**, *66*, 265–275. [CrossRef]
24. Prasanth Ram, J.; Rajasekar, N. A novel flower pollination based global maximum power point method for solar maximum power point tracking. *IEEE Trans. Power Electron.* **2017**, *32*, 8486–8499. [CrossRef]
25. Guo, L.; Meng, Z.; Sun, Y.; Wang, L. A modified cat swarm optimization based maximum power point tracking method for photovoltaic system under partially shaded condition. *Energy* **2018**, *144*, 501–514. [CrossRef]
26. Mokhtari, Y.; Rekioua, D. High performance of Maximum Power Point Tracking Using Ant Colony algorithm in wind turbine. *Renew. Energy* **2018**, *126*, 1055–1063. [CrossRef]
27. Yang, B.; Yu, T.; Shu, H.; Zhu, D.; An, N.; Sang, Y.; Jiang, L. Energy reshaping based passive fractional-order PID control design and implementation of a grid-connected PV inverter for MPPT using grouped grey wolf optimizer. *Sol. Energy* **2018**, *170*, 31–46. [CrossRef]

28.	Nowdeh, S.A.; Moghaddam, M.J.H.; Nasri, S.; Abdelaziz, A.Y.; Ghanbari, M.; Faraji, I. *A New Hybrid Moth Flame Optimizer-Perturb and Observe Method for Maximum Power Point Tracking in Photovoltaic Energy System. Modern Maximum Power Point Tracking Techniques for Photovoltaic Energy Systems*; Springer: Berlin/Heidelberg, Germany, 2020; pp. 401–420.

29.	Sundareswaran, K.; Vigneshkumar, V.; Sankar, P.; Simon, S.P.; Nayak, P.S.R.; Palani, S. Development of an Improved P&O Algorithm Assisted Through a Colony of Foraging Ants for MPPT in PV System. *IEEE Trans. Ind. Inform.* **2016**, *12*, 187–200. [CrossRef]

30.	Seyedmahmoudian, M.; Rahmani, R.; Mekhilef, S.; Oo, A.M.T.; Stojcevski, A.; Soon, T.K.; Ghandhari, A.S. Simulation and Hardware Implementation of New Maximum Power Point Tracking Technique for Partially Shaded PV System Using Hybrid DEPSO Method. *IEEE Trans. Sustain. Energy* **2015**, *6*, 850–862. [CrossRef]

31.	Mohanty, S.; Subudhi, B.; Ray, P.K. A Grey Wolf-Assisted Perturb & Observe MPPT Algorithm for a PV System. *IEEE Trans. Energy Convers.* **2017**, *32*, 340–347. [CrossRef]

32.	Ishaque, K.; Salam, Z. A Deterministic Particle Swarm Optimization Maximum Power Point Tracker for Photovoltaic System under Partial Shading Condition. *IEEE Trans. Ind. Electron.* **2012**, *60*, 3195–3206. [CrossRef]

33.	Thongpron, J.; Kirtikara, K. Effects of low radiation on the power quality of a distributed PV-grid connected system. *Sol. Energy Mater. Sol. Cells* **2006**, *90*, 2501–2508. [CrossRef]

34.	Javed, M.Y.; Murtaza, A.F.; Ling, Q.; Qamar, S.; Gulzar, M.M. A novel MPPT design using generalized pattern search for partial shading. *Energy Build.* **2016**, *133*, 59–69. [CrossRef]

35.	Castaner, L.; Silvestre, S. *Modeling Photovoltaic Systems Using PSpice*; Wiley: Hoboken, NJ, USA, 2002.

36.	Mohan, N.; Robbin, W.P.; Undeland, T. *Power Electronics: Converters, Applications, and Design*, 2nd ed.; Wiley: New York, NY, USA, 1995.

37.	Salhi, M.; El-Bachtiri, R.; Matagne, E. The development of a new maximum power point tracker for a PV panel. *Int. Sci. J. Altern. Energy Ecol. (ISJAEE)* **2008**, *62*, 138–145.

38.	Hasaneen, B.M.; Mohammed, A.A.E. Design and simulation of DC/DC boost converter. In Proceedings of the 2008 12th International Middle-East Power System Conference, Aswan, Egypt, 12–15 March 2008; pp. 335–340.

39.	Yin, W.; Ma, Y. Research on three-phase PV grid-connected inverter based on LCL filter. In Proceedings of the 2013 IEEE 8th Conference on Industrial Electronics and Applications (ICIEA), Melbourne, VIC, Australia, 19–21 June 2013; pp. 1279–1283.

40.	Wang, L. *PID Control System Design and Automatic Tuning Using MATLAB/Simulink*; Wiley-IEEE Press: Hoboken, NJ, USA, 2020; ISBN 978-1-119-46940-7.

 *sustainability*

Article

Optimized Economic Load Dispatch with Multiple Fuels and Valve-Point Effects Using Hybrid Genetic–Artificial Fish Swarm Algorithm

Abdulrashid Muhammad Kabir [1,†], Mohsin Kamal [2,†], Fiaz Ahmad [3,†], Zahid Ullah [4,*,†], Fahad R. Albogamy [5,†], Ghulam Hafeez [6,†] and Faizan Mehmood [7,*,†]

1 Department of Electrical Engineering, Kebbi State University of Science and Technology, Aliero 863104, Nigeria; abdulrashidmkabir@yahoo.com
2 KIOS Research and Innovation Center of Excellence, University of Cyprus, Nicosia 2109, Cyprus; kamal.mohsin@ucy.ac.cy
3 Department of Electrical and Computer Engineering, Air University Islamabad, Islamabad 44000, Pakistan; fiaz.ahmad@mail.au.edu.pk
4 Department of Electrical Engineering, Sialkot Campus, University of Management and Technology (Lahore), Sialkot 51310, Pakistan
5 Computer Sciences Program, Turabah University College, Taif University, P.O. Box 11099, Taif 21944, Saudi Arabia; f.alhammdani@tu.edu.sa
6 Department of Electrical and Computer Engineering, COMSATS University Islamabad, Islamabad 44000, Pakistan; ghulamhafeez@uetmardan.edu.pk
7 Department of Electrical Engineering, University of Engineering and Technology, Taxila 47050, Pakistan
* Correspondence: zahid.ullah@skt.umt.edu.pk (Z.U.); mehmood.faizan@ucy.ac.cy (F.M.)
† These authors contributed equally to this work.

Citation: Kabir, A.M.; Kamal, M.; Ahmad, F.; Ullah, Z.; Albogamy, F.R.; Hafeez, G. and Mehmood, F. Optimized Economic Load Dispatch with Multiple Fuels and Valve-Point Effects Using Hybrid Genetic–Artificial Fish Swarm Algorithm. *Sustainability* **2021**, *13*, 10609. https://doi.org/10.3390/su131910609

Academic Editors: Sheraz Aslam, Herodotos Herodotou and Nouman Ashraf

Received: 9 August 2021
Accepted: 20 September 2021
Published: 24 September 2021

Publisher's Note: MDPI stays neutral with regard to jurisdictional claims in published maps and institutional affiliations.

Abstract: Economic Load Dispatch (ELD) plays a pivotal role in sustainable operation planning in a smart power system by reducing the fuel cost and by fulfilling the load demand in an efficient manner. In this work, the ELD problem is solved by using hybridized robust techniques that combine the Genetic Algorithm and Artificial Fish Swarm Algorithm, termed the Hybrid Genetic–Artificial Fish Swarm Algorithm (HGAFSA). The objective of this paper is threefold. First, the multi-objective ELD problem incorporating the effects of multiple fuels and valve-point loading and involving higher-order cost functions is optimally solved by HGAFSA. Secondly, the efficacy of HGAFSA is demonstrated using five standard generating unit test systems (13, 40, 110, 140, and 160). Finally, an extra-large system is formed by combining the five test systems, which result in a 463 generating unit system. The performance of the developed HGAFSA-based ELD algorithm is then tested on the six systems including the 463-unit system. Annual savings in fuel costs of $3.254 m, $0.38235 m, $2135.7, $9.5563 m, and $1.1588 m are achieved for the 13, 40, 110, 140, and 160 standard generating units, respectively, compared to costs mentioned in the available literature. The HGAFSA-based ELD optimization curves obtained during the optimization process are also presented.

Keywords: artificial fish swarm algorithm; economic load dispatch; genetic algorithm; hybrid genetic–artificial fish swarm algorithm; multi-objective optimization; sustainable power generating system

1. Introduction

Modern power systems around the world are becoming increasingly complex, with interconnections and varying load demands. There is an emergent need for power systems to be sustainable, reliable, low-cost, smarter, and cleaner, which would allow the broad participation of end users for energy generation and consumption and the management of loads by intelligent devices [1]. With this changing outlook, Economic Load Dispatch (ELD) is needed due to the lack of energy resources, increased power generation costs, and environmental concerns. In the actual scenario, the power plants are not equidistant from the load and there is no similar fuel cost function. Therefore, in order to provide cheaper power, loads must be distributed to various power plants to minimize power generation costs. A practical economic dispatch (ED) problem has a highly nonlinear objective function with equality and inequality constraints. The ELD problem is solved using conventional methods such as lambda iteration, gradient methods, and non-conventional methods (heuristic methods). However, these technologies may not provide optimal solutions to find a global optimal solution as they require a piecewise linear and monotonically increasing incremental fuel cost curve. Optimal economic operation and planning of sustainable power generation systems is a very important pillar in the power industry [2]. ELD also refers to the operation of a power generation facility that produces energy at the lowest cost in order to recognize operational limitations of power generation facilities and provide reliable services to consumers. ELD schedules the output of available power generation units at specific times to minimize overall production costs while satisfying equality and inequality constraints [3]. Before 1973, due to the oil embargo measures that led to a sharp rise in fuel prices, utility companies spent around 20% of their total revenue on fuel for the generation of electric energy [4]. By 1980, this figure had increased to over 40% of the total revenue. In the five years following 1973, the fuel cost for electric utilities in the United States increased by 25% per year. Due to the fact that fuel is an irreplaceable natural resource, the efficient use of available fuels is of increasing importance [5].

Table 1 shows the parameters and assumptions for a moderately large power system [2]. The idea of size of money was obtained by considering the annual operation cost of a large utility for purchasing fuel. This cost reveals the direct requirement for customers' income to be an average of 3.15 cents/kWh, targeted for recovering fuel costs. Savings in the operation of a small part of the system refer not only to the amount of fuel consumed but also to the substantial reduction in operating costs. Therefore, this field has gained tremendous attention from engineers for many years. However, a regular change in the basic fuel price level plays the role of emphasizing the problem and enhancing its economic significance. Inflation also poses problems in developing economic operational techniques, methods, and examples of power generation systems [6]. Moreover, the rapid increase in the size and power demand of power systems resulted in reduced operating costs while maintaining the thermal limitations of voltage security and transmission line branching. Many mathematical programming and artificial intelligence techniques, such as GA-based ELD, Particle Swarm Optimization (PSO)-based ELD [7,8], ELD based on dynamic programming and evolutionary programming, and hybrid GA–PSO-based ELD, are applied to solve the aforementioned problem. In the most common formulation, the ELD problem is modeled as a large-scale, non-convex, nonlinear, static optimization problem in both discrete and continuous control variables [9].

Table 1. Parameters and assumptions for total annual fuel cost.

Parameters		Assumptions	
Annual peak load and load factor	10,000 MW and 60 %	Annual energy produced	10^7 MW $\times$ 8760 h/year $\times$ 0.60 $= 5.256 \times 10^{10}$ kWh
Average annual heat rate for converting fuel to electric energy	10,550.56 KJ/kWh	Annual fuel consumption	10,550.56 KJ/kWh $\times$ 5.256 $\times$ 10^{10} kWh $= 55.45 \times 10^{13}$ KJ
Annual fuel cost (corresponds to oil price at \$18/bbl)	\$3.00/1.055 GJ	Annual fuel cost	$55.45 \times 10^{13} \times 3/1.055 \times$ 10^{-9} \$/J \$1.5767 million

Many researchers [6–9] have modeled the nonlinear, convex nature of ELD problems using pure quadratic functions, with the quadratic coefficients (a, b, and c) defined at the start of the solution search process. Meanwhile, research work [10] developed realistic models to incorporate the effect of multiple fuel cost functions and valve-point loading into the formulation of the ELD problem. For example, the authors in [8] employed PSO with the BAT algorithm for solving ELD problems considering inspired acceleration coefficients. In [9], the authors analyzed the generating unit profiles of various distributed generation systems of different technologies. The energy loss was minimized for a distribution system with a mix of renewable energy resources using optimization techniques [10]. The authors in [11] proposed dynamic ELD in a sustainable power system using an accurate forecasting model and improved salp-swarm optimizer considering PV, energy storage, the power system, and various constraints of generating units. In [12], the authors proposed a combined dispatch strategy (load following and cycle charging) for the energy management and optimal operation of a hybrid energy system. A gradient-based optimizer inspired by the Newton technique was used for solving the ELD problem considering the valve-point effect [13]. The authors in [14] proposed a teaching–learning optimization for dynamic ELD of wind energy and load demand uncertainties considering various operation constraints. Most of the research works utilized various optimization techniques to minimize the fuel costs, reduce environmental concerns, and ensure dynamic ELD. In this regard, this work proposes a hybrid optimization technique to achieve the aforementioned objectives.

The key contribution of this work is to address the ELD problem by developing a more realistic model and by considering the effects of valve-point loading and multiple fuel cost functions. A multi-objective optimization problem is formulated that minimizes the fuel cost of generating units and the amount of nitrogen-bearing (NOx) gases emitted by the generating units during their operation. Therefore, there is an emergent need to employ robust techniques to provide a reliable solution to the aforementioned complex optimization problem. Some have used heuristic techniques in an attempt to solve the abovementioned optimization technique to some extent, but this did not provide guaranteed efficient solutions. In order to provide a reliable and efficient solution, a hybridization of two conventional heuristic techniques (GA and AFSA), called HGAFSA, is proposed in order to solve the complex multi-objective optimization problem. HGAFSA is applied to solve a multi-objective ELD problem considering the effects of multiple fuel cost functions and valve-point loading. The effectiveness of the proposed approach is demonstrated using five standard generating unit test systems (13, 40, 110, 140, and 160) and a 463 generating unit system formed by the combination of the five systems, and the results are compared with the best results presented in the literature. The choice of GA and AFSA is based on the following factors: (a) GA is a heuristic technique with a well-defined set of search equations that is effective in solving problems such optimal sizing and location of capacitor banks and distributed generators [15], optimal power flow [16], optimal location of tie and sectionalizing switches in distribution systems, and optimal network expansion [17], and (b) AFSA is a relatively new heuristic technique based on well-refined and sophisticated solution search equations and is widely applied in controller design, optimal PID tuning, and objective function minimization/maximization [18].

The major contributions of this work to the existing body of knowledge are summarized as follows:

1. HGAFSA, capable of solving a higher-order ELD, is developed and used to solve several higher-order ELD problems, including 13, 40, 110, 140, 160, and 463 unit systems.
2. An ELD encoder algorithm is developed and linked to the developed HGAFSA to form a HGAFSA-based ELD algorithm that minimizes any ELD cost function better than every algorithm mentioned in the available literature.
3. The effectiveness of the developed HGAFSA is demonstrated on six ELD systems, including the 463-unit system. Annual savings in fuel costs of $3.254 m, $0.38235 m, $2135.7, $9.5563 m, and $1.1588 m for the 13, 40, 110, 140, and 160 units, respectively, for the first five systems are achieved, compared to costs reported in the available literature.

The remainder of the paper is organized as follows. Section 2 formulates the ELD problem. The formulation of the proposed HGAFSA is presented in Section 3. Section 4 provides the performance validation and simulation settings of the proposed system. The paper concludes with a brief summary and future directions in Section 5.

2. Formulation of ELD

The purpose of the ELD problem is to find an optimal combination of power generation that meets the constraints of equality and inequality while minimizing the total power generation cost. The fuel cost curve for any unit is an approximation of the quadratic function segment of the generator's active power output by assumption [19].

2.1. The Cost Function

Cost function is a financial term used by a company's economists and managers as a way to express how different costs differ under different circumstances. This shows how to display monetary output. Changes in the level of activities related to these outputs will change rates and fees from overhead and operating expenses [19]. The linear cost function has three basic types:

1. Functions of fixed cost (defined by a straight line with a zero (0) gradient).
2. Functions of variable cost (defined by a straight line with positive gradient and having no intercept).
3. Functions of mixed cost (defined by a line having single or multiple gradient(s) and intercept(s)).

In mixed environments, costs are fixed to specific points, which can be changed based on related activities. Analysts use this type of function to make important predictions about the market and to inform various decision-making tasks [19]. For a given network of power generating units, a set of cost functions are usually defined based on the mixed function types to account for the operational cost of each generating unit. Conventionally, the quadratic form of the cost functions is most widely used [20]. A quadratic cost function is a mixed cost function that partly comprises a single fixed cost function (represented by a fixed coefficient, e.g., a) and two variable cost functions (represented by products of coefficients b and c, and functions of output power P and P^2). For a network of n generating units, the overall cost function (F_T) of the system is a summation of the n individual cost functions $[F(P_1), F(P_2)..., F(P_{n-1}), F(P_n)]$ of the various units in the system. This can be simply represented by Equation (1). An ELD problem is an optimization problem that is aimed at minimizing the F_T subject to a set of operating constraints [21].

$$F_T = \sum_{i-1}^{n} F(P_i) = \sum_{i=1}^{n} a_i + b_i P_i + c_i P_i^2 . \tag{1}$$

In general, F_T is used to represent the fuel or emission cost (\$/hr) depending on the values of the coefficients (a_i, b_i and c_i) used [21]. In order to ensure proper operation of the network of generating units, it is essential to ensure that a set of useful constraints are satisfied. These constraints include the generator capacity (inequalities) and active power balance (equality) constraints given in Equations (2) and (3), respectively [21].

$$P_{i,min} \leq P_i \leq P_{i,max} \quad for \ i = 1, 2, ..., n , \tag{2}$$

where $P_{i,min}$ and $P_{i,max}$ are the minimum and maximum power output of the ith unit.

$$P_D = \sum_{i=1}^{n} P_i - P_{Loss} . \tag{3}$$

where P_D is the total power demand and P_{Loss} is the total transmission loss. The transmission loss P_{Loss} can therefore be calculated by using the B matrix technique and is defined by Equation (4) as follows [21]:

$$P_{Loss} = \sum_{i=1}^{n} \sum_{j=1}^{n} P_i B_{ij} P_j , \tag{4}$$

where $B_{ij}s$ are the elements of loss coefficient matrix B. The cost function defined by F_T in Equation (1) assumes a smooth quadratic fuel cost function without the valve-point loadings of the generating units (where the valve-point effects are ignored). The generating units with multi-valve steam turbines exhibit greater variation in the fuel cost functions. Since the valve point results in ripples, a cost function with higher-order nonlinearity will result. Therefore, the function $F(P_i)$ in Equation (1) can be replaced by Equation (5) to account for the valve-point effects [22]. Conventionally, sinusoidal functions are often added to the quadratic cost function to account for the valve-point effect, as given in Equation (5).

$$F(P_i) = a_i + b_i P_i + c_i P_i^2 + |e_i \times sin(f_i \times (P_{i,min} - P_i))| \, . \tag{5}$$

where ei and f_i are the cost coefficients of the ith unit with valve-point effects. In general, the cost coefficients ei and f_i are introduced as in Equation (5) to model the valve-point loadings. Similarly, Equation (5) can be used to represent either the fuel or emission cost ($/hr) depending on the values of the coefficients (a_i, b_i, c_i, e_i, and f_i) that are used. Finally, Equation (5) also represents the proposed higher-order cost function considering the valve-point loading effect. In practical situations, generating units are made up of subunits. These subunits combine to give rise to the overall installed capacity of the unit. Most units are designed to operate using more than one fuel type, particularly in the case in which there is great fluctuation in the price and availability of the dominant fuel types [19]. In the case of moderately large units, a combination of the available fuel types may be used to cover the power demand over the specified period of time. This type of scenario introduces greater nonlinearity into the overall fuel cost function. Therefore, the fuel cost function of such a system can be modeled using a multiple fuel cost function, which is only defined for a particular range of power output within the specified maximum and minimum power generation. Considering both the valve-point loading effect and multiple fuels, the cost function of the system may easily be represented using Equation (6) [19].

$$F(P_i) =$$
$$\begin{cases} a_{i1} + b_{i1} P_i^2 + c_{i1} P_i^2 + |e_{i1} \times sin(f_{i1} \times (P_{i1,min} - P_{i1})| Fuel1 : P_i^m in \leq P_i \leq P_{i1} \\ a_{i2} + b_{i2} P_i^2 + c_{i2} P_i^2 + |e_{i2} \times sin(f_{i2} \times (P_{i2,min} - P_{i2})| Fuel2 : P_i^m in \leq P_i \leq P_{i2} \\ a_{ik} + b_{ik} P_i^2 + c_{ik} P_i^2 + |e_{ik} \times sin(f_{ik} \times (P_{ik,min} - P_{ik})| Fuel3 : P_{ik-1}^{min} \leq P_i \leq P_{ik}^{max} \, . \end{cases} \tag{6}$$

2.2. Artificial Fish Swarm Algorithm

AFSA guarantees a global optimum in solution search problems [23], which is of great importance in artificial intelligence to perform behavioral modeling. Consider a swarm consisting of N artificial fishes and a state vector $X = (x_1, x_2, ...x_n)$, where n states or attributes of the artificial fish are to be optimized via the AFSA algorithm. In addition, suppose that $Y = f(X)$ represents the objective function giving the food concentration of the artificial fish at the current position, and let $Dij = ||X_i - X_j||$ be used to describe the distance between artificial fishes i and j. Other important parameters for the artificial fish, including its vision field, maximum step for motion, the congestion factor, and maximum attempts in each praying, are also taken into account and are expressed as visual, step, δ, and try number, respectively. For better results, the congestion factor is used to constrain the size of the artificial swarm [5]. The behavior of the artificial fish is described next as praying, swarm, and chasing.

2.2.1. Praying

If the artificial fish is currently in state X_i, in order to carry out praying, then it must select another state, e.g., X_j, that is located within its visual field. Afterwards, the search for a minimal solution is continued until $Y_i \geq Y_j$; if this is the case, praying is completed by moving one step in the direction taken. However, if $Yi \leq Yj$, another state X_j must be reselected from the visual field randomly to analyze whether it can move forward based on a certain forwarding condition. This procedure is repeated for try-number times, and if the forward motion condition is still not satisfied, it will take one step in a random direction. Mathematically, this can be expressed as given in Equation (7);

$$x_{i-next-k} = x_{i \to k} + \frac{x_{jk} - x_{ik}}{\|X_j - X_i\|} * random(step) \quad Y_j > Y_i \,,$$

$$x_{i-next-k} = x_{i \to k} + random(step) \quad Y_j \leq Y_i \,,$$

(7)

where $k = 1, 2 \ldots n$, x_{ij} represents the kth element of X_i, which is the current state of the artificial fish; x_{jk} is the kth element of X_j, which is the state of the artificial fish after random movement, and $x_{i-next-k}$ represents the kth element of x_{i-next}, which is the next state of the artificial fish. Similarly, Y_i and Y_j are the values of the objective function of the current state and that after a random movement, respectively, and random(step) represents a random number selected from the range defined by [0 step].

2.2.2. Swarm

In the swarming process, the fish has the natural ability to share food and avoid any distraction that is encountered. Suppose that the current state of the artificial fish is given by X_j, and the total number of other fishes in its vision domain is denoted by n. Now, if $n_f = 0$, this should mean that the visual domain of the given artificial fish is empty, so it is time to implement praying. However, if $n_f \geq 0$, this means that there are other companion fishes present in its vision domain, and it must start searching the central position X_c (i.e., center between the present fishes) of its companions according to Equation (8) [24].

$$X_{ck} = \frac{(\sum_{j=1}^{n_f} x_{jk})}{n_f} \,,$$

(8)

where X_c represents the central position of the artificial fish among other fishes, X_{ck} gives the kth element of X_c, and X_{jk} denotes the kth element of the vector of jth companion j = (1, 2, ... , n). The calculation of the food concentration of the artificial fish at the central position, given by Y_c, is the objective function with the constraint of $Y_c \frac{n_f}{Y_i} > 1$. If the central position is less congested and safer, the artificial fish must move towards this position using Equation (9); otherwise, praying is implemented [24].

$$x_{i-next-k} = x_{i \to k} + \frac{x_{ck} - x_{ik}}{\|X_c - X_i\|} * random(step) \,.$$

(9)

2.2.3. Chasing

In an artificial fish swarm, when fishes are in search of food, neighboring partners have the natural ability to trace and reach food more quickly. Suppose that X_i denotes the current state of the artificial fish and n denotes the total number of companions in its visual field. Now, if $n_f = 0$, this shows that the visual field of the artificial fish is empty; therefore, praying should be implemented. However, if $n_f \geq 1$, this indicates that some companions do exist in its visual field; therefore, it should search and find a companion with a minimum value of the corresponding function X_{max}. Then, the constraint is checked, i.e., $Y_{max} \frac{n_f}{Y_i} > 1$; if it is valid, this means that the fitness value of the corresponding companion is small and it is not congested; thus, Equation (10) is implemented; otherwise, praying is implemented [24].

$$x_{i-next-k} = x_{i \to k} + \frac{x_{max,k} - x_{ik}}{\|X_{max} - X_i\|} * random(step) \,.$$

(10)

where $x_{max,k}$ gives the kth element of state vector X_{max}.

2.3. Genetic Algorithm

Genetic algorithm (GA) is an optimization algorithm that simultaneously works on several solutions (also called population), as opposed to other optimization methods that work on one solution at a time [25,26]. It is an iterative optimization algorithm and comprises several steps, briefly described below.

2.3.1. Reproduction

The foremost operation on a population is called reproduction, which establishes a mating pool by the selection of good strings from a population. The mating pool is fed with duplicate copies of the good strings, i.e., above average strings. Proportionate selection of strings from the present population is the most common operation in the reproduction process, where each string is selected based on its fitness probability. Hence, an ith string is selected with a fitness probability of ϵ_i. The cumulative probability of all the strings in a population is always '1' because the population size in GA is normally fixed. The fitness probability of ith string is given as $\frac{f_i}{\sum_{j=1}^{N} f_j}$. N represents the size of the population. In [27], the authors have presented a method for achieving a proportionate selection using a roulette wheel, where the circumference for every string is marked exactly according to its fitness.

2.3.2. Crossover

Crossover is the next operation applied to the string of the mating pool after the reproduction operation. In this operation, two strings are selected randomly from the mating pool and some of their portions are exchanged. For instance, in a single point crossover operation, two new strings are produced by swapping the right-side portions of two strings after cutting these at arbitrary places, as presented in [15].

A better child string can be produced by combining good sub-strings from either parent if a suitable site is selected, which is usually selected randomly since the suitability of a site is not always known [15]. However, it must be noted that the random selection of a site does not make the search process random. If a single-point crossover is applied to two 1-bit strings from either parent, at most, different strings can be found as a solution in the given search space. With the selection of a random site, 2^i children strings are produced. These strings may or may not contain good sub-strings from parent strings, which depends on the selection of an appropriate site. This aspect is of little interest as, if the crossover operation fails to produce good strings, the reproduction operation will produce more copies in the following mating pool. In a similar way, with a two-point crossover operation, two sites are chosen randomly. A multi-point crossover operation can be carried out in a similar fashion, and this extension is usually called the uniform crossover operator.

For a case of binary strings, the uniform crossover operation is applied by selecting from either parent every bit with a probability of 0.5 [15]. The major purpose of the crossover operation is the search of the parameter space and the preservation of the information from parent strings since these are labeled as good strings after due selection by the reproduction operation. Maximum information is transferred or preserved from parent to child strings with a single-point crossover operator search as opposed to that of the uniform crossover operator, where the search is extensive but the information preservation form parent to child is minimal. For a crossover probability of P_c, the crossover operation is applied to $100P_c\%$ of the strings in the population and the remaining strings, i.e., $100(1-P_c)\%$ are transferred to a new population [15].

2.3.3. Mutation

In genetic algorithms, search procedures are normally carried out with the crossover operator. However, a mutation operation may also be used sometimes for this purpose, which uses certain mutation probability P_m, to change a 1 to 0 and vice versa. The detailed process of mutation is explained in [15] where a new solution is created after a change in the value of the fourth gene. The mutation operation is necessary to diversify the population, which can be demonstrated with the above example, where a long string of zeros occurs, and a 1 is required to obtain a new solution that is more optimal or near to optimal. Mutation is also helpful in improving a local solution.

3. Formulation of Hybrid Genetic–Artificial Fish Swarm Algorithm

The proposed HGAFSA is designed based on the available parameter dredging steps present in the conventional GA and AFSA. However, each of the separate algorithms (GA and AFSA) is assumed to be composed of three major steps, as described below.

1. GA

    (a) Reproduction;
    (b) Crossover;
    (c) Mutation.

2. AFSA

    (a) Praying;
    (b) Swarming;
    (c) Chasing.

The mathematical formulation of these steps has been described in Section 2. The proposed HGAFSA is a logical combination of the six steps listed above. It is worth noting that the GA uses a binary operation on a set of binary codes known as chromosomes, which further comprise gins, whereas the AFSA uses real numbers ranging between zero and one as the parameters of the so-called artificial fish. As such, a decoder function and an encoder function are required to serve as converters from GA to AFSA and vice versa. These functions are intended to decode binary code into real numbers and later encode real numbers into binary.

3.1. Decoder Function

The decoder function takes in four parameters as inputs and generates a decoded version of the main parameter as the output. Here, the main parameter is the chromosome (X), whereas the remaining three parameters are:

1. $x_{min} \rightarrow$ Lower boundary of the desired decoded output;
2. $x_{max} \rightarrow$ Higher boundary of the desired decoded output;
3. $N_{bits} \rightarrow$ Number of bits per parameter.

The steps involved in decoding a single chromosome can be described using Algorithm 1.

Algorithm 1: The decoder function

---

    input $\ : X, x_{min}, x_{max} \, and \, N_{bits}$
    output$: X_{decoded}$

1 *Evaluate:* $N = number \ of \ bits \ in \ X$;
2 *Evaluate:* $p = N * (N_{bits})^{-1}$ *//number of parameters*;
3 *Evaluate:* $q = 0.5^{[1,2,...N_{bits}]}$ *//quantization levels*;
4 *Evaluate:* $q_{norm} = q * (\sum_{i=1}^{N_{bits}} q_i)^{-1}$ *//quantization level normalization*;
5 *Evaluate:* $X_{decoded}(i) = [q(1) * X(j+1)q(2) * X(j+2)...q(N_{bits} * X(j + Nbits))]$;
6 **for** $i \leftarrow 1$ **to** p *and* $j \leftarrow (i-1) * N_{bits}$ **do**
7     $X_{decoded} = [X_{decoded}(1)X_{decoded}(2)...X_{decoded}(p)] * (x_{max} - x_{min}) + x_{min}$;

Using the decoder function, a vector X with $p \times N_{bits}$ elements is decoded into a vector $X_{decoded}$ with p elements. As an illustration, consider $X = [110110110111011110111011111011]$, if $x_{min} = 0$ and $x_{max} = 1$, let $N_{bits} = 6$. Algorithm 1 yields $= [0.8571 \ 0.8730 \ 0.4762 \ 0.9365 \ 0.9365]$. In the proposed HGAFSA, once any of the GA steps is executed, the resulting output/population must be decoded before their respective fitnesses can be evaluated. Meanwhile, the resulting output/population from the AFSA steps are directly evaluated using the fitness function without being decoded.

3.2. Encoder Function

The encoder function is also written to counteract the effect of the decoder function presented in Algorithm 1. However, a reversed procedure is adopted based on Algorithm 1 (moving from step 7 to 1). Here, the decoder function is intended to generate chromosome (X) given its decoded version $X_{decoded}$, x_{min}, x_{max} and N_{bits}. However, the encoder function formulation is omitted from this manuscript for brevity. Generally, it can be said that the decoder function converts a chromosome into a fish, whereas the encoder function converts a fish back into a chromosome. This can be further described using Figure 1. Furthermore, it is worth noting that both X and $X_{decoded}$ are kept for reference during the optimization process using the proposed HGAFSA. However, either X or $X_{decoded}$ is later discarded depending on which of the GA or AFSA steps perform better at a given generation and at a given step in the HGAFSA dredging process. At first, the entire population is stored as chromosomes. However, each chromosome is either left as a X or transformed into a fish $X_{decoded}$ depending on which of the HGAFSA steps (GA step or AFSA step) performs better.

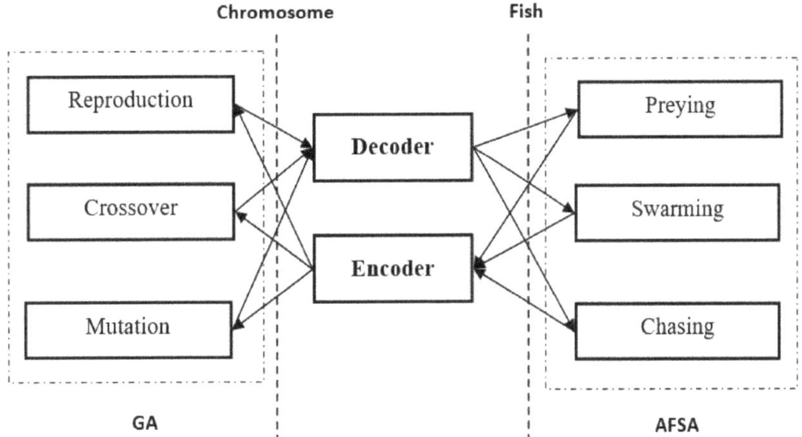

Figure 1. Illustration of chromosome to fish conversion and vice versa.

3.3. Population Update

The proposed HGAFSA is composed of two unique algorithms (GA and AFSA) with completely different parameter dredging procedures. When one of the GA steps is executed on an encoded fish (chromosome), the resulting chromosome might be of poor fitness compared to the result if an AFSA step had been directly performed on the fish itself. However, this consequence might be reversed. Therefore, the population must be carefully updated for optimality and an improved convergence rate. Algorithm 2 further describes the population update procedure.

Algorithm 2: Population updated

    input ://list of steps to be executed in chronological order)
              combined population of fish and chromosomes
    output: $P(k)$

1 *Define:* $\Omega =;$
2 **for** $k \leftarrow 1$ **to** *popsize* **do**
3   **for** $s \leftarrow 1$ **to** N_{steps} **do**
4   **if** $P(k) = fish$ & $S(s)$ *AFSA* **then**
5   *Evaluate:* $f = F_{obj}(P(k));$
6   *Execute:* $P(k) \rightarrow S(s) \rightarrow P(k)_{new};$
7   *Evaluate:* $f_{new} = F_{obj}(P(k)_{new});$
8   **if** f_{new} *is better than* f **then**
9   *Store:* $[f_{new}P(k)_{new}] \rightarrow \Omega;$
10   **else**
11   *Store:* $[fP(k)] \rightarrow \Omega;$
12   **else if** $P(k) = fish$ & $S(s)GA$ **then**
13   *Evaluate:* $f = F_{obj}(P(k));$
14   *Execute:* $P(k) \rightarrow Encoder \rightarrow P(k)_{Encoded};$
15   *Execute:* $P(k)_{Encoded} \rightarrow S(s) \rightarrow P(k)_{Encoded,new};$
16   *Evaluate:* $f_{new} = F_{obj}(P(k)_{Encoded,new});$
17   **if** f_{new} *is better than* f **then**
18   *Store:* $[f_{new}P(k)_{Encoded,new}] \rightarrow \Omega;$
19   **else**
20   *Store:* $[fP(k)] \rightarrow \Omega;$
21   **else if** $P(k) = chromosome$ & $S(s)AFSA$ **then**
22   *Execute:* $P(k) \rightarrow Decoder \rightarrow P(k)_{Decoded};$
23   *Evaluate:* $f = F_{obj}(P(k)_{Decoded});$
24   *Execute:* $P(k)_{Decoded} \rightarrow S(s) \rightarrow P(k)_{Decoded,new};$
25   *Evaluate:* $f_{new} = F_{obj}(P(k)_{Decoded,new});$
26   **if** f_{new} *is better than* f **then**
27   *Store:* $[f_{new}P(k)_{Decoded,new}] \rightarrow \Omega;$
28   **else**
29   *Store:* $[fP(k)] \rightarrow \Omega;$
30   **else if** $P(k) = chromosome$ & $S(s)GA$ **then**
31   *Execute:* $P(k) \rightarrow Decoder \rightarrow P(k)_{Decoded};$
32   *Evaluate:* $f = F_{obj}(P(k)_{Decoded});$
33   *Execute:* $P(k) \rightarrow S(s) \rightarrow P(k)_{new};$
34   *Execute:* $P(k) \rightarrow Decoded \rightarrow P(k)_{new,Decoded};$
35   *Evaluate:* $f_{new} = F_{obj}(P(k)_{new,Decoded});$
36   **if** f_{new} *is better than* f **then**
37   *Store:* $[f_{new}P(k)_{new,Decoded}] \rightarrow \Omega;$
38   **else**
39   *Store:* $[fP(k)] \rightarrow \Omega;$

40 *Sorting:* Rearrage the vectors in Ω (in the order of optimality);

3.4. Model of the Economic Load Dispatch Problem

To solve the ELD problem using the proposed HGAFSA, a function is required to convert the random numbers generated by it into electrical power demand scheduled to the set generating units. Let P be a set of power to be generated by the available generating units forming the ELD problem. Let P_{max} and P_{min} be the maximum and minimum power allowable for each of the units, respectively. N_G is the number of generating units. Then,

$$P = [P_1 P_2 P_3 ... P_{N_G-2} P_{N_G-1} P_{N_G}], \tag{11}$$

$$P_{max} = [P_{max,1} P_{max,2} ... P_{max,N_G-1} P_{max,N_G}], \tag{12}$$

$$P_{min} = [P_{min,1} P_{min,2} ... P_{min,N_G-1} P_{min,N_G}]. \tag{13}$$

The next most important parameter of ELD problem formulation is the total power (P_T) to be generated by the generating units to meet both the power demand and the power losses along the network. This can be described using the equality constraint as in Equation (14).

$$P_T = P_D + P_L \, . \tag{14}$$

where P_D is the total power demanded by the consumers and P_L is the total power losses in the network. P_T can also be expressed using Equation (15).

$$P_T = \sum_{i}^{N_G} P_i \, . \tag{15}$$

As described earlier, to evaluate the fitness of a population generated by HGAFSA, the population must be decoded into a fish ($X_{decoded}$). However, this fish must be further converted into a real power demand P. To achieve this, let $X_{decoded}$ be replaced by χ having p elements, such that Equation (16) holds.

$$\chi = x_1 x_2 x_3 ... x_{p-2} x_{p-1} x_p \, . \tag{16}$$

An ELD encoder function is developed to transform χ into an equivalent P. The overall process can be described using Figure 2.

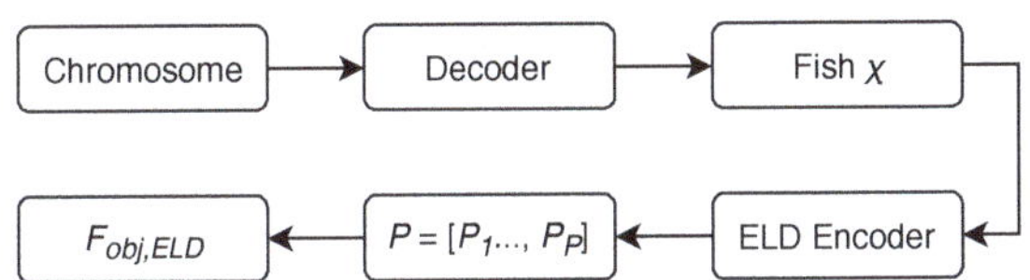

Figure 2. Evaluating the fitness of a chromosome/fish.

An ELD encoder function is developed to transform χ into an equivalent P. The overall process can be described using Figure 2. The function $F_{obj,ELD}$ is the objective function of the ELD problem. The ELD encoder function is given in Algorithm 3. The proposed ELD encoder has the advantage that no generating unit can generate below its minimum allowable generating limit. However, its generation may exceed the allowable maximum. To prevent this, steps 11 to 14 are added to Algorithm 3 to enforce the maximum limit constraint. Furthermore, in the developed ELD encoder, the power generation at any unit cannot be lower than the minimum allowable generation at that unit. This is because, in line 10 of the ELD encoder Algorithm, the addition of $P_{min}(h)$ enforces the lower limit constraint. However, the parameter γ can be greater than the upper limit at a particular generating unit due the presence of a rational function at the first part. The rational part of γ usually results in approximation, pushing its value beyond the maximum allowable limit. This phenomenon can be avoided by replacing γ with $P_{max}(h)$ when the upper limit is exceeded or using line 11 to 14 of Algorithm 3.

However, it can be observed from step 16 of Algorithm 3 that the proposed ELD encoder performs optimization during the encoding process. Therefore, it could be termed an optimal ELD encoder of order p. To further increase the performance of the proposed ELD encoder, the dimension of X can be extended for every given χ generated by the proposed HGAFSA. This extension is modeled in Equation (17).

$$\chi_{ext} = [\chi, \chi|(sin(2\pi\chi)|, \chi|(cos(2\pi\chi))|] \, . \tag{17}$$

With this extension, a new optimal ELD encoder of order 3×p can be formed by modifying steps 6 and 14 of the former optimal ELD encoder (order p) described earlier. This can be achieved using Equations (18) and (19), respectively.

$$par = [1, 2, 3, ..., 3 \times p] \, , \tag{18}$$

$$s = [1, 2, 3, ..., 3 \times p] .\qquad(19)$$

Equation (17) can be used to boost the performance of HGAFSA by further exploring the solution search space. This is because, for every new set of possible solutions χ, two additional solution sets are also searched, i.e., $\chi_1 = \chi|(sin(2\pi\chi))|$ and $\chi_2 = \chi|(cos(2\pi\chi))|$. Therefore, the probability that the optimal solution will lie within the three search spaces (χ, χ_1, and χ_2) must always be greater than or equal to the probability of the obtained optimal solution being within χ alone. As such, the performance of the original ELD encoder (which makes use of χ alone) will definitely improve. In the other hand, the ELD encoder of order $3 \times p$ is of a higher rate of convergence than the original ELD encoder (with order p).

Algorithm 3: ELD Encoder

---

    input :$\chi, p, N_G, P_{max}, P_{min}$ and P_T
    output: P

1   *Creat:* $\zeta = []$;
2   **for** $g = [1, 2, 3, ..., N_G]$ **do**
3      $\partial = (P_{max}(g) - P_{min}(g)) * \chi / max(\chi);$
4      $\zeta = \begin{bmatrix} \zeta \\ \partial \end{bmatrix};$

5   *Create:* $\varepsilon = [];$
6   **for** $par = [1, 2, 3, ..., p]$ **do**
7   **for** $g = [1, 2, 3, ..., N_G]$ **do**
8        $\rho(g) = \zeta(g, par);$

9   **for** $h = [1, 2, 3, ..., N_G]$ **do**
10        $\gamma = \dfrac{(P_T - \sum_{g}^{N_G} P_{min}(g)) * \rho}{(\sum_{g}^{N_G} \rho(g))} + P_{min}(h);$

11   **while** $max(\gamma - P_{max}) > 0$ *(enforce the upper limit constraint for each generating unit)* **do**
12        $\rho = rand(1, N_G). * \rho + rand(1, N_G);$
13   **repeat**
14   for loop at step 9;
15   **until**;
16      $\varepsilon = \begin{bmatrix} \varepsilon \\ \gamma \end{bmatrix};$

17   **for** $s = [1, 2, 3, ..., p]$ **do**
18   *Evaluate:* $Fit(s) = F_{obj,ELD}([\varepsilon(s, 1)\ \varepsilon(s, 2)\ \varepsilon(s, 3) ... \varepsilon(s, N_G - 1)\ \varepsilon(s, N_G)]);$

19   *Determine:* $K \leftarrow Fit(K) = min(Fit);$
20   *Assign:* $P = [\varepsilon(K, 1)\ \varepsilon(K, 2)\ \varepsilon(K, 3) ... \varepsilon(K, N_G - 1)\ \varepsilon(K, N_G)];$

3.5. The Proposed HGAFSA-Based Higher-Order ELD Algorithm

The ELD problem is one of the power system analysis problems with a large number of possible solutions. However, such solutions form a set of local optimums. As the number of generating units increases, the possible solutions increase exponentially. As such, an algorithm that can deeply search into the solution domain is required to locate the global optimum solution. In this work, the HGAFSA optimization algorithm with high computation capability and a fast rate of convergence is developed for complex ELD problem solving. The flow chart for the proposed HGAFSA-based higher-order ELD problem solver is shown in Figure 3.

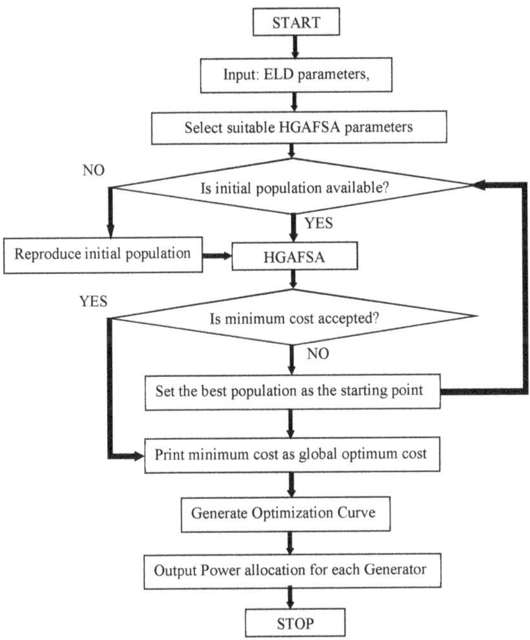

Figure 3. HGAFSA-based ELD algorithm.

4. Performance Validation

To demonstrate the effectiveness of the developed HGAFSA-based ELD problem solving algorithm, six test systems are designed. The developed algorithm is programmed in the MATLAB 2016a environment on a setup with 8 GB RAM and a 2.3 GHz Core I3 processing Computer running Windows 10.1.

A set of suitable parameters chosen for simulation analysis are listed in Table 2. To achieve the desired objective of outperforming all existing ELD algorithms in the literature, the best solution generated by HGAFSA is set as the starting point if it does not meet the desired goal. This process is repeated as far as the optimum cost is higher than the best cost presented in the literature so far.

Table 2. HGAFSA simulation parameter settings.

Parameter	Algorithm	Abbreviation	Value
Population Size	GA, AFSA	Psize	64
Number of Parameters	GA, AFSA	NoP	32
Visual Distance	AFSA	VD	0.875 to 1
Crowdness Factor	AFSA	CF	0.09 to 0.5
Step Size	AFSA	Ssize	0.00125 to 0.1
Max. Iteration	GA, AFSA	Max_Iter	10,000
Mutation Rate	GA	MR	0.4 to 0.75
Selection Probability	GA	SProb	0.375 to 0.5
Number of Bits	GA	NoB	8

4.1. Test System 1

The proposed HGAFSA algorithm is tested for a 13 generating units (Gen. units) test system having a non-smooth fuel cost function with valve-point effect, and its effectiveness is demonstrated compared to that of the oppositional grey wolf optimization (OGWO) algorithm [28] for the mentioned test system. The system data, such as fuel cost coefficients and active power limit thresholds for different generators, are adopted from [29] and the power demand assumed for the given test system is 2.52 GW. The simulation results of the 13-unit system are comparatively analyzed in Table 3 for our proposed HGAFSA and other optimization algorithms, such as improved coordinated aggregation-based PSO [30], shuffled differential evolution (SDE), oppositional real-coded chemical reaction optimization (ORCCRO) [31], biogeography-based optimization (BBO), hybrid differential evolution-based BBO (DE/BB) [32], and oppositional–invasive weed optimization (OIWO) [33]. The performance of HGAFSA is measured in terms of fuel cost ($/h) and power loss (MW), compared to the mentioned optimization algorithms, in Table 3. The fuel cost of HGAFSA is lower (24,141.26 $/h) compared to other algorithms. Comparatively, HGAFSA reduces the fuel cost of the generation system under consideration. Similarly, SDE and ORCCRO result in the same power loss (40.09 and 40.11) as HGAFSA (40.22), but lower than GWO, OGWO, and OIWO. Therefore, the reduced fuel cost and minimum power loss shown in Table 3 prove the efficacy of our proposed HGAFSA. Figure 4a shows the HGAFSA-based optimization curve. Figure 4a reveals that the algorithm converges at the thirteenth generation system to a cost of 24,141.2687 $/h. This results in annual savings of $3.253 m. Figure 4b presents the cumulative power generated by the units. In ELD, the power generated by each unit must lie within its maximum and the minimum allowable power output (upper and lower limits). Therefore, the power allocations must lie within this limit and the optimum power allocated to each unit (P_i) must lie within its limits. Finally, the cost functions of the various generating units influence the optimum power allocated the units and, thus, define the pattern of the 'optimum P_i' curves for each of the six test systems in this work. In general, the peak of the cumulative power curves can be described as follows.

Table 3. Comparison of results for 13-unit system.

No. of Gen. Units	HGAFSA (MW)	OGWO (MW)	GWO (MW)	OIWO (MW)	SDE (MW)	ORCCRO (MW)
1	628.32	628.29	628.16	628.31	628.32	628.32
2	299.20	299.18	298.92	299.19	299.20	299.20
3	299.20	297.50	298.22	299.19	299.20	299.20
4	159.73	159.72	159.72	159.73	159.73	159.73
5	159.73	159.73	159.72	159.73	159.73	159.73
6	159.73	159.72	159.72	159.73	159.73	159.73
7	159.73	159.73	159.71	159.73	159.73	159.73
8	151.73	159.73	159.67	159.73	159.73	159.73
9	148.02	159.73	159.66	159.73	144.74	144.72
10	114.79	77.39	77.39	77.39	113.12	112.14
11	95.58	114.74	114.60	113.10	92.40	92.40
12	92.40	92.39	92.38	92.35	92.40	92.40
13	92.40	92.37	92.35	92.39	92.40	92.40
Fuel Cost ($/h)	24,141.26	24,512.72	24,514.47	24,514.83	24,514.90	24,513.91
Power Loss (MW)	40.11	40.28	40.29	40.36	40.09	40.11

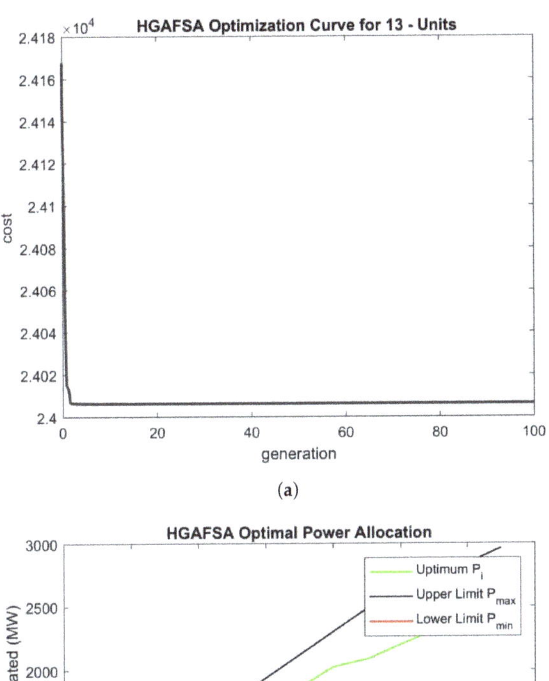

(a)

(b)

Figure 4. HGAFSA optimization and optimal power allocation curves for 13 units. (**a**) Optimization curve; (**b**) cumulative power generated.

1. Optimum P_i Curve: Defines the total power generated (Demand + Losses) by the system of generating units;
2. Upper Limit P_{max} Curve: Defines the maximum power that can be generated by the system of generating units;
3. Lower Limit P_{min} Curve: Defines the minimum power generated by the system of generating units.

In the case of Figure 4b, $P_i^{peak} = 25{,}604$, $P_{min}^{peak} = 550$, and $P_{max}^{peak} = 2960$.

4.2. Test System 2

The performance of the proposed algorithm, HGAFSA, and OGWO is validated for a 40 generating unit test system, where the data for valve-point coefficients and 40 generating units are adopted from [34]. A total power demand of 10.5GW is assumed for the given system. The real power generation output and fuel cost are calculated for the 40-unit test system by various methods, such as HGAFSA, GWO, OGWO [28], SDE [31], ORC-CRO, quasi-oppositional teaching–learning-based optimization (QOTLBO) [35], hybrid ant colony–genetic algorithm (GAAPI) [36], and krill herd algorithm (KHA) [37], and are tabulated in Table 4. The HGAFSA appraoch results in a reduced fuel cost (136,396.9 \$/h) and minimum power loss (957.29 MW) compared to the approaches mentioned in Table 4. Figure 5a shows the HGAFSA-based optimization curve, which converges at the fifty-first generation to a cost of 136,396.9727 \$/h. This results in annual savings of \$382,350. Furthermore, Figure 5b shows that all generators satisfy their inequality constraints.

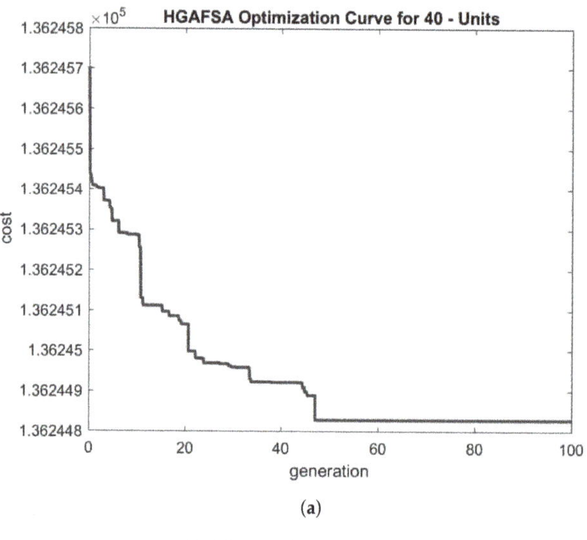

(a)

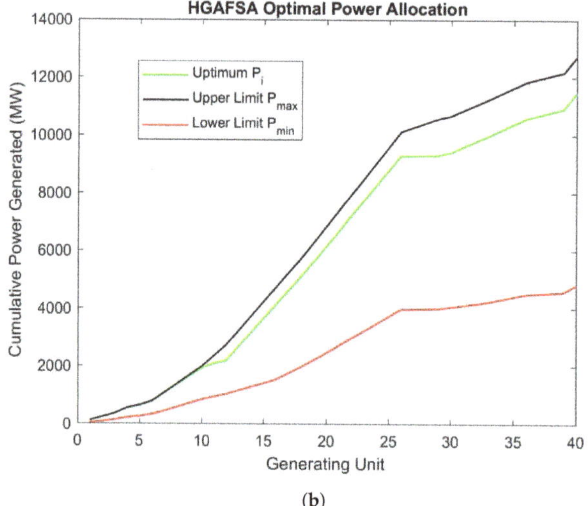

(b)

Figure 5. HGAFSA optimization and optimal power allocation curves for 40 units. (**a**) Optimization curve; (**b**) cumulative power generated.

Table 4. Comparison of results for 40-unit system.

No. of Gen. Units	HGAFSA (MW)	OGWO (MW)	GWO (MW)	OIWO (MW)	SDE (MW)	ORCCRO (MW)	GAAPI (MW)	QOTLBO (MW)	KHA (MW)
1	113.96	114	114	113.9908	110.06	111.68	114	114	114
2	113.69	114	114	114	112.41	112.16	114	114	114
3	120	120	120	119.99	120	119.98	120	107.82	120
4	179.74	183.57	181.04	182.51	188.72	182.18	190	190	190
5	96.97	87.81	87.83	88.42	85.91	87.28	97	88.37	88.59
6	140	140	140	140	140	139.85	140	140	105.51
7	300	300	300	299.99	250.19	298.15	300	300	300
8	284.8	300	300	292.06	290.68	286.89	300	300	300
9	289.02	300	300	299.88	300	293.38	300	300	300
10	279.65	279.72	279.97	279.70	282.01	279.34	205.25	211.20	280.67
11	168.81	243.61	243.62	168.81	180.82	162.35	226.3	317.27	243.53
12	94	94.17	94.14	94	168.74	94.12	204.72	163.76	168.80
13	484.04	484.27	484.45	484.07	469.96	486.44	346.48	481.57	484.11
14	484.05	484.33	484.23	484.04	484.17	487.02	434.32	480.54	484.16
15	484.04	484.04	484.24	484.03	487.73	483.39	431.34	483.76	485.23
16	484.08	484.07	484.03	484.08	482.3	484.51	440.22	480.29	485.06
17	489.28	489.21	489.62	489.28	499.64	494.22	500	489.24	489.45
18	489.3	489.26	489.32	489.29	411.32	489.48	500	489.55	489.30
19	511.32	511.33	511.46	511.32	510.47	512.2	550	512.54	510.71
20	511.33	511.49	511.49	511.33	542.04	513.13	550	514.29	511.30
21	549.94	523.47	523.47	549.94	544.81	543.85	550	527.08	524.46
22	549.94	546.64	547.68	549.99	550	548	550	530.10	535.57
23	523.3	523.38	523.37	523.28	550	521.21	550	524.29	523.37
24	523.32	523.33	523.13	523.32	528.16	525.01	550	524.65	523.15
25	523.27	523.40	523.34	523.58	524.16	529.84	550	525.05	524.19
26	523.28	523.30	523.35	523.58	539.1	540.04	550	524.46	523.54
27	10.01	10.01	10.06	10.01	10	12.59	11.44	10.89	10.12
28	10.01	10.01	10.63	10.01	10.37	10.06	11.56	17.43	10.18
29	10.01	10.06	10.51	10.01	10	10.79	11.42	12.78	10.02
30	96.96	87.80	87.80	87.86	96.1	89.7	97	88.81	87.81
31	190	190	190	190	185.33	189.59	190	190	190
32	190	190	190	189.99	189.54	189.96	190	190	190
33	190	190	190	190	189.96	187.61	190	190	190
34	199.99	200	200	199.99	199.9	198.91	200	200	200
35	200	200	200	200	196.25	199.98	200	168.08	164.91
36	169.2	164.89	164.83	164.82	185.85	165.68	200	165.50	164.97
37	110	110	110	110	109.72	109.98	110	110	110
38	109.99	110	110	109.99	110	109.82	110	110	110
39	110	110	110	110	95.71	109.88	110	110	110
40	550	511.85	511.54	550	532.43	548.5	550	511.53	512.06
Fuel Cost ($/h)	136,396.9	136,440.6	136,446.8	136,452.7	138,157	136,855.1	139,865	137,329.8	136,670
Power Loss (MW)	957.29	973.12	973.28	957.29	974.43	958.75	1045.06	1008.96	978.92

4.3. Test System 3

This system comprises 110 generating units with quadratic cost function characteristics. A load demand of 15GW is assumed for this system, and other system data (fuel valve coefficients and active power thresholds) are obtained from [38]. The minimum fuel cost and generation capacity computed using HGAFSA and OIWO are tabulated in Table 5. The results show that HGAFSA provides efficient and cheap power generation compared to OIWO and other optimization algorithms mentioned in the literature. Figure 6a shows the HGAFSA-based optimization curve. The algorithm is truncated at the hundredth generation and at a cost of 197,988.892 $/h. This results in annual savings of $2135.69. The saving in cost is lower because of the narrow margins for some of the generating units, especially units 1 to 9, and the high cost of generation. Furthermore, Figure 6b shows that all generators satisfy their inequality constraints.

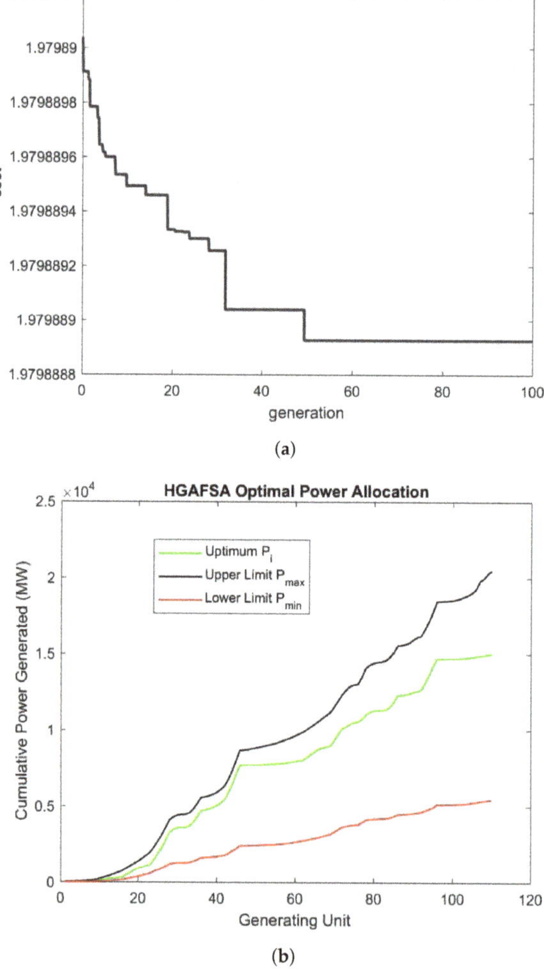

Figure 6. HGAFSA optimization and optimal power allocation curves for 110 units. (**a**) Optimization curve; (**b**) cumulative power generated.

Table 5. Comparison of results for 110 unit system.

No. of Gen. Units	HGAFSA (MW)	OIWO (MW)	No. of Gen. Units	HGAFSA (MW)	OIWO (MW)	No. of Gen. units	HGAFSA (MW)	OIWO (MW)
1	2.4	2.4	38	69.99	69.98	75	89.99	89.99
2	2.40	2.40	39	99.99	99.99	76	49.99	49.99
3	2.40	2.40	40	120	120	77	160	160.01
4	2.4	2.4	41	157.18	156.8	78	295.76	291.36
5	2.4	2.4	42	220	220	79	175.05	177
6	4.01	4.01	43	440	440	80	98.01	97.75
7	4	4	44	560	560	81	10.01	10.01
8	4	4	45	660	660	82	12.01	12.30
9	4	4	46	616.43	619.53	83	20.01	20.04
10	64.39	63.05	47	5.40	5.40	84	199.98	199.99
11	62.16	59.27	48	5.4	5.4	85	324.99	324.51
12	36.29	35.65	49	8.40	8.40	86	439.99	439.99
13	56.62	57.43	50	8.4	8.4	87	14.42	18.86
14	25	25	51	8.4	8.4	88	24.32	23.33
15	25	25	52	12	12	89	82.44	84.40
16	25	25	53	12	12	90	89.25	91.9
17	155	155	54	12.01	12.01	91	57.61	58.29
18	155	155	55	12	12	92	99.99	98.07
19	155	155	56	25.2	25.2	93	440	440
20	155	155	57	25.2	25.2	94	499.99	499.97
21	68.9	68.9	58	35	35	95	600	600
22	68.9	68.9	59	35.01	35	96	471.47	469.27
23	68.9	68.9	60	45.01	45.01	97	3.6	3.6
24	350	350	61	45.01	45.01	98	3.6	3.6
25	400	400	62	45	45	99	4.4	4.4
26	400	400	63	184.99	185	100	4.40	4.40
27	500	500	64	185	184.99	101	10.01	10.01
28	500	500	65	185	185	102	10.01	10.01
29	200	199.99	66	184.99	185	103	20.01	20.01
30	100	100	67	70	70	104	20.01	20.01
31	10.01	10.01	68	70	70	105	40	40
32	19.99	19.99	69	70.01	70.01	106	40.01	40.01
33	79.99	79.48	70	359.99	360	107	50	50
34	250	250	71	400	400	108	30	30
35	360	360	72	400	400	109	40	40
36	400	399.99	73	104.96	107.83	110	20	20
37	39.99	39.99	74	191.49	188.81	Fuel Cost ($/h)	197,988.8	197,989.1

4.4. Test System 4

In this scenario, a power system comprising 140 generating units is considered and the performance of the proposed HGAFSA algorithm is compared to OGWO, SDE, and OIWA. These results validate that HGAFSA outperforms all other methods in producing cheap power. The performance of both HGAFSA and OGWO is satisfactory but HGAFSA is significantly better than OGWO. The fuel costs are computed as 1,558,619 $/h for HGAFSA and 1,559,710 $/h and 1,559,953 $/h for OGWO and GWO, respectively. Figure 7a shows the HGAFSA-based optimization curve. HGAFSA converges at the sixty-sixth generation to a cost of 1,559,710 \$/h, resulting in annual savings of \$9.55 m. Furthermore, Figure 7b shows that all generators satisfy their inequality constraints.

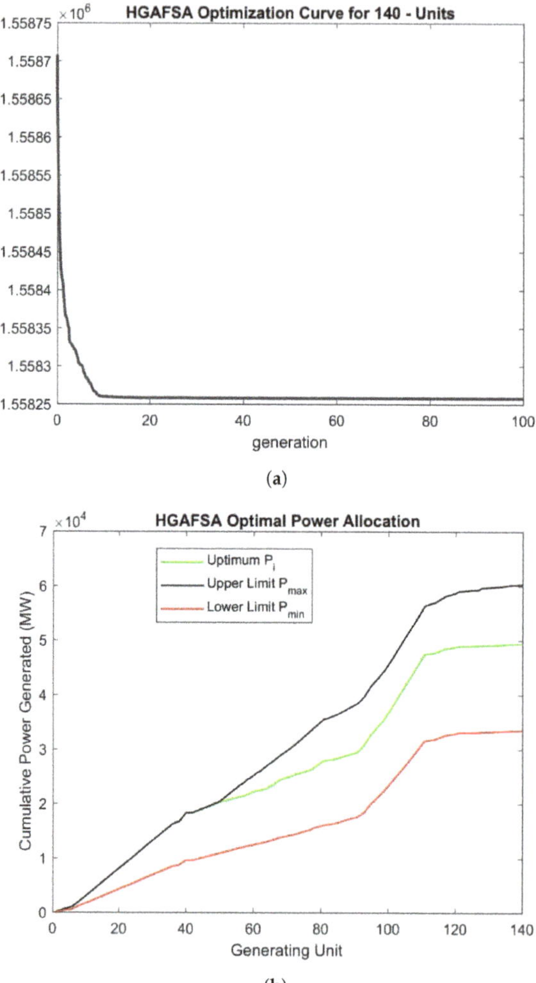

Figure 7. HGAFSA optimization and optimal power allocation curves for 140 units. (a) Optimization curve; (b) cumulative power generated.

4.5. Test System 5

The proposed algorithm is also tested for a test system consisting of 160 generating units with non-smooth valve-point cost functions. In order to show the effectiveness of the proposed technique for solving a large-scale ELD problem, transmission losses are ignored. The fuel costs are computed as 9612.8 \$/h for HGAFSA and 9745.1 \$/h and 9813.3 \$/h for OGWO and GWO, respectively. These results indicate that the total production cost is lower than that of all other methods mentioned in this work. Figure 8a shows the HGAFSA-based optimization curve. The HGAFSA algorithm converges at the eighth generation to a cost of 9612.8295 \$/h, providing annual savings of \$1.158 m. Furthermore, Figure 8b shows that all generators satisfy their inequality constraints.

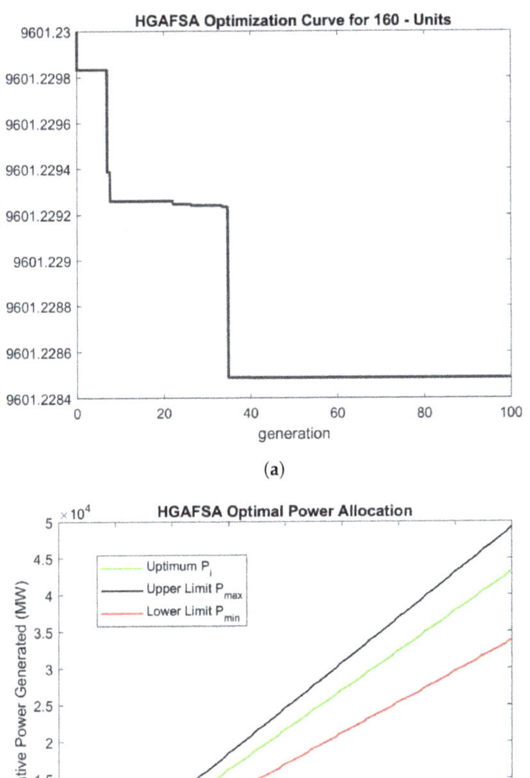

Figure 8. HGAFSA optimization and optimal power allocation curves for 160 units. (**a**) Optimization curve; (**b**) cumulative power generated.

4.6. Test System 6

The test system comprises a 463-unit system formed by combining test systems 1 to 5 (Gen. Units of 13 + 40 + 110 + 140 + 160). The coefficient of the valve point is assumed to be zero if not available for any generating unit. The simulation study is carried out under valve-point loading multi-fuel cost and emotion. A total power demand of 120,000 MW is used for the dispatch problem. The power loss components are assumed to be negligible. The problem is solved using the proposed HGAFSA. The fuel cost is computed as 1,645,338.7 \$/h. Figure 9a also shows the HGAFSA-based optimization curve. The HGAFSA algorithm converges at the eighty-sixth generation to a cost of 1,645,338.7385 \$/h. Furthermore, Figure 9b shows that all generators satisfy their inequality constraints.

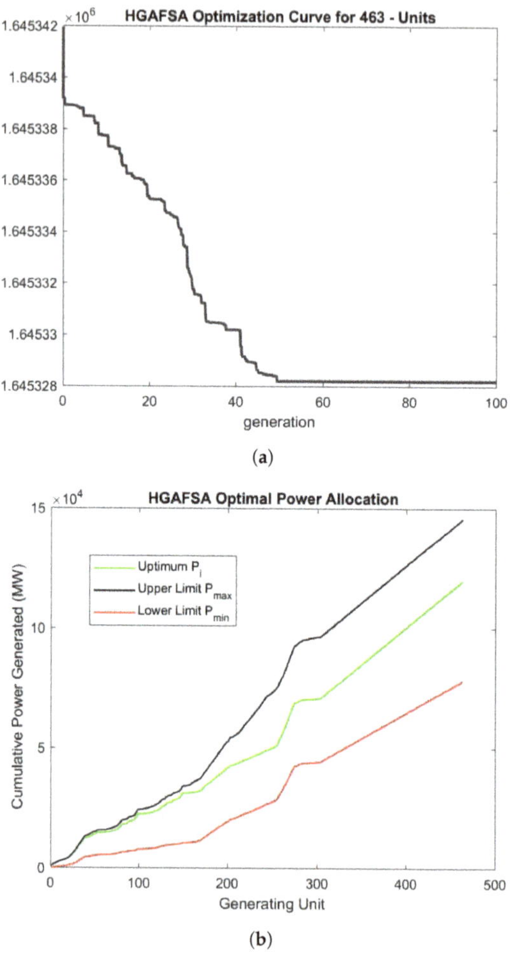

Figure 9. HGAFSA optimization and optimal power allocation curves for 463 units. (**a**) Optimization curve; (**b**) cumulative power generated.

In general, the results of the overall simulation study are summarized and presented in Table 6, which provides a comparative analysis of the developed HGAFSA and the best-performing algorithms mentioned in the literature. Table 6 reveals that HGAFSA outperforms all other algorithms for all simulation scenarios, with the best performance obtained in the 140-unit test system. This, however, results in relatively high overall annual savings. This work also provides a new benchmark test system of 463 units. In order to further demonstrate the optimization parameter sensitivity of the proposed HGAFSA, a range of sensitivity analysis is performed and is reported in the next subsection.

4.7. Sensitivity Analysis

Sensitivity analysis is carried out to provide an insight towards the selection of the HGAFSA-based optimization parameter settings (a) to achieve the desired optimization trade-off, (b) to boost the capability of replicating this work, and (c) to ease future advancement in this area of research. Tables 7–13 present the optimization results obtained by varying the most sensitive HGAFSA parameters over a number of Monte Carlo (NMC) simulation trials. In the analysis, the NMC is maintained at 10 trials and the optimization result is averaged over the 10 trials. In each of the tables, the convergence optimum cost, the number of function evaluations before convergence (NFEBC), the number of generations/iterations before convergence (NGBC), and the CPU time are averaged (over the 10 trials/NMC) and recorded/presented.

The convergence/optimization curve reaches a minimum value at a point referred to as the 'point of convergence'. The number of optimization trials/generation before the point of convergence, also known as the number of generations before convergence (NGBC), is used to compare the performance of HGAFSA for different cases of optimization. Contrarily, the NFEBC gives an idea of the computational intensity/expensiveness of any given optimization algorithm. In Table 7, the Psize is randomly varied between 2 and 500 individual/candidate solutions, and it is observed that the best performance (i.e., the lowest cost and a relatively lower NFEBC, NGBC, and CPU time) is achieved with 64-candidate solutions. Furthermore, for Psize greater than 64-candidate solutions, HGAFSA yields the optimum cost but with higher NFEBC and CPU time. However, the NGBC is observed to decrease with an increase in Psize.

Table 6. Summary of results for the comparison of HGAFSA and OGWO/OIWO.

Test System	Number of Gen. Units	HGAFSA ($)	Best Cost in Literature ($)	Annual Savings ($)	Total Power (MW)	CPU (s)
1	13	24,141.26	24,512.72	3,253,957.2	2560.36	5.02
2	40	136,396.97	136,440.62	382,350.35	11,457.29	10.11
3	110	197,988.89	197,989.14	2135.68	15,000	104.3
4	140	1,558,619.09	1,559,710	9,556,337	49,342	47.12
5	160	9612.82	9745.11	1,158,803.5	43,200	10.23
6	463	1,645,338.73	none	none	120,000	

Table 7. Effect of population size (Psize) on HGAFSA convergence.

NoP = 32; VD = 1; CF = 0.5; Ssize = 0.005; Max.Iter = 100; MR = 0.5; Sprob = 0.5; NoB = 8; NMC = 10					
Test Sys.	Psize	Cost ($)	NFEBC	NGBC	CPU (s)
1	2	1,558,959	44,544	87	5.823
2	16	1,558,651	319,488	78	41.77
3	64	1,558,619	360,448	22	47.12
4	100	1,558,619	460,800	18	60.24
5	500	1,558,619	1,536,000	12	200.8

Table 8. Effect of number of parameters (NoP) on HGAFSA convergence.

Psize = 64; VD = 1; CF = 0.5; Ssize = 0.005; Max.Iter = 100; MR = 0.5; Sprob = 0.5; NoB = 8; NMC = 10					
Test Sys.	NoP	Cost ($)	NFEBC	NGBC	CPU (s)
1	2	1,559,930	51,200	100	6.693
2	16	1,558,700	729,088	89	95.31
3	32	1,558,619	360,448	22	47.12
4	64	1,558,619	622,592	19	81.39
5	124	1,558,619	1,079,296	17	141.1

Table 9. Effect of visual distance (VD) on HGAFSA convergence.

Psize = 64; NoP = 32; CF = 0.5; Ssize = 0.005; Max.Iter = 100; MR = 0.5; Sprob = 0.5; NoB = 8; NMC = 10					
Test Sys.	**VD**	**Cost ($)**	**NFEBC**	**NGBC**	**CPU (s)**
1	0.25	1,559,130	1,638,400	100	71.39
2	0.5	1,559,009	1,638,400	100	71.39
3	0.75	1,558,909	1,490,944	91	64.97
4	1	1,558,619	360,448	22	47.12
5	1.25	1,558,619	491,520	30	50.25
6	1.2	1,559,658	1,638,400	100	71.39

Table 10. Effect of crowdedness factor (CF) on HGAFSA convergence.

Psize = 64; NoP = 32; VD = 0.875; Ssize = 0.005; Max.Iter = 100; MR = 0.5; Sprob = 0.5; NoB = 8; NMC = 10					
Test Sys.	**CF**	**Cost ($)**	**NFEBC**	**NGBC**	**CPU (s)**
1	0.03	1,559,992	1,638,400	100	71.39
2	0.06	1,559,986	1,556,480	95	67.82
3	0.09	1,559,930	1,490,944	91	64.97
4	0.5	1,558,619	360,448	22	47.12
5	0.6	1,558,666	851,968	52	55.69
6	0.7	1,558,802	1,343,488	82	58.54

Table 11. Effect of step size (Ssize) on HGAFSA convergence.

Psize = 64; NoP = 32; VD = 0.875; CF = 0.5; Max.Iter = 100; MR = 0.5; Sprob = 0.5; NoB = 8; NMC = 10					
Test Sys.	**Ssize**	**Cost ($)**	**NFEBC**	**NGBC**	**CPU (s)**
1	0.00125	1,558,619	360,448	22	70.68
2	0.005	1,558,619	360,448	22	47.12
3	0.01	1,558,689	524,288	32	68.54
4	0.025	1,558,679	1,097,728	67	71.75
5	0.05	1,558,692	1,343,488	82	58.54
6	0.1	1,558,629	1,507,328	92	65.68

Step size (Ssize) is another parameter of the HGAFSA that randomly affects its characteristics. In general, the smaller the value of Ssize, the better the performance, as presented in Table 11. An Ssize of 0.005 is used in this work. An important parameter of the HGAFSA is the selection probability (Sprob). Sprob also randomly affects the characteristics of HGAFSA, as shown in Table 12. Maintaining the value of Sprob at 50% (0.5) can work for most optimization problems. Similar to Sprob, the mutation rate (MR) also randomly affects the behavior of the developed HGAFSA, as shown in Table 13. A 50% mutation rate (MR = 0.5) is sufficient to handle most optimization problems. Finally, it is noted that, despite the effect of parameter variation, the HGAFSA outperforms most of the existing algorithms, regardless of its parameter settings.

Table 12. Effect of selection probability (Sprob) on HGAFSA convergence.

Psize = 64; NoP = 32; VD = 0.875; CF = 0.5; Max.Iter = 100; MR = 0.5; Ssize = 0.005; NoB = 8; NMC = 10					
Test Sys.	**Sprob**	**Cost ($)**	**NFEBC**	**NGBC**	**CPU (s)**
1	0.1	1,558,629	442,368	27	48.19
2	0.2	1,558,621	507,904	31	55.33
3	0.3	1,558,650	475,136	29	51.76
4	0.5	1,558,619	360,448	22	47.12
5	0.7	1,558,629	622,592	38	67.82
6	0.8	1,558,622	491,520	30	64.25

Table 8 shows that an increase in the number of parameters (NoP) results in earlier convergence but with higher computational expensiveness and convergence time. It is also found out that a choice of 32 parameters is sufficient to achieve a relatively good convergence characteristic. Table 9 presents the effect of the visual distance (VD) on the convergence characteristics of HGAFSA. The choice of VD alters the value of NGBC, and an improper selection of VD may result in a large value of NGBC or even lead to local minimum/premature convergence. A trail and error procedure can be used to obtain a suitable VD for any given optimization problem. However, it is found that a VD = 1 is most suitable for solving the ELD problem. The effect of the crowdedness factor (CF) is demonstrated in Table 10. Even though the results reported are the average of 10 trials, the effect of CF on HGAFSA is observed to be random. However, with CF = 0.5, the least/optimum cost is obtained for all scenarios.

Table 13. Effect of mutation rate (MR) on HGAFSA convergence.

Psize = 64; NoP = 32; VD = 0.875; CF = 0.5; Max.Iter = 100; Ssize = 0.005; Sprob = 0.5; NoB = 8; NMC = 10					
Test System	**MR**	**Cost ($)**	**NFEBC**	**NGBC**	**CPU (s)**
1	0.1	1,558,629	540,672	33	58.9
2	0.25	1,558,621	638,976	39	69.61
3	0.5	1,558,619	360,448	22	47.12
4	0.75	1,558,650	524,288	32	57.12
5	1	1,558,629	573,440	35	62.47

5. Conclusions and Future Directions

HGAFSA is developed using a hybridization of the conventional binary-coded GA and real-coded AFSA using a decoder (which converts a GA 'chromosome' into an AFSA 'fish') and an encoder (which performs the reverse operation of the decoder). An ELD encoder algorithm is also developed and integrated with the HGAFSA to form a more robust, efficient, and reliable technique (with a guaranteed high rate of convergence) for solving complex ELD problems. A sensitivity analysis is carried out on the optimization parameter settings of the developed HGAFSA. It is concluded that a trial and error method can be used to select a suitable set of parameters in order to optimize an ELD problem. The convergence of HGAFSA is randomly dependent on most of its parameters, and despite the limitation of the parameter choice, its solutions would always be better than most of the existing algorithms. The choice of encoder/decoder function further enhances the optimization process. This feature of HGAFSA makes it robust and almost insensitive to the choice of optimization parameter setting and enables it to support a wide range of optimization parameter settings. The effectiveness of the HGAFSA is demonstrated through the simulation of solutions to a multi-objective ELD problem with higher-order cost functions using 13, 40, 110, 140, 160, and 463 generating unit test systems. Outstanding performance is demonstrated by the proposed HGAFSA-based ELD approach for most of the test systems. The future directions include the integration of the proposed ELD encoder with other meta-heuristics techniques in order to increase the efficiency and performance and its comparison with the proposed HGAFSA. The computational expensiveness of HGAFSA can also be further reduced by minimizing the number of repetitive steps in the algorithm while still introducing a powerful function that could compensate for their solution search ability. Another possible improvement is to modify the solution search procedure by introducing a set of steps that could adjust the parameter settings of HGAFSA based on the iteration counter to further enhance its rate of convergence.

Author Contributions: Conceptualization, A.M.K., M.K. and F.A.; methodology, F.A. and F.M.; formal analysis, M.K., F.A. and F.M.; validation, A.M.K. and F.A.; investigation, F.A. and Z.U.; funding acquisition, F.R.A. and G.H.; resources, F.R.A. and G.H.; data curation, M.K. and F.M.; project administration, Z.U., F.R.A. and G.H.; writing—original draft preparation, A.M.K., and M.K.; writing—review and editing, Z.U., F.R.A. and G.H.; supervision, F.M., F.A. All authors have read and agreed to the published version of the manuscript.

Funding: The APC is funded by Taif University Researchers Supporting Project Number(TURSP-2020/331), Taif University, Taif, Saudi Arabia.

Acknowledgments: The authors would like to acknowledge the support from Taif University Researchers Supporting Project Number(TURSP-2020/331), Taif University, Taif, Saudi Arabia.

Conflicts of Interest: The authors declare no conflict of interest.

References

1. Li, Q.; Wang, J.; Zhang, Y.; Fan, Y.; Bao, G.; Wang, X. Multi-Period Generation Expansion Planning for Sustainable Power Systems to Maximize the Utilization of Renewable Energy Sources. *Sustainability* **2020**, *12*, 1083. [CrossRef]
2. Chouhdry, Z.U.R.; Hasan, K.M.; Raja, M.A.Z. Design of reduced search space strategy based on integration of Nelder–Mead method and pattern search algorithm with application to economic load dispatch problem. *Neural Comput. Appl.* **2018**, *30*, 3693–3705. [CrossRef]
3. Zhang, Q.; Zou, D.; Duan, N.; Shen, X. An adaptive differential evolutionary algorithm incorporating multiple mutation strategies for the economic load dispatch problem. *Appl. Soft Comput.* **2019**, *78*, 641–669. [CrossRef]
4. Babatunde, O.; Munda, J.; Hamam, Y. Power system flexibility: A review. *Energy Rep.* **2020**, *6*, 101–106.
5. Azad, M.A.K.; Rocha, A.M.A.; Fernandes, E.M. Improved binary artificial fish swarm algorithm for the 0–1 multidimensional knapsack problems. *Swarm Evol. Comput.* **2014**, *14*, 66–75. [CrossRef]
6. Iqbal, S.; Sarfraz, M.; Ayyub, M.; Tariq, M.; Chakrabortty, R.K.; Ryan, M.J.; Alamri, B. A Comprehensive Review on Residential Demand Side Management Strategies in Smart Grid Environment. *Sustainability* **2021**, *13*, 7170. [CrossRef]
7. Mehmood, F.; Ashraf, N.; Alvarez, L.; Malik, T.N.; Qureshi, H.K.; Kamal, T. Grid integrated photovoltaic system with fuzzy based maximum power point tracking control along with harmonic elimination. *Trans. Emerg. Telecommun. Technol.* **2020**, e3856. [CrossRef]
8. Ellahi, M.; Abbas, G.; Satrya, G.B.; Usman, M.R.; Gu, J. A Modified Hybrid Particle Swarm Optimization With Bat Algorithm Parameter Inspired Acceleration Coefficients for Solving Eco-Friendly and Economic Dispatch Problems. *IEEE Access* **2021**, *9*, 82169–82187. [CrossRef]
9. Bernardon, D.; Mello, A.; Pfitscher, L.; Canha, L.; Abaide, A.; Ferreira, A. Real-time reconfiguration of distribution network with distributed generation. *Electr. Power Syst. Res.* **2014**, *107*, 59–67. [CrossRef]
10. Atwa, Y.; El-Saadany, E.; Salama, M.; Seethapathy, R. Optimal renewable resources mix for distribution system energy loss minimization. *IEEE Trans. Power Syst.* **2010**, *25*, 360–370. [CrossRef]
11. Mahmoud, K.; Abdel-Nasser, M.; Mustafa, E.; Ali, Z.M. Improved Salp–Swarm Optimizer and Accurate Forecasting Model for Dynamic Economic Dispatch in Sustainable Power Systems. *Sustainability* **2020**, *12*, 576. [CrossRef]
12. Aziz, A.S.; Tajuddin, M.F.N.; Adzman, M.R.; Ramli, M.A.M.; Mekhilef, S. Energy Management and Optimization of a PV/Diesel/Battery Hybrid Energy System Using a Combined Dispatch Strategy. *Sustainability* **2019**, *11*, 683. [CrossRef]
13. Deb, S.; Abdelminaam, D.S.; Said, M.; Houssein, E.H. Recent Methodology-Based Gradient-Based Optimizer for Economic Load Dispatch Problem. *IEEE Access* **2021**, *9*, 44322–44338. [CrossRef]
14. Alanazi, M.S. A Modified Teaching—Learning-Based Optimization for Dynamic Economic Load Dispatch Considering Both Wind Power and Load Demand Uncertainties With Operational Constraints. *IEEE Access* **2021**, *9*, 101665–101680. [CrossRef]

15. Dhebar, Y.; Deb, K. A computationally fast multimodal optimization with push enabled genetic algorithm. In *Proceedings of the Genetic and Evolutionary Computation Conference Companion*; ACM: New York, NY, USA, 2017; pp. 191–192.

16. Fang, N.; Zhou, J.; Zhang, R.; Liu, Y.; Zhang, Y. A hybrid of real coded genetic algorithm and artificial fish swarm algorithm for short-term optimal hydrothermal scheduling. *Int. J. Electr. Power Energy Syst.* **2014**, *62*, 617–629. [CrossRef]

17. Holland, J.H. *Adaptation in Natural and Artificial Systems: An Introductory Analysis with Applications to Biology, Control, and Artificial Intelligence*; MIT Press: Cambridge, MA, USA, 1992.

18. Jubril, A.M.; Komolafe, O.A.; Alawode, K.O. Solving multi-objective economic dispatch problem via semidefinite programming. *IEEE Trans. Power Syst.* **2013**, *28*, 2056–2064. [CrossRef]

19. Dharamjit, D. Load flow analysis on IEEE 30 bus system. *Int. J. Sci. Res. Publ.* **2012**, *2*, 1–6.

20. Sivagnanam, G.; Mauryan, K.C.; Maheswar, R.; Subramanian, R. An Efficient ITA Algorithm to Solve Economic Load Dispatch with Valve Point Loading Effects. *Int. J. Print. Packag. Allied Sci.* **2017**, *2017*, 121–128.

21. Khoa, T.; Vasant, P.; Singh, B.; Dieu, V. Solving Economic Dispatch By Using Swarm Based Mean-Variance Mapping Optimization (MVMOS). *Glob. Technoloptim* **2015**, *6*, 184.

22. Al-Bahrani, L.; Seyedmahmoudian, M.; Horan, B.; Stojcevski, A. Solving the Real Power Limitations in the Dynamic Economic Dispatch of Large-Scale Thermal Power Units under the Effects of Valve-Point Loading and Ramp-Rate Limitations. *Sustainability* **2021**, *13*, 1274. [CrossRef]

23. Jiang, C.; Wan, L.; Sun, Y.; Li, Y. The application of PSO-AFSA method in parameter optimization for underactuated autonomous underwater vehicle control. *Math. Probl. Eng.* **2017**, *2017*, 7482. [CrossRef]

24. Costa, M.F.P.; Rocha, A.M.A.; Fernandes, E.M. An artificial fish swarm algorithm based hyperbolic augmented Lagrangian method. *J. Comput. Appl. Math.* **2014**, *259*, 868–876. [CrossRef]

25. Hussain, I.; Ali, S.; Khan, B.; Ullah, Z.; Mehmood, C.; Jawad, M.; Farid, U.; Haider, A. Stochastic wind energy management model within smart grid framework: A joint Bi-directional service level agreement (SLA) between smart grid and wind energy district prosumers. *Renew. Energy* **2019**, *134*, 1017–1033. [CrossRef]

26. Ali, S.M.; Ullah, Z.; Mokryani, G.; Khan, B.; Hussain, I.; Mehmood, C.A.; Farid, U.; Jawad, M. Smart grid and energy district mutual interactions with demand response programs. *IET Energy Syst. Integr.* **2020**, *2*, 1–8. [CrossRef]

27. Chatterjee, S.; Laudato, M.; Lynch, L.A. Genetic algorithms and their statistical applications: An introduction. *Comput. Stat. Data Anal.* **1996**, *22*, 633–651. [CrossRef]

28. Li, M.; Huang, X.; Liu, H.; Liu, B.; Wu, Y.; Xiong, A.; Dong, T. Prediction of gas solubility in polymers by back propagation artificial neural network based on self-adaptive particle swarm optimization algorithm and chaos theory. *Fluid Phase Equilibria* **2013**, *356*, 11–17. [CrossRef]

29. Bhattacharya, A.; Chattopadhyay, P.K. Hybrid differential evolution with biogeography-based optimization for solution of economic load dispatch. *IEee Trans. Power Syst.* **2010**, *25*, 1955–1964. [CrossRef]

30. Vlachogiannis, J.G.; Lee, K.Y. Economic load dispatch—A comparative study on heuristic optimization techniques with an improved coordinated aggregation-based PSO. *IEEE Trans. Power Syst.* **2009**, *24*, 991–1001. [CrossRef]

31. Reddy, A.S.; Vaisakh, K. Shuffled differential evolution for large scale economic dispatch. *Electr. Power Syst. Res.* **2013**, *96*, 237–245. [CrossRef]

32. Ellahi, M.; Abbas, G. A Hybrid Metaheuristic Approach for the Solution of Renewables-Incorporated Economic Dispatch Problems. *IEEE Access* **2020**, *8*, 127608–127621. [CrossRef]

33. Roy, P.K.; Bhui, S. Multi-objective quasi-oppositional teaching learning based optimization for economic emission load dispatch problem. *Int. J. Electr. Power Energy Syst.* **2013**, *53*, 937–948. [CrossRef]

34. Mandal, B.; Roy, P.K.; Mandal, S. Economic load dispatch using krill herd algorithm. *Int. J. Electr. Power Energy Syst.* **2014**, *57*, 1–10. [CrossRef]

35. Wang, Y.; Li, B.; Weise, T. Estimation of distribution and differential evolution cooperation for large scale economic load dispatch optimization of power systems. *Inf. Sci.* **2010**, *180*, 2405–2420. [CrossRef]

36. Chiang, C.L. Improved genetic algorithm for power economic dispatch of units with valve-point effects and multiple fuels. *IEEE Trans. Power Syst.* **2005**, *20*, 1690–1699. [CrossRef]

37. Muralikrishnan, N.; Jebaraj, L.; Rajan, C.C.A. A Comprehensive Review on Evolutionary Optimization Techniques Applied for Unit Commitment Problem. *IEEE Access* **2020**, *8*, 132980–133014. [CrossRef]

38. Vishwakarma, K.K.; Dubey, H.M.; Pandit, M.; Panigrahi, B.K. Simulated annealing approach for solving economic load dispatch problems with valve point loading effects. *Int. J. Eng. Sci. Technol.* **2012**, *4*, 60–72. [CrossRef]

MDPI

St. Alban-Anlage 66

4052 Basel

Switzerland

www.mdpi.com

Sustainability Editorial Office

E-mail: sustainability@mdpi.com

www.mdpi.com/journal/sustainability